Structure and Function of
Invertebrate Oxygen Carriers

Serge N. Vinogradov   Oscar H. Kapp
Editors

Structure and Function of Invertebrate Oxygen Carriers

With 140 Illustrations

Springer-Verlag
New York Berlin Heidelberg London
Paris Tokyo Hong Kong Barcelona

Serge N. Vinogradov
Biochemistry Department
Wayne State University
School of Medicine
Detroit, MI 48201 USA

Oscar H. Kapp
Enrico Fermi Institute
University of Chicago
Chicago, IL 60637 USA

Proceedings of the Satellite Symposium of the Xth International Biophysics Congress held in Whistler, British Columbia, Canada, July 24–27, 1990

Cover: Space-filled model of the crystal structure of *Glycera dibranchiata* carbonmonoxyhemoglobin rendered on an IBM RS-6000 workstation. Crystal structure provided by Professor Warner E. Love of Johns Hopkins University. Workstation provided to O. Kapp by IBM Corporation as part of their Workstation Technology Program. Ryan, M.J. and Kapp, O.H. (1991) *SPIE Proceedings* **1396**: 335–339.

Library of Congress Cataloging-in-Publication Data
Structure and function of invertebrate oxygen carriers / Serge N.
  Vinogradov, Oscar H. Kapp, editors.
    p. cm.
    "Based predominately on . . . the Satellite Symposium 'Structure and
  Function of Invertebrate Oxygen-Binding Proteins,' held in Whistler,
  British Columbia, Canada, July 24–27, 1990, the week prior to the
  Xth International Biophysics Congress in Vancouver"—Introd.
    Includes bibliographical references and index.
    ISBN 0-387-97585-3. — ISBN 3-540-97585-3
    1. Hemoproteins—Congresses.   2. Invertebrates—Physiology—
  Congresses.   3. Oxygen—Physiological transport—Congresses.
  I. Vinogradov, Serge N., 1933- .   II. Kapp, Oscar H.
  III. Satellite Symposium "Structure and Function of Invertebrate
  Oxygen-Binding Proteins" (1990 : Whistler, B.C.)   IV. International
  Biophysics Congress (10th : 1990 : Vancouver, B.C.)
  QP552.H46S77   1991
  592'.012—dc20   91-14654

Printed on acid-free paper.

Camera-ready copy provided by the editors.
Printed and bound by Edwards Brothers, Inc., Ann Arbor, MI.
Printed in the United States of America.

9 8 7 6 5 4 3 2 1

ISBN 0-387-97585-3 Springer-Verlag New York Berlin Heidelberg
ISBN 3-540-97585-3 Springer-Verlag Berlin Heidelberg New York

Introduction

The content of this volume is based predominantly on the lectures and posters presented by the participants of the Satellite Symposium "Structure and Function of Invertebrate Oxygen-Binding Proteins," held in Whistler, British Columbia, Canada, July 24–27, 1990, the week prior to the Xth International Biophysics Congress in Vancouver, British Columbia. Because not everyone who wished to attend this symposium was able to do so, we decided to also include in this volume contributions from several non-participants.

This volume is dedicated to the memories of Eraldo Antonini, Bernt Linzen and Robert Terwilliger, whose untimely deaths in recent years have tragically deprived all of us working in the area of oxygen-binding proteins of esteemed colleagues and personal friends. We would like to thank Maurizio Brunori, Emilia Chiancone, Heinz Decker, Jürgen Markl and Joseph Bonaventura for contributing the memorial essays.

We would like to thank the following for their support: Dr. G.T. Heberlein, Vice President for Research and Dean of the Graduate School, and Dr. R.J. Sokol, Dean of the School of Medicine, Wayne State University, the Center for Imaging Science of the University of Chicago, IBM Corporation, the National Institutes of Health and the U.S. Department of Energy. In particular, we would like to acknowledge our debt to Ms. Virginia M. Boyce for the careful editing and preparation of the manuscripts for publication.

S.N. Vinogradov
O.H. Kapp

In Memoriam:

An Homage to Eraldo Antonini

Maurizio Brunori and Emilia Chiancone

Department of Biochemistry and CNR Center of Molecular Biology,
University of Rome "La Sapienza," 00185 Rome, Italy

When Eraldo Antonini died on 19 March 1983, he was one of the most famous scientists of our time. Many of his papers are still classics not only in the fields of biochemistry and molecular biology, but also in related areas of biomedical research, such as pharmacology and biotechnology. Indeed, the *Citation Index* indicates that he was one of the most cited scientists in the period spanning 1965 to 1978.

Antonini was born in San Piero à Sieve near Florence on 20 April 1931. He graduated from the University of Rome in 1954 with a degree in Medicine and Surgery. At that time, under the guidance of Professor Alessandro Rossi-Fanelli, the Director of the Institute of Biochemistry, he was already deeply engaged in active research on the structural and functional properties of hemoglobin and myoglobin. In 1967, Antonini was named Full Professor of Molecular Biology at the University of Camerino, and from 1970 until his death he was Full Professor of Chemistry in the Faculty of Medicine at the University of Rome. He was a founding member of the European Molecular Biology

Organization (EMBO), and a member of many Italian and international scientific societies, such as the Royal Society of Medicine. Antonini was also the recipient of a number of prestigious awards, such as the Feltrinelli Prize for Biology and Medicine awarded to him by the Accademia Nazionale dei Lincei in 1971, and the special Saint Vincent Prize for Medicine, awarded *ad memoriam* in 1983.

Antonini's scientific interest was initially, and for many years thereafter, the study of the functional properties of human hemoglobin and myoglobin. His original contributions and seminal findings in this area of biochemistry are innumerable. During the first years of his scientific activity he accurately determined the oxygen association curves of hemoglobin and myoglobin by applying a simple tonometric method which is still in use today. One of his many outstanding contributions was the preparation of native globin from human hemoglobin and the *in vitro* reconstitution of the functionally active protein by the addition of natural and modified hemes. This brilliant result demonstrated that the necessary "information" for the reconstitution of the native protein is intrinsic to the globin fold, an impressive achievement. In the late fifties, he also developed new approaches for studying the effect of the porphyrin and metal structure on the functional modulation of hemoglobin. While continuing his work on the purification and characterization of myoglobins and hemoglobins from several species in the sixties, Antonini engaged in the study of the influence of chemical modifications on oxygen-binding, and of the variables which control the Bohr effect. At this time, he also attacked the problem of the reversible dissociation of human hemoglobin into subunits and the relationship between dissociation and ligand binding. It was this work that aroused the curiosity of Jeffries Wyman in 1961 and prompted his collaboration with the Rome group to prove that the puzzlingly high cooperativity, observed under conditions in which the oxygenated form of the molecule is largely dimeric, was due to oxygen-linked dissociation. Wyman, who had planned to stay in Rome for only a few months, became so engaged by the remarkable intuition and the warm personality of Eraldo that he happily and fruitfully spent 25 years of his life working at the "Regina Elena" and the Biochemistry Institute. One of his first discoveries, which originated from this unique and "cooperative" collaboration, was the proteolytic removal of the C-terminal residues of the α- and β-chains of hemoglobin that resulted in the loss of the ligand-linked conformational changes and, thus, of the characteristic functional features of hemoglobin (the heme-heme interactions and the Bohr effect). This illuminating information about the crucial role of the C-terminal residues of the polypeptide chains in the regulation of hemoglobin function was exploited later by M. F. Perutz on the basis of the crystal structures of oxy(met) and deoxyhemoglobin.

With the participation of a growing number of enthusiastic young scientists, the Rome group entered into a productive era which attracted a large number of scientists from abroad, the first of whom was Rufus Lumry who initiated a series of informal discussions known as "La Cura Conferences." Among the most significant contributions of the late sixties was the preparation and characterization of artificial symmetrical hybrid molecules, such as $\alpha_2 X_2 \beta_2 Y_2$ (with X = CN, NO and Y = O$_2$, CO). The acid-base equilibria of hemoglobin, the Bohr effect and its dependence on temperature, the oxidation-reduction equilibria and the relationship between these properties and oxygen binding were also investigated, one idea and one experiment leading to another. All of these results contributed significantly to our understanding of human hemoglobin and to the formulation of the allosteric theory. Antonini's thorough knowledge of hemoglobin was summarized in a classic review article written in 1964 for *Advances in Protein Chemistry*, when the three-dimensional structure of hemoglobin was already available, and later in the volume *Hemoglobin and Myoglobin in their Reactions with Ligands*, which is still widely used around the world 28 years after its publication.

Although human hemoglobin held Antonini's attention, he continued to study respiratory pigments from different sources, searching for a general interpretation of the regulation phenomena in these systems. His contributions in this specific area, which is directly relevant to the proceedings of the *Satellite Symposium of the Xth Biophysics Congress* (1990), span from the purification and characterization of *Aplysia* myoglobin (the first example of a myoglobin that lacks the distal His) and the pioneering work on *Spirographis spallanzanii* chlorocruorin and other giant extracellular hemoglobins, to the study of the cooperative hemoglobins from the mollusc *Scapharca* that are characterized by a unique assembly of globin folds. The discovery of a cooperative dimer in *Scapharca*, in particular, represented an exciting discovery that has led to new, unexpected developments.

Antonini's scientific stature has hopefully emerged from this brief account of his achievements in the study of hemoglobin and myoglobin; his human qualities are more difficult to put into words. However, those who were fortunate to meet him at work and at home will never forget his generosity, vitality, warm personal approach and enthusiasm for new ideas which made him an extremely successful scientist and a rare human being.

In Memoriam:

An Homage to Bernt O. Linzen

Heinz Decker[a] and Jürgen Markl[b]

[a]Zoological Institute, University of Munich, Luisenstrasse 14, Munich, Federal Republic of Germany
[b]Zoological Institute, University of Würzburg, Röntgenring 10, Würzburg, Federal Republic of Germany

In early August 1988, the sudden death of one of our most respected colleagues and teachers shocked the scientific community. The biochemist and animal physiologist Prof. Dr. Bernt Linzen was a revered teacher and loyal friend: we are grateful for this opportunity to pay homage to his memory.

Those who had the privilege to meet Bernt Linzen were fascinated by his engaging discussions and stimulating examinations of biological phenomena. He captured our attention and inspired some of us to even change our own fields of research (the authors left inorganic chemistry (J. M., 1974) and high energy physics (H. D., 1975) in order to join his group). Bernt Linzen had a profound ability to fascinate others in his zoological pursuits, but he did not limit himself to biological themes. He could spend hours discussing classical music or fine art. His respect and love for the arts live on in our memories of him playing the piano, accompanied by singing colleagues from around the world at his home during the Tutzing Conference in 1985.

Those who worked in our laboratory were members of a research family headed by Bernt Linzen. He concerned himself not only with our research, but also with our personal welfare. Though he would arrive late in the morning, he would work with us until well after midnight in his office at the end of a long corridor on the top floor of the Zoological Institute in Munich. When he was taken from us on the fifth of August, nineteen hundred eighty-eight, we lost both a valued scientific advisor and a compassionate, fatherly friend.

Bernt O. Linzen was born in Cali, Columbia in 1931. During a visit to Hamburg, Germany, he and his parents were surprised by the beginning of the Second World War, which forced them to stay. Bernt grew up in Germany and began studying biology and chemistry in Hamburg and Tübingen. Though attracted at a very early age by biochemical questions, he worked for several months in 1953 as a bird watcher on the small coastal island of Mellum. Thus, his long list of publications begins with a small paper about a 27 year-old oyster fisherman (1).

Bernt Linzen then studied under Nobel Laureate Adolf Butenandt at the Max-Planck-Institute for Biochemistry in Tübingen, and received his doctorate in 1957 for his thesis, *Über Ommine und das Auftreten von Ommochromen im Tierreich*. He spent the next 15 years pursuing his studies in this field, first with Butenandt at the Max-Planck-Institutes in Tübingen and Munich and, after 1968, independently at the Zoological Institute of the University of Munich. These studies are summarized in a review article entitled *The Tryptophan-Ommochrome Pathway in Insects* (2). From 1960 to 1962 he was a postdoctoral fellow in Gerald Wyatt's laboratory at Yale University, where he studied nucleic acid metabolism and the regulation of mitosis in insects (3). In 1975, he was appointed to the chair of General Biology at the Zoological Institute of the University of Munich, where he became responsible for the curricula in biochemistry and metabolic physiology.

In 1968 he discovered that a colleague was keeping large spiders in the cellar of the institute. Aware that spider chemistry was an as yet undiscovered field, he persuaded Renate Loewe, a student at the time, to analyze the blood components of these animals (4). This study drew his attention to the blue hemolymph pigment that was soon to dominate his scientific life: the respiratory protein hemocyanin. During the next 20 years he interested more than 50 students and scientists in the study of the structure, function and evolution of hemocyanin, and in the examination of physiological aspects involving hemocyanin, notably in the tarantula *Eurypelma californicum*. A number of co-workers continued this field of study for many years: Renate Loewe, Dieter Angersbach, Walter Schartau, Hans-Jürgen Schneider, Anette Savel-Niemann, Rüdiger Paul, Renate Voit and the two authors.

Bernt Linzen and his group deduced that tarantula hemocyanin is composed of seven distinct subunits, which they then succeeded in isolating (5). Prof. Dr. Linzen was particularly successful in collaborating with other research groups, often with potential competitors. He worked with 16 authors from various laboratories to publish a comparison of all known hemocyanin sequences with the three-dimensional model of spiny lobster hemocyanin that had been determined by Wim Hol's group in Groningen (6). In collaboration with Prof. Lontie in Leuven, Linzen was the first to address the question of a possible relationship between arthropod and molluscan hemocyanins by sequencing the functional domain of the hemocyanin of a molluscan species (7). This work also confirmed a presumed structural relationship between hemocyanin and tyrosinase. Collaborations with Ernst van Bruggen's group in Groningen resulted in the elucidation of the quaternary structure of tarantula hemocyanin (8). Using his knowledge of such molecular details, Bernt Linzen began to study how the manifestation of the native oxygen-binding properties of tarantula hemocyanin is related to its quaternary structure (9). His ultimate goal was to understand everything about spider respiratory physiology, from the regulation of hemocyanin gene expression to the adaptation of the whole animal to ecophysiological constraints. Thus, his central studies on the structure and function of hemocyanin were flanked by two additional approaches: on one side was the structural analysis of hemocyanin genes (10), and on the other, the physiological analysis of book lung function, blood circulation, oxygen supply and energy metabolism (11–13).

Although Bernt Linzen was a continual stimulating and coordinating force in the research activities of his group, he always supported the attempts of his co-workers to establish independent projects such as the "nesting" model of allosteric interaction (14), the comparative analysis of a large variety of hemocyanins (15) and the identification of the site of hemocyanin biosynthesis in spiders (16).

Despite his many research activities and heavy teaching load, he periodically took additional responsibilities as Head of the Institute, Dean of the Faculty, President of the German Zoological Society and Managing Editor of the *Journal of Comparative Physiology B*. In early 1988, Bernt Linzen officially retired from these duties to devote himself completely to research. He planned to write a book on invertebrate respiratory physiology and to study the genes of molluscan hemocyanins. Although his life was too short to realize these plans, his scientific enthusiasm, deep insight, fine humor and friendly advice will always be remembered.

References

1. Linzen, B. (1954) *Die Vogelwarte* 17: 43–49.
2. Linzen, B. (1974) *Adv. Insect. Physiol.* 10: 117–246.
3. Linzen, B. (1965) *Biochim. Biophys. Acta* 103: 588–600.

4. Linzen, B. and Lowe, R. (1969) *Z. Vergl. Physiol.* **55**: 27–34.
5. Schneider, H.-J., Markl, J. Schartau, W. and Linzen, B. (1977) *Hoppe-Seyler's Z. Physiol. Chem.* **358**: 1382–1409.
6. Linzen, B., Soeter, N.M., Riggs, A.F., Schneider, H.-J., Schartau, W., Moore, M.D., Yokota, E., Behrens, P.Q., Nakashima, H., Takagi, T., Nemoto, T., Vereijken, J.M., Bak, H.J., Beintema, J.J., Volbeda, A., Gaykema, W.P.J. and Hol, W.G.J. (1985) *Science* **229**: 519–524.
7. Drexel, R., Siegmund, S., Schneider, H.-J., Linzen, B., Gielens, C., Préaux, G., Lontie, R., Kellermann, J. and Lottspeich, F. (1987) *Hoppe-Seyler's Z. Physiol. Chem* **368**: 617–635.
8. Markl, J., Kempter, B., Linzen, B., Bijlholt, M.M.C. and Van Bruggen, E.F.J. (1981) *Hoppe-Seyler's Z. Physiol. Chem.* **362**: 1631–1641.
9. Savel-Niemann, A., Markl, J. and Linzen, B. (1988) *J. Mol. Biol.* **204**: 385–395.
10. Voll, W. and Voit, R. (1990) *Proc. Natl. Acad. Sci. U.S.A.* **87**: 5312–5316.
11. Angersbach, D. (1978) *J. Comp. Physiol.* **123**: 113–125.
12. Paul, R., Fincke, T. and Linzen, B. (1987) *J. Comp. Physiol.* **157**: 209–217.
13. Paul, R., Fincke, T. and Linzen, B. (1989) *J. Comp. Physiol.* **159**: 409–418.
14. Robert, C.H., Decker, H., Richey, B., Gill, S.J. and Wyman, J. (1987) *Proc. Natl. Acad. Sci. U.S.A.* **84**: 1891–1895.
15. Markl, J. (1986) *Biol. Bull.* **171**: 90–115.
16. Kempter, B. (1983) *Naturwissenschaften* **70**: 255–256.

In Memoriam:

An Homage to Robert C. Terwilliger

Joseph Bonaventura

Marine Biomedical Center, Duke University Marine Laboratory,
Pivers Island, Beaufort, NC 28516, USA

> *We shall not cease from exploration*
> *And the end of all our exploring*
> *Will be to arrive where we started*
> *And know the place for the first time.*
> *Through the unknown, remembered gate*
> *When the last of earth left to discover*
> *Is that which was the beginning;*
> *At the source of the longest river*
> *The voice of the hidden waterfall*
> *And the children in the apple tree*
> *Not known, because not looked for*
> *But heard, half-heard, in the stillness*
> *Between two waves of the sea.*
> *T.S. Eliot*

Bob Terwilliger. I can't write it like that. It's Bob and Nora. Of course, everyone knows they are two people; but, as with many scientific couples, they are often thought of as a unit. So it is with Bob and Nora, and the indefatigable style they developed together is continued by Nora in the exploration of the wonders of oxygen-carrying proteins.

In 1974, after a lecture at the C. Ladd Prosser "retirement" symposium, I was approached by a scientist from Charleston, Oregon, who politely asked if I would spend a few minutes talking with him. We found a nice spot outside in the warm December sun of Tucson, and for the next two hours Bob Terwilliger and I carried on a fascinating (at least to me) conversation about a wide variety of topics. First and foremost, he mentioned Nora and the rest of his family, how much he cared for them and how important it was for them all to take time to be together, to read books aloud and, in general, to enjoy the spirit of living. That conversation led to an invitation to visit and then an enjoyable nine-month stay by the Terwilliger family in Beaufort.

Bob's interest in oxygen-binding proteins began early in his post-doctoral graduate years at Boston University. He and his collaborator, Ken Read, began an extensive investigation of invertebrate myoglobins and hemoglobins, picking up, to a certain extent, where Clyde Manwell had left off. Bob turned the technique of Sephadex chromatography into a powerful analytical and preparative tool in the characterization of both cellular and extracellular hemoglobins. His first paper on the subject (1) set the stage for many future discoveries. His collaboration with Nora (other than as husband and father) began with studies on *Pista pacifica* (2). The studies continued with a focus on invertebrate hemoglobins; not, however, to the exclusion of vertebrate hemoglobins and the hemocyanins and hemerythrins! Some of the highlight discoveries made by Bob and Nora include the following.

Pearls-on-a-String Hemoglobins. The hemoglobins from Planorbid snails are assembled from multidomain subunits, each domain being globin-like and of *ca.* 15 kDa (3, 4). This characteristic of molluscan hemoglobins is reminiscent of the multidomain nature of molluscan hemocyanins. These hemoglobins are the largest ones that have ever been found, having a molecular weight in the range of 8–12,000 kDa.

Vacation Hemoglobins. While on vacation in Baha, California, the ever-inquisitive Terwilligers walked and lounged on the beaches, occasionally cracking open an unsuspecting invertebrate. This vacation activity led to the discovery of the marvelous hemoglobin system of the clam *Cardita affinis* (5). The Terwilligers characterized this system functionally as well as by electron microscopy. The system is now under extensive molecular biological investigation in the laboratory of Austen Riggs.

Rain Dance Hemoglobin. In the deserts and high plains of the Pacific Northwest, rain occurs only occasionally, frequently in the springtime. Hence, vernal pools appear at the slightest downpour, and they literally explode with an incredible variety of life forms, all of which are required to carry out their full life cycles in brief time intervals. The tadpole shrimp, *Lepidurus bilobatus*, is one of these wonderful organisms that quickly appears in vernal pools. The Terwilligers' study of this creature led to finding a most bizarrely shaped hemoglobin having a two-domain, 33 kDa subunit structure (6).

Garlic Worm Hemoglobin. These organisms, when dredged up by shrimpers who called them "passion plugs," smelled like a cross between mercaptoethanol and garlic. These interesting creatures were found to have three kinds of oxygen-binding protein: polymeric vascular hemoglobin, a dimeric coelomic hemoglobin and a dimeric body-wall myoglobin (7). The organisms also have an apt name: *Travisia foetida*!

Bi-Color Hemoglobin. Serpulids from Friday Harbor, *Serpula vermicularis*, have a vascular globin containing both heme and chloroheme. Accordingly, the vascular fluid has a most peculiar color. In oxygen-binding experiments, the spectra of the two forms develop independently and arise from the two chromophores that appear to be a part of the same protein (8).

Barnacle Bill Hemoglobins. The chances are good that essentially all investigators interested in invertebrate oxygen-binding proteins had looked at the blood and tissues of barnacles in search of an oxygen-binding pigment. These look-sees, negative with respect to finding any major pigment, were usually not published. Getting hemoglobin from a barnacle was probably thought to be something like getting blood from a turnip! There is a most unusual barnacle, however, in which the Terwilligers did find hemoglobin. This barnacle, *Briarosaccus callosus*, a parasite that lives in crabs, was found to have an extremely interesting hemoglobin system. It is unlike other arthropod hemoglobins in that it has a single domain subunit of 15 kDa and an aggregate size of around 8,000 kDa (9).

Bob was not only a devoted researcher; his academic life was also filled with inspired teaching and mentorship of both graduate and undergraduate students. Both on the Eugene campus and on the Oregon coast, Bob taught students about the fascinating invertebrates that so intrigued him. He showered the students with enthusiasm and, in an allosteric way, their positive reactions led to positive cooperativity. I find it difficult to imagine a professor more devoted to his students and their well-being than was Bob. I can easily imagine him jumping up on a desk and shouting to the students *"carpe diem"* in the way Robin Williams did in the film *Dead Poet's Society*. They were his students and were given the appropriate attention and rigorous training that students need. In addition, Bob and his students often became life-long friends. Bob's interest and concern about teaching and research at marine stations are beautifully illustrated in his 1988 *American Zoologist* article which describes the dynamic interplay that makes first-class science so exciting (10).

Best of all, I'll never forget his whimsical smile, the sparkle of his eyes, the high-pitched oscillating whistle (sounding like the approach of an alien spacecraft), and the ready puns and jokes that continuously came from Bob's mouth. He was a delight to be around and lived with great joy and appreciation for the wonders of nature. We all miss him greatly.

References

1. Terwilliger, R.C. and Read, K.R.H. (1969) *Comp. Biochem. Physiol.* **29**: 551–560.
2. Terwilliger, R.C., Terwilliger, N.B. and Roxby, R. (1975) *Comp. Biochem. Physiol.* **50**B: 225–232.
3. Terwilliger, R.C. and Terwilliger, N.B. (1977) *Comp. Biochem Physiol.* **58**B: 283–289.
4. Terwilliger, R.C., Terwilliger, N.B., Bonaventura, C. and Bonaventura, J. (1977) *Biochim. Biophys. Acta* **494**: 416–425.
5. Terwilliger, R.C., Terwilliger, N.B. and Schabtach, E. (1978) *Comp. Biochem. Physiol.* **59**B: 9–14.
6. Dangott, L.J. and Terwilliger, R.C. (1979) *Biochim. Biophys. Acta* **579**: 452–461.
7. Terwilliger, R.C., Garlick, R.L. and Terwilliger, N.B. (1980) *Comp. Biochem. Physiol.* **66**B: 261–266.
8. Terwilliger, R.C. (1978) *Comp. Biochem. Physiol.* **61**B: 463–469.
9. Terwilliger, R.C., Terwilliger, N.B. and Schabtach, E. (1986) In *Invertebrate Oxygen Carriers*, ed. B. Linzen, 125–127. Berlin: Springer-Verlag.
10. Terwilliger, R.C. (1988) *Am. Zool.* **28**: 27–34.

Contents

Part II.   Structure and Function

Part III.   Amino Acid and cDNA Sequences

Part IV.   Gene Structure and Physiological Role

Part I
Quarternary and
Three-Dimensional Structure

1

The Architecture of 4x6-meric Arachnid Hemocyanin

Nicolas Boisset, Jean-Christophe Taveau and Jean Lamy

François Rabelais University and CNRS URA 1334, 2 bis Boulevard Tonnellé, 37042 Tours, France

Introduction

Arthropod Hcs have highly heterogeneous M_ws ranging from approximately 450 to 3,600 kDa. For example, 1x6-mers of 75 kDa polypeptide chains have been found in crustaceans, 2x6-mers in crustaceans and spiders, 4x6-mers in most chelicerate groups and in the thalassinid shrimps, 6x6-mers in centipedes, and 8x6-mers in merostoms.

The elucidation of arthropod Hc architecture follows the development of electron microscopy. The first micrographs of an arthropod Hc, the horseshoe crab *Limulus polyphemus*, were taken in 1942 by Stanley and Anderson (1), though much had been done before on the characterization of the Hc function (2, 3). Stanley and Anderson only observed that the molecule was essentially spherical. Twenty years later, Levin (4) published impressive micrographs of negatively stained *Limulus* Hc showing most of the views of the 8x6-mer and the side view of the 4x6-mer, but they were completely misinterpreted and the structure remained unknown.

Great progress in the understanding of the structure of arthropod Hcs was accomplished in 1966. Fernandez-Moran *et al.* (5) used electron microscopy to identify the ring view of the 8x6-mer, and Wibo (6, 7) described the side view of the arachnid 4x6-mer and the cross and bow tie views of the 8x6-mer. He also mentioned the existence of a pentagonal view in *Limulus* Hc. In addition, he understood that *Limulus* Hc was an 8x6-mer resulting from the juxtaposition of two 4x6-mers of the cheliceratan type. Finally, he correctly interpreted the structural difference between the crustacean and the cheliceratan 2x6-mers, and suggested that all arthropod Hcs might be composed of a universal hexameric building block. Unfortunately, this pioneer work was only published as a thesis (6) with a restricted circulation (the nomenclature of the various EM views is described in detail in reference 8).

In the 1970s, after the original work of Sullivan *et al.* (9) on the fractionation of dissociated *Limulus* Hc into five chromatographic zones, the subunit heterogeneity of the main cheliceratan Hcs was established by physical and immunological methods (10–12).

The publication by Van Heel and Frank (13) of an image processing method based on the multifactorial method of correspondence analysis considerably improved the results of electron microscopy, and facilitated the recognition of different faces of a molecule from slightly different views. The two pentagonal views of the 8x6-mer (8), the 45° view of the 4x6-mer (14) and the opposite faces of the 2x6-mer (15) were identified by this method. The same year, the intramolecular location of the 24 subunits of *Androctonus australis* Hc was resolved by immunoelectron microscopy (16), and comparable results were obtained thereafter on the Hcs of *Eurypelma californicum* (17) and *Limulus* (18).

Between 1983 and 1985, it was debated whether the dodecameric building block of the chelicerate Hcs was of the left or right isomeric type, as originally suggested by Van Heel *et al.* (14). Other landmarks in understanding arthropod Hc structure were the determination of the first complete amino acid sequence by Schneider *et al.* (19), promptly followed by several others, and the crystallographic determination of the hexameric Hc of *Panulirus interruptus* at 3.2 Å resolution by Gaykema *et al.* (20) (for an up-to-date description see reference 21).

The model built by Lamy *et al.* in 1985 (22) was in agreement with these disparate pieces of information. However, as it was based mainly on indirect arguments collected from different species, often evolutionarily very distant (such as *Panulirus* and *Limulus*), a more direct verification was required. The publication in 1988 of the single exposure, random conical tilt series method of three dimensional reconstruction by Radermacher (23) provided the opportunity for such direct verification. The present paper summarizes the main results of the first 3D reconstruction of a 4x6-meric Hc from randomly oriented EM views (24) and their implication to 4x6-mer architecture.

Results and Discussion

The Disposition of the Subunits in the Model. The model (22) was based on four main hypotheses: 1) the 4x6-mer is composed of four hexameric units of the *Panulirus* type; 2) the dimeric subunit ($Aa3C$–$5B$) is involved in both interhexamer and the interdodecamer contacts, as suggested by the crucial role of this subunit in reassembly experiments; 3) the right isomeric type of dodecamer is preferred to the left because it better explains the stain exclusion pattern of the

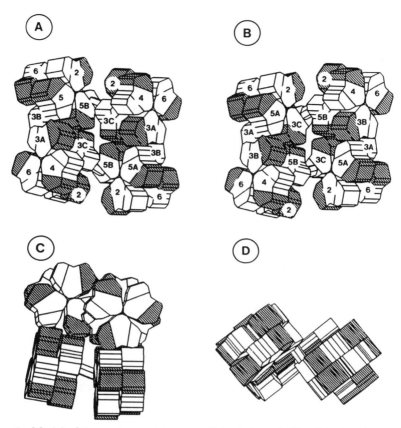

Figure 1. Model of the quaternary structure of the 4x6-meric Hc of *Androctonus*: (A) top view flop; (B) top view flip; (C) side view; (D) 45° view. The hatched zone of each subunit roughly corresponds to that of domain 1.

45° view (the left portion of the molecule is more deeply embedded in the stain than the right one, when the hexamers with rosette contours are located in the top of the picture); 4) the respective positions of the two dodecamers in the 4x6-mer were postulated to explain the rocking effect observed in the top view (the rocking axis passing through the upper left and lower right hexamers when the interdodecamer cleft is oriented top to bottom in the picture). The four main views of the model as well as the topological disposition of the subunits deduced by immunoelectron microscopy are shown in Figure 1.

The 3D Reconstruction of Androctonus *Hc*. Three-dimensional reconstructions of native 4x6-meric Hc were carried out using the single-exposure, random conical tilt series method (23). The molecules were first negatively stained using the double carbon layer technique (25). Then, each field was successively exposed in the electron microscope at 50° and 0° tilt angles. After digitization of the negatives, the images were selected, windowed and sorted into three groups containing the top, side and 45° views, respectively, and each group was studied separately. Homogeneous image subsets were obtained by submitting the aligned untilted-specimen images to correspondence analysis (13) and automatic classification (26), then the reconstruction was performed on the corresponding tilted images (for a detailed description of the materials and methods, the reader is directed to reference 24). Finally, the consistency of the structural features produced by the reconstruction was studied by comparing the 3D volumes computed from the top, side and 45° views with those of the model.

Solid Body Surface Representation. Figure 2A depicts the solid body surface representation resulting from the reconstruction of the top view flip (in the flip view the right dodecamer is shifted up). In this orientation, the 3D volume is not very informative. However, if we rotate it around a horizontal axis in the plane of the picture (Figure 2B), it appears that the two bridges are perpendicularly oriented exactly as predicted by the 1985 model (Figure 2C).

However, a flattening of the structure is visible on the solid-body surface representation of Figure 2. Similar compressions were observed in the 3D reconstruction volumes obtained from the side and 45° views. The double carbon film method of negative staining seems to be responsible for this plastic deformation. Indeed, the flattening is constantly observed in a direction perpendicular to the plane of the support grid.

Sections through the 3D Volumes. Figure 3 shows sections cut at the same level in the 3D volume (Figures 3B1–4), and in the 1985 model (Figures 3C1– 4) reconstructed from the 45° views. Figure 3A1 portrays the level of the four sections, and Figure 3A2 the surface of the model in its 45° orientation for

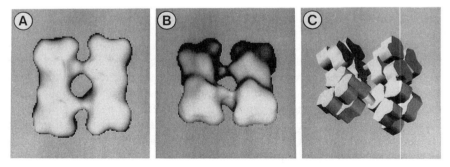

Figure 2.   Visualization of the interdodecameric bridges of *Androctonus* Hc reconstructed by the random conical tilt series method: (A) solid-body surface representation of the 3D reconstruction from the top view flip (original orientation), resolution 1/49 Å^{-1}; (B) after a 70° rotation around an horizontal axis in the image plane; (C) model of Lamy *et al.* (22), in the same orientation as in B.

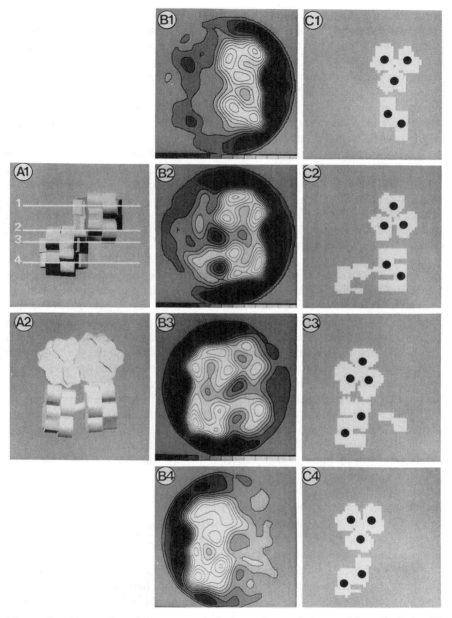

Figure 3. Comparison between equivalent sections of the model and of the 3D reconstruction volume oriented as in the 45° view: (*A*1) level of four sections; (*A*2) view of the model in an orientation directly comparable to the sections of rows *B* and *C*; (*B*1–*B*4) sections *N*°1–4 from the reconstruction volume of the 45° views (resolution 1/22 Å^{-1}, thickness of the voxel = 5.5 Å); (*C*1–*C*4) sections *N*°1–4 from the model.

comparison. When passing from section 1 (Figures 3*B*1 and 3) to section 4 (Figures 3*B*4 and 3*C*4), the disposition of the main nuclei of stain exclusion changes. In section 1, the three main nuclei of the upper hexamer form a triangle pointing down exactly as do the corresponding three subunits of the model. In the lower hexameric area, the main nucleus on the left side is shifted up farther than is the one on the right side. This disposition, in perfect agreement with a dodecamer of the right isomeric type, excludes the possibility of a left enantiomer (according to the nomenclature in (22)). The same topology appears in the model. In the other three sections, there is also an excellent agreement between the stain exclusion patterns of the 3D volume and the model. For example, the positions of the lower bridge in Figures 3*B*2–3 and 3*C*2–3 are quite similar (the upper bridge, perpendicular to the plane of the picture, is not visible in this orientation). It is also remarkable that sections 1 and 4, on the one hand, and sections 2 and 3 on the other, appear to be mirror-inverted both in the model and in the 3D volume. This pattern unequivocally demonstrates that the two dodecamers are facing each other.

Unresolved Problems. The 3D reconstruction described above definitely clarifies the relative positions of the hexamers within the dodecameric half-molecule, as well as the positions of the half-molecules in the 4x6-mer. However, the resolution of the method is insufficient to show the contacts between the subunits in the interhexamer and interdodecamer areas. The number of contact points between the hexamer of each dodecamer is still unknown. There is at least one, and at most four, intersubunit contacts between the two hexamers of each dodecamer, and it is very likely that only two interdodecamer contacts correspond to the two bridges. At present, the only clear information is that the *Aa*3*C* and *Aa*5*B* subunits are present in dissociated Hc as a stable heterodimer and that they are capable of self-association to produce two-dimensional crystals. As their topological locations in the center of the molecule are known, it is very likely that these subunits are involved both in the interhexamer and interdodecamer contacts. In this case, the cohesion of the 4x6-mer would very likely result from a pair of *Aa*3*C*–*Aa*5*B* interactions. With respect to other interhexamer contacts, the only available arguments are that in reassembly experiments, the *Aa*3*B* and *Aa*5*A* subunits are required to obtain a 4x6-mer. These subunits are located in the area between the hexamers, and could also be involved in the cohesion of the 2x6-mer.

Another problem linked to the position of the dimeric subunit is the nature of the intersubunit contacts involved in the free, stable heterodimer. Actually, as they are positioned in the center of the 4x6-mer, there must be two types of contacts between the *Aa*3*C* and *Aa*5*B* subunits. As shown in Figure 1, one of these contacts involves domain 3 (*D*3) of the two subunits; and the other, domain 1 (*D*1). In the model, the *D*1–*D*1 contact is an interhexamer contact and the *D*3–*D*3 contact is an interdodecamer contact. At present, the information needed to decide whether the free stable heterodimer is a *D*1–*D*1 or a *D*3–*D*3 heterodimer is lacking.

Understanding the structural roles of the various subunits raises important questions. For example, it is known that certain subunits are required to obtain given levels of aggregation in reassembly experiments. Thus, the dimeric subunit (*Aa3C–Aa5B*) is needed to obtain a 4x6-mer; but we do not know the reason for this requirement in terms of intersubunit contacts. Similarly, the fact that the Hill coefficient reaches a value of 9.25 at pH 7.8 (27) suggests that the allosteric unit of the 4x6-mer is at least the dodecamer and that the interhexamer contacts are involved in the transmission of the information. However, information about the role of the interdodecamer contacts in cooperativity is lacking.

Acknowledgements

We are indebted to Prof. Joachim Frank (State University of New York) and his collaborators, and especially to Drs. Michael Radermacher and Terry Wagenknecht for their contribution to a long and fruitful collaborative work. In particular, Figures 2 and 3 are adapted from reference 24.

References

1. Stanley, W. and Anderson, T. (1942) *J. Biol. Chem.* **146**: 25–33.
2. Svedberg, T. (1933) *J. Biol. Chem.* **103**: 311–325.
3. Erickson-Quensel, I. and Svedberg, T. (1936) *Biol. Bull.* **71**: 498–547.
4. Levin, Ö. (1963) *Arkiv Kemi* **21**: 29-35.
5. Fernandez-Moran, H., Van Bruggen, E.F.J. and Ohtsuki, M. (1966) *J. Mol. Biol.* **16**: 191–207.
6. Wibo, M. (1966) Ph.D. Dissertation. University of Louvain.
7. Wibo, M., Baudhuin, P. and Berthet, J. (1966) *Arch. Int. Physiol. Biochem.* **74**: 945–947.
8. Lamy, J., Sizaret, P.Y., Frank, J., Verschoor, A., Feldman R.J. and Bonaventura, J. (1982) *Biochemistry* **21**: 6825–6833.
9. Sullivan, B., Bonaventura, J. and Bonaventura, C. (1974) *Proc. Natl. Acad. Sci. U.S.A.* **71**: 2558–2562.
10. Lamy, J., Lamy, J. and Weill, J. (1979) *Arch. Biochem. Biophys.* **193**: 140–149.
11. Lamy, J., Lamy, J., Weill, J., Bonaventura, J., Bonaventura, C. and Brenowitz, M. (1979) *Arch. Biochem. Biophys.* **196**: 324–329.
12. Markl, J., Savel-Niemann, A. and Linzen, B. (1981) *Z. Physiol. Chem.* **362**: 1255–1262.
13. Van Heel, M. and Frank, J. (1981) *Ultramicroscopy* **6**: 187–194.
14. Van Heel, M., Keegstra, W., Schutter, W. and Van Bruggen, E. F. J. (1983) *Life Chem. Rep. Suppl. Ser.* **1**: 69–73.
15. Van Heel, M. and Keegstra, W. (1981) *Ultramicroscopy* **7**: 113–130.
16. Lamy, J., Bijlholt, M.M.C., Sizaret, P.Y., Lamy, J.N. and Van Bruggen, E.F.J. (1981) *Biochemistry* **20**: 1849–1856.

17. Markl, J., Kempter, B., Linzen, B., Bijlholt, M.M.C. and Van Bruggen, E.F.J. (1981) *Z. Physiol. Chem.* **362**: 1631–1641.

18. Lamy, J., Lamy, J., Sizaret, P.Y., Billiald, P., Jolles, P., Jolles, J., Feldman, R.J. and Bonaventura, J. (1983) *Biochemistry* **22**: 5573–5583.

19. Schneider, H.J., Drexel, R., Feldmaier, G., Linzen, B., Lottspeich, F. and Henschen, A. (1983) *Z. Physiol. Chem.* **364**: 1357–1381.

20. Gaykema, W.P.J., Hol, W.G.J., Vereijken, J.M., Soeter, N.M., Bak, H.J. and Beintema, J.J. (1984) *Nature* **309**: 23–29.

21. Volbeda, A. and Hol, W.G.J. (1989) *J. Mol. Biol.* **206**: 531–546.

22. Lamy, J., Lamy, J., Billiald, P., Sizaret, P.Y., Cave, G., Frank, J. and Motta, G. (1985) *Biochemistry* **24**: 5532–5542.

23. Radermacher, M. (1988) *J. Elec. Microsc. Tech.* **9**: 359–394.

24. Boisset, N., Taveau, J.C., Lamy, J., Wagenknecht, T., Radermacher, M. and Frank, J. (1990) *J. Mol. Biol.* **216**: 743–760.

25. Tischendorf, G.W., Zeichardt, H. and Stoffler, G. (1974) *Mol. Gen. Genet.* **134**: 187–208.

26. Lebart, L., Morineau, A. and Warwick, K.M. (1984) *Multivariate Descriptive Statistical Analysis*, 117–143. New York: John Wiley and Sons.

27. Lamy, J., Lamy, J., Bonaventura, J. and Bonaventura, C. (1980) *Biochemistry* **19**: 3033–3039.

2
Static and Kinetic Studies on the Dissociation of *Limulus polyphemus* Hemocyanin with Solution X-Ray Scattering

Kazumoto Kimura,[a] Yoshihiko Igarashi,[b] Akihiko Kajita,[b] Zhi-Xin Wang,[c] Hirotsugu Tsuruta,[d] Yoshiyuki Amemiya[e] and Hiroshi Kihara[f]

[a]Division of Medical Electronics, Dokkyo University School of Medicine, Tochigi 321-02, Japan
[b]Department of Biochemistry, Dokkyo University School of Medicine, Tochigi 321-02, Japan
[c]Laboratory of Molecular Enzymology, Institute of Biophysics, Academia Sinica, Beijing 100080, China
[d]Department of Materials Science, Faculty of Science, Hiroshima University, Hiroshima 730, Japan
[e]Photon Factory, National Laboratory for High Energy Physics, Tsukuba 305, Japan
[f]School of Nursing, Jichi Medical School, Tochigi 329-04, Japan

Introduction

Limulus polyphemus Hc (3,600 kDa) consists of 48 subunits which are assembled to build eight submultiples. The submultiple which comprises six subunits is also called the basic hexameric unit or the hexameric building block (1). Ultracentrifugal studies by Brenowitz *et al.* have shown that, in the absence of Ca^{2+}, the native Hc (48-mer) dissociates into 24-mers at physiological pH and into 12-mers at pH 5.0 (2). This fact is of special interest from either a structural or a functional point of view since native Hcs having similar sizes have been found among the arthropods, for example, in *Androctonus australis* (24-mer), *Astacus leptodactylus* (12-mer), *Panulirus interruptus* (6-mer) and so forth (1). Brenowitz *et al.* also used the stopped-flow light scattering method to study the kinetics of the dissociation of the Hc into smaller fragments. However, they obtained the pseudo-first order rate constants (k') for the initial 10% of the reaction, regardless of the increase in k' during the later part of the dissociation (2).

Although unique models for the quaternary structure of arthropod Hc occurring in nature have been proposed on the basis of molecular immunoelectron microscopic studies by Lamy *et al.* (1), the steric arrangement of their submultiples is still controversial. It seems pertinent to investigate the structure of native Hc and the structure of the dissociated fragments, in particular, by a different approach.

Solution x-ray scattering (SXS) provides a useful tool for gaining insight into the gross conformation of macromolecules (3-7). In fact, its usefulness has been proven in the analysis of the quaternary structure of Hcs from *A. leptodactylus* (4) and *Helix pomatia* (5), and of Hbs from *Helisoma trivalvis* (6) and *Tylorrhynchus heterochaetus* (7). Recently, Kihara *et al.* developed a stopped-flow x-ray scattering (SFXS) method and used it in studies of the dissociation of aspartate transcarbamylase and phosphorylase (8, 9).

The aim of the present study is to investigate the gross conformation of the Hc of *L. polyphemus* and its dissociated fragments by the SXS method. Furthermore, the kinetic process of the dissociation was also followed using the SFXS technique and the results obtained are discussed in relation to the dissociation mechanism.

Materials and Methods

The hemolymph of the horseshoe crab was provided by Dr. H. Sugita (Tsukuba University). The Hc was purified at 4°C according to the method of Brenowitz *et al.* (2). The 60 S Hc was obtained from the hemolymph in pellet form by ultracentrifugation at 225,000 x g for 60 minutes and was dissolved in 50 mM Tris-HCl, pH 7.4, containing 10 mM $CaCl_2$. Specimens of the whole molecule (48-mer), half molecule (24-mer), quarter molecule (12-mer) and monomer were prepared as described previously (10). From the preliminary experiments all of the specimens were thought to consist of a single ultracentrifugal component, despite minor contamination of the quarter molecule.

X-ray scattering experiments were performed at the beam line 15A1 of the Photon Factory, National Laboratory for High Energy Physics. The camera and detection systems were reported in detail elsewhere (11). The sample solution was irradiated with monochromatic x-rays (1.504 Å), and scattered x-ray intensities were recorded on a position-sensitive proportional counter (512 channels) with camera length at 2412 mm and channel width at 0.368 mm. The scattering data were normalized with respect to exposure time and protein concentration. Normalized data were also subjected to background subtraction. In the present study, x-ray scattering data are expressed in terms of $h = 4\pi$ (sin θ)/λ (λ: wavelength, 2θ: scattering angle), and the R_g value was obtained from the slope of the Guinier plots (12) prior to the model analysis. Details of the experimental technique and of the evaluation procedure are described elsewhere (8–11). Model analyses were achieved on the basis of the program developed by Furuno *et al.* (13).

A stopped-flow apparatus fabricated especially for x-ray scattering was used in the experiment. Characteristics of the stopped-flow apparatus were reported elsewhere (9). The dead time of the mixing was 10 ms. SFXS experiments were carried out by mixing the Hc sample (90 mg/ml Hc, Tris-HCl ($I = 0.15$), 10 mM $CaCl_2$) with the dissociation buffer (Tris-HCl ($I = 0.15$), 28.5 μM EDTA) at a ratio of 1 : 3.5. Minimal sample volume was 180 μl.

Results and Discussion

Equilibrium Solution X-Ray Scattering. SXS data of the Hc were obtained at pH 7.0, in the presence and absence of 20 mM EDTA. Figures 1 and 2 (full lines) show the final scattering curves of the whole Hc (48-mer) and half Hc (24-mer) molecules, obtained after extrapolation to zero concentration. Two minima at $h = 0.034$ Å^{-1} and $h = 0.078$ Å^{-1}, and three maxima at $h = 0$ Å^{-1}, $h = 0.045$ Å^{-1} and $h = 0.108$ Å^{-1} were observed in the Hc 48-mer scattering curve. The prominent subsidiary maxima of the Hc 48-mer (Figure 1) reflect the high symmetry of the macromolecule in solution. Figure 2 shows that the first minimum and the second maximum seen in the 48-mer (Figure 1) changed to a shoulder in the Hc 24-mer, while the second minimum and the third maximum remained unchanged, indicating some preservation of the symmetry of the protein. The SXS pattern of the quarter molecule (12-mer) was also recorded at pH 5.0 in the presence of 20 mM EDTA. The SXS pattern of the monomer fragment at pH 9.0 in the presence of 10 mM EDTA was also obtained at a protein concentration of 27 mg/ml.

The radius of gyration R_g, at zero concentration of the Hc and at zero angle intensity, I_o, was determined from Guinier plots of the normalized scattering curve at zero concentration of the protein in the region of $hR_g < 1$. The I_o and R_g thus obtained for each specimen were as follows: I_o (count/s)/R_g(Å) for Hc 48-mer, (4119 ± 79)/(110.67 ± 1.24); for Hc 24-mer, (2128 ± 39)/(91.32 ± 1.42); for Hc 12-mer, (1275 ± 34)/(77.32 ± 2.36). The cubes of the R_g values should be in ratios of 1 : 0.5 : 0.25, when the three molecular species of Hc are assumed to be spheres. However, the ratios obtained from the experimental curve were found to be 1 : 0.56 : 0.34. The significant discrepancy observed for

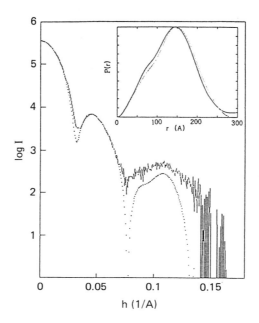

Figure 1. Comparison of the x-ray scattering pattern of the whole molecule (48-mer) of *L. polyphemus* Hc (----) with the calculated x-ray scattering pattern of model *B* (....). λ = 1.504 Å, exposure time: 100 s. Intensity is normalized at 100 nÅ. The inset shows the *p(r)* functions of the whole molecule calculated from the experimental results (----) and according to model *B* (....).

both R_g, and, especially, for the Hc 12-mer indicates a definite inadequacy of the above-mentioned spherical approximation and suggests the necessity of a more detailed model analysis. The R_g value of the monomer estimated from the Guinier plot was 36.5. The ratio of R_g^3 of the 48-mer to that of the monomer was 0.036. This confirms that the protein dissociates mainly into monomer at pH 9.0 in the presence of EDTA. The *p(r)* functions of the four molecular species were calculated and plotted in the insets of Figures 1 and 2. These *p(r)* functions will be discussed later.

Model Analyses. Model analyses were performed on the assumption that the native Hc (48-mer) and its dissociated fragments (24-mer and 12-mer) are composed of eight, four and two submultiples, respectively. The molecular assembly of the whole molecule (48-mer) was approximated by models A and B, both of which consist of eight submultiples arranged in two tetragonal layers, assuming that each submultiple is a sphere.

In model A, the four submultiples of the upper layer are positioned directly above the four submultiples of the lower layer (not shown). Figure 3 illustrates model B, in which the upper layer is rotated 45° relative to the lower layer. From the electron micrographic appearance of the native *L. polyphemus* Hc molecule reported by Lamy *et al.* (1), we measured the radius of a submultiple (hexameric building block) to be 60–70Å. Then, the scattering patterns for models A and B were calculated by changing the radius of submultiple R_o. Figure 1 shows the scattering pattern for model B using $R_o = 58$ Å, which provided the best fit to the experimental results. The R_g value was calculated from model B to be 106 Å, which is compatible with the experimental result (110.7 Å). With regard to the 24-mer, the scattering patterns were also

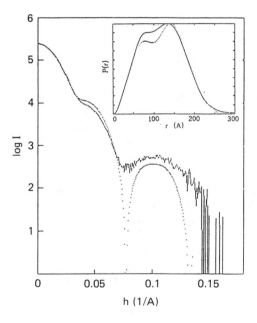

Figure 2. Comparison of the x-ray scattering pattern of the half molecule (24-mer) of *Limulus* Hc in the presence of 20 mM EDTA at pH 7 (----) with the calculated x-ray scattering pattern according to model D (····). Conditions are as those referred to in Figure 1, except for the pH and EDTA concentration. The inset shows the $p(r)$ function of the half molecule (24-mer) calculated from the experimental results (----), and according to model D (····).

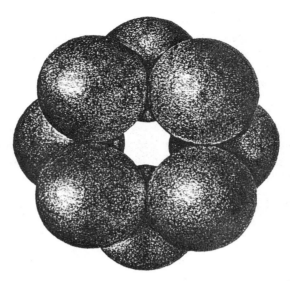

Figure 3. Model B for the whole molecule (48-mer) of *L. polyphemus* Hc.

calculated according to the following two models: 1) model C, a tetrahedron composed of four submultiples, and 2) model D, in which four submultiples are located in a layer as is presumed for the upper or lower layers of model B. The profile of the scattering curve from model D fits the experimental data better than that of model C, with respect to the shoulder around $h = 0.04$ Å^{-1}, in particular. It appears that the R_g value of model D (93.5 Å) is almost equal to the experimental value (91.3 Å).

The simulated $p(r)$ functions for models B and D together with the corresponding experimental data are shown in the insets of Figures 1 and 2. In the figures, the simulated curves agree well with the experimental profiles. By introducing models B and D one may postulate the following mechanism for the dissociation process: upon the removal of Ca^{2+} by EDTA, the native Hc 48-mer composed of two layers, having four submultiples each, dissociates into two 24-mers composed of a single layer like model D. Model fittings for the 12-mer (model E: pH 5.0 in the presence of EDTA) were made with the assumption that two submultiples ($R_o = 58$ Å) are in contact with each other. Although the shoulder of the experimental curve appeared to be shallower, the simulation curve reproduced the experimental data fairly well (not shown).

From models B, D and E, the ratios of R_g^3 values were calculated to be $1 : 0.67 : 0.33$, compared to the experimental values $1 : 0.56 : 0.34$. This fact would imply that four submultiples of the 24-mer are arranged in a layer of two 12-mers, and are possibly twisted somewhat from the plane. At pH 9 in the presence of 20 mM EDTA, R_g of the monomer fragment was experimentally obtained as 36.5 Å, which is smaller than the R_g of a submultiple (44.9 Å), indicating that the submultiples are dissociated into subunits consistent with the sedimentation analysis. The correlation between the calculated and experimental curves suggested that the Hc dissociated mainly into the constituent monomers, which had volumes one-sixth of the submultiple. The results obtained from sedimentation analyses on the native Hc and the dissociated fragments (2) confirm these conclusions. In addition, our model for the native Hc is compatible with that of Lamy *et al.* (1) insofar as the steric arrangement of the submultiple is concerned, and may explain their five electron micrographic views as a single entity observed from five different angles. The shape of the submultiple, however, is somewhat different.

Stopped-Flow X-Ray Scattering Study. Dissociation of the native Hc 48-mer to the 24-mer through the removal of calcium was monitored by SFXS at pH 7.0. Values of zero angle intensity, I_o, and of the radius of gyration, R_g, were estimated from the scattering patterns by utilizing Guinier plots of each time frame, and then were plotted as a function of time, as shown in Figure 4. An

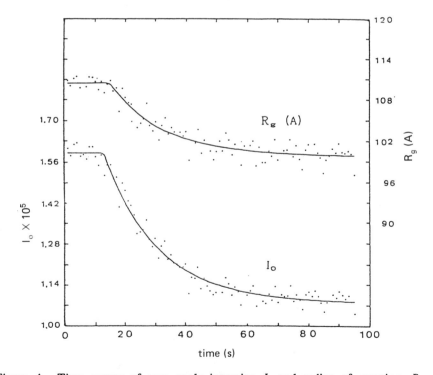

Figure 4. Time course of zero angle. intensity, I_o and radius of gyration, R_g, following the removal of Ca^{2+} by EDTA at pH 7. The experimental conditions are the same as those in the legend of Figure 1. Full lines denote the simulated curves according to equations (1) and (2), respectively.

initial lag followed by a single exponential decay was observed for both curves. A similar decay in the light scattering intensity due to Hc dissociation was observed by Brenowitz *et al.* (2). They estimated the pseudo-first order rate constant (k') using the first 10% of the reaction. However, considering the lag time, we attempted to calculate I_o and R_g from the following equations:

$$I_o = I_{of} + (I_{oi} - I_{of}) \, e^{-k_{app}(t-td)} \tag{1}$$

$$R_g = R_{gf} + (R_{gi} - R_{gf}) \, e^{-k_{app}(t-td)} \tag{2}$$

where I_{oi} and I_{of} denote initial ($t = td$) and final ($t = \infty$) values of zero angle intensities (I_o), respectively, R_{gi} and R_{gf} refer to initial ($t = td$) and final ($t = \infty$) values of the radius of gyration, respectively, and td represents lag time. As is clear from the figure, curves simulated from equations (1) and (2) for I_o and R_g, respectively, fit the data well. In addition, the decrease in these parameters reflects the dissociation of a high M_w protein into a low M_w fragment.

Figure 5 demonstrates the effect of uncomplexed EDTA concentration on the dissociation which represents the remaining portion of the chelator that is not combined with Ca^{2+} after mixing. The lag time in either the I_o or R_g curves decreased with increasing concentration of EDTA, while k_{app} increased.

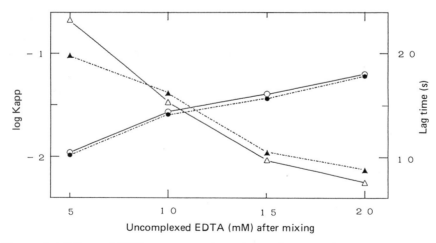

Figure 5. Effect of EDTA concentration of the 48-mer to 24-mer dissociation. Conditions are as referred to in Figure 1, with the exception of EDTA concentration. Uncomplexed EDTA concentration is the remaining portion of the chelator which is not complexed with calcium ions after mixing. Lag time of I_o: (Δ); R_g: (▲); k_{app} of I_o: (O); R_g: (●).

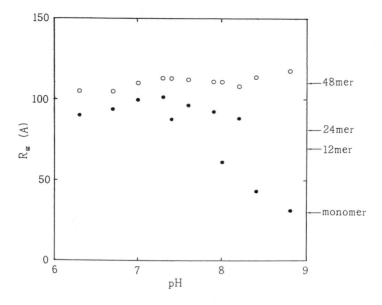

Figure 6. pH dependence of R_g at initial (O) and final (●) state. The assignments at the right-hand side are calculated values based on the assumption that the molecules are spheres in the native Hc and its constituent fragments. R_g of 48-mer was selected as a standard.

Dependence of lag time and k_{app} on the protein concentration (between 2.5 and 20 mg/ml) was not observed in the presence of 20 mM EDTA. The effects of pH on the lag time and on k_{app} were also evaluated. Lag time was decreased and k_{app} increased with decrease in pH. This indicates that dissociation occurs preferentially in the acidic pH range.

Figure 6 shows the pH dependence of the R_g value at the initial (lag) and final state. The R_g values in the initial state corresponded to the 48-mer independent of pH, whereas the values at the final state were found to be those of the 24-mer at a pH range between 6.3 and 7.9. Above pH 8, as may be seen in the same figure, the 48-mer dissociated into the monomer. These results are in accordance with the sedimentation analysis by Brenowitz *et al.* (2), confirming the change in molecular species through the dissociation process. The dissociation process was also investigated by use of the fluorescence stopped-flow technique: the results confirmed those obtained in the SFXS study.

References

1. Lamy, J.N., Lamy, J., Billiad, E., Sizaret, P.Y., Taveau, J.C., Boisset, N., Frank, J. and Motta, G. (1986) In *Invertebrate Oxygen Carriers*, ed. B. Linzen, 185–201. Berlin: Springer-Verlag.

2. Brenowitz, M., Bonaventura, C. and Bonaventura, J. (1984) *Biochemistry* **233**: 879–888.
3. Pilz, I. (1982) In *Small Angle X-Ray Scattering*, eds. O. Glatter and O. Kratky, 239–293. London: Academic Press.
4. Pilz, I., Goral, K., Hoylaerts, M., Witters, R. and Lontie, R. (1980) *Eur. J. Biochem.* **105**: 539–543.
5. Berger, J., Pilz, I., Witters, R. and Lontie, R. (1977) *Eur. J. Biochem.* **80**: 79–82.
6. Pilz, I., Schwarz, E., Tsfadia, Y. and Daniel, E. (1988) *Int. J. Biol. Macromol.* **10**: 353–355.
7. Pilz, I., Schwarz, E., Suzuki, T. and Gotoh, T. (1988) *Int. J. Biol. Macromol.* **10**: 356–360.
8. Nagamura, T., Kurita, K., Tokikura, E. and Kihara, H. (1985) *J. Biochem. Biophys. Methods* **11**: 277–286.
9. Tsuruta, H., Nagamura, T., Kimura, K., Igarashi, Y., Kajita, A., Wang, Z.-X., Wakabayashi, K., Amemiya, Y. and Kihara, H. (1989) *Rev. Sci. Instr.* **60**: 2356–2358.
10. Kimura, K., Igarashi, Y., Kajita, A., Wang, Z.-X., Tsuruta, H., Amemiya, Y. and Kihara, H. (1990) *Biophys. Chem.* **38**: 23–32.
11. Amemiya, Y., Wakabayashi, K., Hamanaka, T., Wakabayashi, T., Matsushita, T. and Hashizume, H. (1983) *Nuclear Instr. Methods* **208**: 471–477.
12. Guinier, A. and Fournet, G. (1955) *Small-Angle Scattering of X-Rays*. New York: John Wiley and Sons.
13. Furuno, T., Ikegami, A., Kihara, H., Yoshida, M. and Kagawa, Y. (1983) *J. Mol. Biol.* **170**: 137–153.

3

The Di-decameric Hemocyanin of the Atlantic Murex Snail, *Muricanthus fulvescens* (Sowerby)

Theodore T. Herskovits,[a] Curley Kieran,[a] Michelle D. Edwards[b] and Mary G. Hamilton[b]

[a]Department of Chemistry, Fordham University, Bronx, NY 10458, USA
[b]Division of Science and Mathematics, Fordham University, New York, NY 10023, USA

Introduction

The sedimentation studies of Svedberg and co-workers (1, 2) of the respiratory proteins of various species of vertebrates and invertebrates, including the Hcs of the gastropods of the phylum Mollusca, have suggested that these Hcs consist largely of a single type of particle characterized by a sedimentation coefficient of about 100 S and having a particle mass close to 9,000 kDa. Only a few Hcs, such as those of the channeled whelk, *Busycon canaliculatum*, and the giant garden slug, *Limax maximus*, were found to have minor amounts of components sedimenting with a higher rate close to 130 S. More recent investigations by our group and others (3) have shown that many gastropod Hcs consist of a mixture of higher aggregates with sedimentation coefficients of about 100 S, 130 S, 150 S, 170 S and even higher, 200 to 230 S, corresponding to di-, tri-, tetra-, penta-, and larger multi-decameric units (4, 5). Scanning transmission electron microscopy (STEM) measurements gave particle masses ranging from

Table 1. Light-scattering M_w data of *M. fulvescens* Hc.

Solvent	Protein conc. range (g.1^{-1})	$(\partial n/\partial c)_\mu$ (cm^{3}g^{-1})	M_w	B (1.mol.g^{-2})
pH 8.0, 0.1 M Tris, 0.05 M Mg^{2+}, 0.01 M Ca^{2+}	0.10–1.9	0.194	8.6 ± 0.6x10^6[a]	3.4 x 10^{-10}[a]
pH 8.5, 0.1 M Tris	0.23–1.1	0.194	5.86 x 10^5	1 x 10^{-8}
pH 10, 0.1 M Bicarb/NaOH	0.21–1.1	0.182	4.87 x 10^5	1.7 x 10^{-8}
pH 11, 0.1 M Bicarb/ NaOH, 0.01 M EDTA	0.20–1.1	0.186	4.05 x 10^5	–7 x 10^{-8}
3.0 M urea, pH 8.0, Tris	0.19–0.9	0.184	7.44 x 10^5	–1.6 x 10^{-7}
8.0 M urea, pH 8.0, Tris	0.26–1.2	0.148	4.17 x 10^5	2.5 x 10^{-7}
2.0 M GdmCl, pH8.0, Tris	0.24–1.0	0.174	7.36 x 10^5	1.7 x 10^{-7}
6.0 M GdmCl, pH 8.0, Tris	0.28–1.1	0.133	3.69 x 10^5	1.8 x 10^{-7}

[a]Based on $K_{diss}^{20,\ 10}$ = 1 x 10^{-8} M and a hard sphere value of the second virial coefficient, B = 3.4 x 10^{-10} 1.mol.g^{-2}.

8,900 for the di-decamer to 43,400 kDa for the largest, 10 unit aggregate of *Busycon contrarium* in the Melongenidae family (4, 5). Three additional groups of Hcs from the Naticidae or moon snail family (3, 6), the opisthobranch or sea-hare group, *Aplysia vaccaria* (unpublished data), *Aplysia limacina* (7), and *Dolabella auricularia* (8), and the recently discovered protobranch bivalves, *Yoldia limatula* and *Yoldia thraciaeformis* (9) show analogous ranges of aggregates, from essentially all di-decamers in *Calinaticina oldroydii* (6), *A. vaccaria*, and *Y. limatula* (9), to higher polymeric aggregates exemplified by the Hcs from *D. auricularia* (8) and *Y. thraciaeformis* (9).

The studies of Savel-Niemann *et al.* (10) on the polymeric Hc of the giant keyhole limpet, *Megathura crenulata* (10, 11) have suggested that the formation of aggregates higher than di-decamers may require the participation of two immunologically distinct subunits. The presence of at least two different Hc chains necessary for the reassociation of the giant Atlantic Murex snail *Muricanthus fulvescens* has been demonstrated (12). We have used light scattering and analytical ultracentrifugation to follow the dissociation behavior of this Hc. The predominantly di-decameric character of this Hc detailed in the present study suggests that factors other than, or in addition to, subunit or chain heterogeneity are required for the polymerization of molluscan Hcs beyond the di-decameric stage.

Results and Discussion

The Hc of *M. fulvescens* sediments as a single boundary at pH 8.0, 0.05 M Mg^{2+}, 0.01 M Ca^{2+}, with a sedimentation coefficient of 103 S. The M_w obtained by light scattering in the same solvent was found to be 8,600 ± 600 kDa (Table 1). Preliminary STEM mass measurements of one specimen gave a lower average particle mass of 8,150 ± 350 kDa. The summary of the light scattering M_w data in Table 1, obtained at high pH and in 8.0 M urea, gives M_w estimates of 405 to 487 kDa for the dissociated Hc subunits that are close to one-twentieth of the parent M_w.

All muricid Hcs which have been investigated to date have a predominantly di-decameric organization. Wood and Peacock (14) found that *Murex trunculus* Hc sediments at 102.7 S and has an M_w of 8,850 kDa based on the combined sedimentation and diffusion coefficients. We find that another Hc, isolated from the hemolymph of the lace murex *Chicoreus florifer dilectus* also sediments as 100 S di-decamers and has a light scattering M_w of 9,000 kDa (unpublished data). It seems that despite the presence of more than one type of chain in at least one of these Hcs (that of *M. fulvescens* (12)), these muricid Hcs form no higher aggregates of the type observed with the *M. crenulata* Hc.

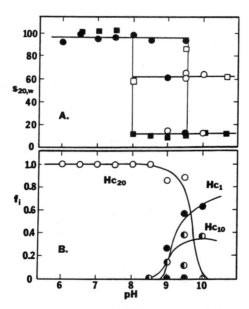

Figure 1. Effects of pH on the sedimentation coefficients ($S_{20^\circ,w}$) and the species distribution (f_i) of *M. fulvescens* Hc in the presence of 0.01 M Mg^{2+} (O,●) and the absence of divalent ions (□,■). Rotor speed, 24,000 rpm; Hc concentration, 0.2–0.3%. In panel *B*, open, half-filled, and filled circles represent di-decameric (Hc$_{20}$), decameric (Hc$_{10}$), and monomeric (Hc$_1$) components, respectively, examined in the presence of 0.01 M Mg^{2+}.

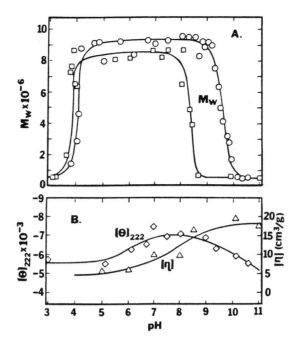

Figure 2. The effects of pH on M_w, the circular dichroism at 222 nm ($[\theta]_{222}$) and the intrinsic viscosity ($[\eta]$) of *M. fulvescens* Hc. Panel *A*: 0.01 M Mg^{2+}; no divalent ions; Hc concentration: 0.010%. Panel *B*: no divalent ions present.

The influence of pH, guanidine hydrochloride (GdmCl) and the different urea and salt concentrations on the subunit dissociation and denaturation of *M. fulvescens* Hc represent a second facet of our present investigation. The sedimentation data of Figure 1 reflect the changes in subunit organization as the pH is increased, with the dissociation of the 100 S di-decamers to decamers, with sedimentation coefficients of 60–65 S, and the smaller, predominantly monomeric, fragments of about 10 to 14 S in the pH region of 9–10 and above. The M_w changes shown in Figure 2 indicate a shift in the midpoints of the dissociation transition from about pH 8.5 in the absence of stabilizing divalent ions, to pH 9.5 in the presence of 0.01 M Mg^{2+}. The changes in intrinsic viscosity ($[\eta]$) and circular dichroism in the conformationally sensitive, far ultraviolet region ($[\theta]_{222}$) (also shown in this figure) are found to be relatively minor. This suggests that the folded regions of the Hc chains housing the functional Cu ions remain largely unaltered in the acidic and alkaline pH regions. The changes in the intrinsic viscosity and circular dichroism accompanying denaturation by GdmCl are much more pronounced, as shown by the data in Figure 3.

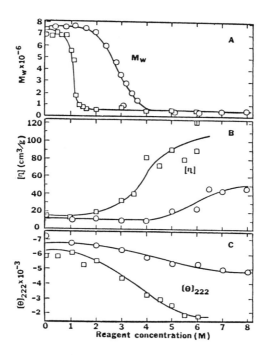

Figure 3. The influence of urea (O) and GdmCl (□) concentration on M_w, the intrinsic viscosity ($[\eta]$) and the circular dichroism at 222 nm ($[\theta]_{222}$) of *M. fulvescens* Hc. Panel *A*: pH 8.5, 0.1 Tris, 0.01 M Mg^{2+}, 0.01 M Ca^{2+}. Panels *B* and *C*: pH 8.5, 0.1 M Tris.

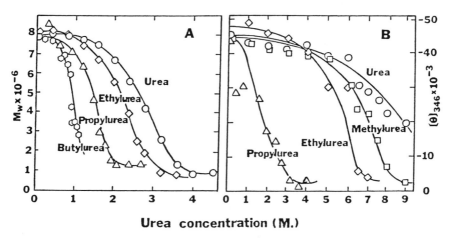

Figure 4. The effects of the urea series of reagents on M_w and the circular dichroism at 346 nm of *M. fulvescens* Hc. Panel *A*: pH 8.5, 0.1 M Tris, 0.01 M Mg^{2+}, 0.01 M Ca^{2+}; concentration = 0.010%. Panel *B*: pH 8.5, 0.1 M Tris.

Figure 4 shows the effects of the hydrophobic urea series of reagents on the subunit dissociation and the solution conformation of *M. fulvescens* Hc reflected by the changes in the light-scattering M_ws and the circular dichroism. The trend in both sets of measurements, reflected by the concentrations of reagent at the midpoints of the transitions, suggests hydrophobic stabilization of both the subunit contacts within the Hc decamers, discussed in earlier papers (3, 13, 15), and the folded functional units of the individual chains. It should be noted that M_w changes and the dissociation of the Hc subunits occur at much lower reagent concentrations than does the unfolding of the subunits.

Acknowledgements

Supported in part by Faculty Research Grants from Fordham University and Grant GM-08055 from the National Institutes of Health, U.S. Public Health Service.

References

1. Eriksson-Quensel, I.-B. and Svedberg, T. (1936) *Biol. Bull.* **71**: 498–547.
2. Svedberg, T. and Pedersen, K.O. (1940) *The Ultracentrifuge.* Oxford: Oxford University Press, 355–372.
3. Herskovits, T.T., Mazzella, L.J. and Villanueva, G.B. (1985) *Biochemistry* **24**: 3862–3870.
4. Herkovits, T.T., Blake, P.A., Gonzalez, J.A., Hamilton, M.G. and Wall, J.S. (1989) *Comp. Biochem. Physiol.* **94**B: 415–421.
5. Hamilton, M.G., Herskovits, T.T., Furcinitti, P.S. and Wall, J.S. (1989) *J. Ultrastruct. Mol. Struct. Res.* **102**: 221–228.
6. Hamilton, M.G., Herskovits, T.T. and Wall, J.S. (1990) *Proc. XIIth Int. Congr. for Electron Microscopy*, 810–811.
7. Ghiretti-Magaldi, A., Salvato, B., Tallandini, B. and Beltramini, M. (1979) *Comp. Biochem. Physiol.* **62**A: 579–584.
8. Makino, N. (1971) *J. Biochem.* **70**: 149–155.
9. Terwilliger, N.B., Terwilliger, R.C., Meyhofer, E. and Morse, M.P. (1988) *Comp. Biochem. Physiol.* **89**B: 189–195.
10. Savel-Niemann, A., Wegener-Strake, A. and Markl, J. (1991) In *Invertebrate Dioxygen Carriers*, ed. G. Préaux. Leuven: Leuven University Press. In press.
11. Senozan, N.M., Landrum, J., Bonaventura, J. and Bonaventura, C. In *Invertebrate Oxygen Binding Proteins: Structure, Active Site, and Function*, eds. J. Lamy and J. Lamy, 703–718. New York: Marcel Dekker.
12. Brouwer, M., Ryan, M., Bonaventura, J. and Bonaventura, C. (1978) *Biochemistry* **17**: 2810–2815.
13. Herskovits, T.T., Blake, P.A. and Hamilton, M.G. (1988) *Comp. Biochem. Physiol.* **90**B: 869–874.
14. Wood, E.J. and Peacocke, A.R. (1973) *Eur. J. Biochem.* **35**: 410–420.
15. Herskovits, T.T. (1988) *Comp. Biochem. Physiol.* **91**B: 597–611.

4
Scanning Transmission Electron Microscopy (STEM) Studies of Molluscan Hemocyanins

Mary G. Hamilton,[a] Theodore T. Herskovits[b] and Joseph S. Wall[c]

[a]Division of Science and Mathematics, Fordham University, New York, NY 10023, USA
[b]Department of Chemistry, Fordham University, Bronx, NY 10458, USA
[c]Biology Department, Brookhaven National Laboratory, Upton, NY 11973, USA

Introduction

In the past four years we have studied some 20 molluscan Hcs of the Polyplacaphoran, Gastropodan and Bivalvian classes by STEM and other physical methods (1–6). With STEM we have been able to measure the masses of individual particles in unstained, freeze-dried specimens, and to examine the arrangement of the cylindrical decameric units within various aggregates. Although the most intensively studied Hcs (*e.g., Helix pomatia*) are di-decameric, many of the gastropodan Hcs that we have studied are multi-decameric assemblies. The appearance of the di-decameric Hc may be represented schematically as a closed box composed of two decameric units facing one another, [x], where x is used to identify such units in the longer assemblies. In the bracket notation of Van Holde and Miller (7) for the decamer,], the closed end represents the collar end formed by the folding over of two of the eight functional units of each of the ten monomeric chains (8). The model for a di-decameric Hc is based primarily on Mellema and Klug's image analyses (9) of negatively-stained transmission electron microscopic (TEM) images of the isoionic (*pI*) polymers of *Kelletia kelletia* Hc that look like stacks of closed boxes: [x][x][x]. We have found that the isoionic type of regular stacking is not seen in multi-decameric Hcs. Rather, as others have also noted (10, 11), there is a polarity in the arrangement of the decameric units: usually only one "Mellema-Klug" di-decamer is present with decamers added in both directions, and with collar ends never facing one another.

Ghiretti-Magaldi *et al.* (10) suggested that the molluscan Hcs may be grouped into three subclasses on the basis of their "molecular dimensions." We would modify this classification in several ways. First, we would propose as subclass one and as the prototype molluscan Hc, as suggested by Terwilliger *et al.* (11), the 60 S Hc of the chitons since it has a single decameric unit with a mass similar to the decameric increment of higher assemblies (1) and appears to have only one collar end in electron micrographs (11). The 50 S Hcs of cephalopods may be considered to be a modified group of this subclass with a lower M_r and a different architecture; in other words, they have collars at both ends of a single decameric unit (12). As subclass two, 100 S di-decameric Hcs with collars at both ends that do not polymerize, we add several examples from the tulip and murex families (4, 13). We would place in subclass three the moon snail Hcs that are predominantly tri-decamers. Finally, as a fourth subclass, we add the Hcs that aggregate to varying degrees to form multi-decamers, such as the Hcs of the sea hare, *Dolabella auricularia* (14), the keyhole limpet, *Megathura crenulata* (15) and some members of the Melongenidae family (2).

Why are the molluscan Hcs found in such a diversity of sizes when they appear to consist of identical morphological units? What promotes or limits the association of the decamers and di-decamers? Mellema and Klug (9) suggested that there might be a cap at the collar end of the di-decameric Hc of *K. kelletia*. They described the polymers of *Busycon canaliculatum* formed at pH 4.4 as "closely-fused" and "suggesting that something has been lost from the collar and/or cap." In our detailed mass analysis of the native polymers of *Busycon contrarium* (1), we found a uniform mass increment of 4,400 kDa that was close to the mass of the non-polymerizing decameric Hcs of the chitons. This suggests that if there is a cap, its mass must be less than 100 kDa to be consistent with our mass measurements. In their extensive TEM studies of *H. pomatia* Hc that supported the Mellema-Klug model for the di-decameric Hc, Siezen and Van Bruggen (16) did not observe a cap. We have analyzed the radial mass distribution of the cylinders in end-views to examine the cap question more directly (17).

Studies of the dissociation of Hcs in response to changes in the ionic composition of the solvent (reviewed in (18)) have shown that the inter-decameric interactions are mainly polar and ionic. Divalent ions are always required except at low pH, where protons may apparently substitute. Since differences in the collars probably determine the state of aggregation, we are comparing a number of Hcs in various polymeric states by ultracentrifugal analysis and STEM so that we may more closely study the structure of the collars themselves. We are trying to prepare stacked polymers such as those formed by *K. kelletia* Hc (19) at pH 4.5. We are also examining the effects of proteolysis on the polymerizing capability of various Hcs in order to study another type of polymer: tubular aggregates such as those formed by the di-decameric Hc of *H. pomatia* when treated with trypsin (20).

Results and Discussion

The images obtained by STEM with two different specimen preparations are illustrated in Figure 1. Figure 1A portrays an unstained, freeze-dried specimen; and Figure 1B, an uranyl acetate-stained, air-dried specimen of the Hc of *Calinaticina oldroydii*. Less detail is observed in the freeze-dried samples. Although the "tiers" are not seen, internal cavities are visible. In fact, the images of freeze-dried specimens more closely resemble those obtained with palladium-shadowed specimens by Van Bruggen *et al.* (8) than do the negatively-stained ones. As noted by Van Bruggen *et al.* (8), the side-views of the di-decameric particle have a depression that represents not the channel through the cylinder, but the facing, non-collar ends of the two decameric units. Van Bruggen *et al.* (8) presented a model that incorporates the tiered appearance and depicts the collar formation. The collar pieces may be clearly seen in the negatively-stained image of *C. oldroydii* Hc shown in Figure 1B.

The advantage of using STEM in the study of Hcs is that the masses of identifiable individual particles (21) may be measured. This is best illustrated by comparing STEM results with the information available from ultracentrifugal analyses of Hcs. Table 1 presents the distribution of sedimenting components calculated from schlieren optical patterns for some representative Hcs. Though we may calculate the sedimentation coefficients of the well-defined peaks, the more rapidly sedimenting components of Hcs, like those of *Busycon carica,* are poorly resolved.

Table 2 gives the STEM analyses of the particle masses for five of these Hcs. For the monodisperse Hcs the STEM values are similar to those obtained by light scattering (1). We may use these data to correlate the sedimenting components with the particle distribution, recognizing that the electron microscopic analysis may not include all particles, whereas the ultracentrifugal pattern will.

Thus, while the sedimentation pattern may indicate that larger species are present, STEM clearly shows that they are discrete multi-decamers. In the best specimens (*e.g.,* those in Figures 1A, C and D), the decamers may not only be counted, but their arrangement within the assembly may be seen. The number of decamers in selected examples are labeled. Although negatively-stained images show these internal arrangements even more clearly, the mass measurements were required to validate the conclusions based on appearance alone.

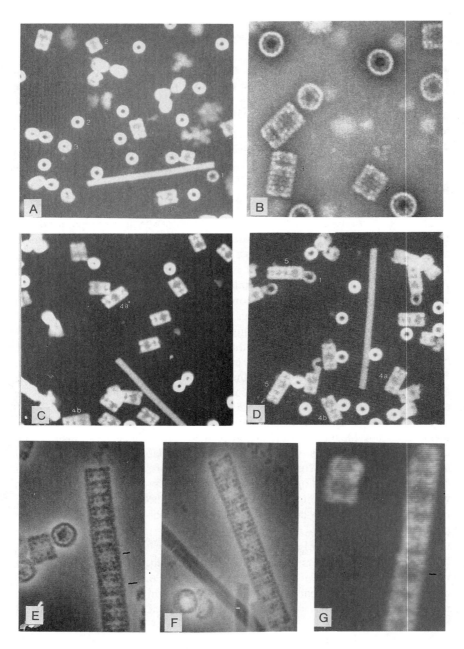

Figure 1. STEM images of various molluscan Hcs. The specimens also contained tobacco mosaic virus (diameter, 18 nm). (A) and (B) C. oldroydii Hc. (C) E. (=Lunatia) heros Hc. (D) B. carica Hc. (E) M. crenulata Hc. (F) N. proxima Hc. (G) B. carica Hc treated with trypsin (see text). (B), (E), (F), negatively-stained with uranyl acetate, air-dried. (A), (C), (D), (G), unstained, freeze-dried. Reproduced with permission from (22) for (C) and (D) and from (17) for (A); copyright by San Francisco Press, Inc. San Francisco, CA 94101, USA.

Table 1. Distribution of sedimenting components of molluscan Hcs.

Component	100 S	130 S	150 S	175 S	200 S	>200 S
P. gigantea	100%					
E. heros		95%	<- 5% ->			
B. canaliculatum	60%	30%	10%			
B. carica	13%	43%	26%	<- 18% ->		
B. contrarium	6%	28%	36%	16%	<- 14% ->	
M. crenulata	22%	10%	26%	13%	12%	17%
N. proxima	55%[a]	<-	35%	->		

[a]10% of 60 S was also present.

Table 2. STEM particle masses (MDa) for some gastropodan Hcs.[a]

Component[b]:	100 S	130 S	156 S	178 S	200 S	220 S	240 S	256 S	275 S
No. decamers:	2	3	4	5	6	7	8	9	10
P. gigantea	9.0								
B. canaliculatum	8.2	13.4	17.4						
E. heros		13.5							
B. carica	8.9	13.4	17.5	22.0	26.6				
B. contrarium	8.7	13.3	17.9	22.1	26.7	31.0	35.1	38.9	43.4

[a]Measurements reproduced from references 1, 2.
[b]Sedimentation coefficients estimated by extrapolation on plots of log S *vs.* log M.

One application of STEM is unmatched by any other technique: the radial mass distribution may be determined. Analysis of the end views of cylinders of one, two, and three decamers, such as those in Figure 1A, showed that the mass falls to zero at the center of all three types of particles, which indicates that no cap is present (17). We may not, however, eliminate the possibility of losing the cap in the specimen deposition at the freeze-drying step or at some other point in the preparation. We are presently refining these analyses to obtain a better picture of the collar ends. We are also examining other polymeric forms.

Let us consider the various polymers that have been observed among the molluscan Hcs; first, at ionic conditions close to physiological that we shall call "native"; and second, at pHs close to the isoionic point ("*pI* polymers").

Native Polymers. With a few exceptions to be noted below, all higher assemblies examined seem to contain a single di-decameric unit per particle. This feature has been noted by others (*e.g.*, Ghiretti-Magaldi *et al.* (10) and

Terwilliger *et al.* (11)). The tri-decameric Hc of *Euspira heros* shown in Figure
1*C* is the most convincing example, since higher assemblies lacking a
"nucleating" di-decamer might represent broken particles. Figure 1*D* depicts
several such examples in the image of *B. carica* Hc where long cylinders have
evidently hit the grid and then toppled over, leaving attached single decameric
units, identified by mass (labeled 1 next to a penta-decamer in Figure 1*D*). The
tri-decamers (Figure 1*C*) of *E. heros* Hc always showed closed ends.
Furthermore, tetra-decamers that were also present showed two possible
positions for the di-decameric unit: [[*x*]] and [[[*x*] (labeled 4*a* and 4*b*,
respectively, in Figure 1*C*). Thus far, we have not seen arrangements in which
the non-collar end faces outward:][*x*][or][*x*]]. Enumeration of the possible
unique arrangements is illustrated in Figure 2. Notice that as the particles get
longer, the number of these "positional isomers" increases. We wonder whether
these subtle differences in structure contribute to the polydispersity observed in
sedimentation.

Exceptions to the Rules of Assembly. First, more than one di-decameric unit
per particle may be seen in the Hc of *M. crenulata* (two are indicated by short
lines in Figure 1*E*), which had the complex sedimentation pattern shown in
Table 1. Second, *pI*-type ([*x*][*x*][*x*]) polymer formation was observed (Figure
1*F*) in the Hc of a primitive bivalve, *Nucula proxima*. In addition to the classic
di-decamers, other long aggregates (not shown) were also seen in this specimen;
the latter were similar to those described as "polar" in the Hc of another bivalve,
that of *Yoldia thraciaeformis* (11).

Limited trypsinolysis of the multi-decameric Hc of *B. carica*, following the
procedures of Van Breemen *et al.* (20), which had generated collarless tubes with
H. pomatia Hc, produced the long, abnormal assemblies shown in Figure 1*G*.
Many of these have two Mellema-Klug di-decameric units, not unlike those of
M. crenulata Hc. Evidently, we succeeded in altering the collars of the end-most
units, but the collars of the inner units were not accessible. Compare these with
the native *B. carica* Hc particles in Figure 1*D*.

pI *Polymers.* We have analyzed the sedimentation patterns of nine Hcs at pH
4.5, and have classified as three groups 1) those that are unaffected (chiton, tulip,
Marisa cornuarietis); 2) those that aggregate into discrete sedimenting
components (*K. kelletia* and *Muricanthus fulvescens*); and 3) those that
precipitate completely (*C. oldroydii, E. heros, B. carica*). Except for the di-
decameric Hc of *C. oldroydii*, there is a correlation between the prevalence of
native higher assemblies and precipitation. Thus far, we have examined only
three of these by electron microscopy. *K. kelletia* Hc formed end-wise stacks
([*x*][*x*][*x*]) as first shown by Condie and Langer (19). *M. fulvescens* Hc also
formed stacks, but some particles showed internal irregularities reminiscent of
the reassembled aggregates of *M. cornuarietis* Hc (5). In addition to the end-
wise stacking seen in the molecules that form discrete aggregates, precipitation

Figure 2. Models for possible arrangements of the decameric units in molluscan Hcs of various sizes assuming one "nucleating" di-decamer per particle. There are $1 \le j = (n/2)$ unique arrangements for each particle containing n decameric units. [= decameric unit with collar at left. [x] indicates "nucleating" di-decamer.

s rate	n	j			
60 S			[		
100 S	2	1	[x]		
130 S	3	1	[x]]		
156 S	4	2	[x]]]	[[x]]	
178 S	5	2	[x]]]]	[[x]]]	
200 S	6	3	[x]]]]]	[[x]]]]	[[[x]]]
220 S	7	3	[x]]]]]]	[[x]]]]]	[[[x]]]]
etc.					

represents a different mode of aggregation with lateral associations. The precipitate of *E. heros* Hc showed such clumps, and also short stacks of tri-decamers aligned endwise in both configurations ([[x][x] and [[x][x]]), as though there is a preference for collar-ends to associate at low pH. As it seems likely that differences in the structure of the collars account for the variety of the states of aggregation of the molluscan Hcs, we are continuing these studies of the various kinds of polymers with STEM and other techniques.

Acknowledgements

This work was supported in part by Faculty Research Grants from Fordham University. Support for the Brookhaven STEM is provided by the National Institutes of Health Grant No. RR 01777 and the U.S. Department of Energy. We thank Martha Simon and Frank Kito for STEM operation.

References

1. Hamilton, M.G., Herskovits, T.T., Furcinitti, P.S. and Wall, J.S. (1989) *J. Ultrastruct. Mol. Struct. Res.* **102**: 221–228.
2. Herskovits, T.T., Blake, P.A., Gonzalez, J.A., Hamilton, M.G. and Wall, J.S. (1989) *Comp. Biochem. Physiol.* **94**B: 415–421.
3. Herskovits, T.T., Rodriquez, R.R. and Hamilton, M.G. (1990) *Comp. Biochem. Physiol.* **97**B: 631–636.
4. Herskovits, T.T., Gonzalez, J.A. and Hamilton, M.G. (1991) *Comp. Biochem. Physiol.* **98**B.
5. Herskovits, T.T., Otero, R.M. and Hamilton, M.G. (1990) *Comp. Biochem. Physiol.* **97**B: 623–629.

6. Herskovits, T.T., Hamilton, M.G., Cousins, C.J. and Wall, J.S. (1990) *Comp. Biochem. Physiol.* **96**B: 497–503.
7. Van Holde, K.E. and Miller, K.I. (1982) *Q. Rev. Biophys.* **15**: 1–129.
8. Van Bruggen, E.F.J., Wiebenga, E.H. and Gruber, M. (1962) *J. Mol. Biol.* **4**: 1–7.
9. Mellema, J.E. and Klug, A. (1972) *Nature* **239**: 146–150.
10. Ghiretti-Magaldi, A., Salvato, B., Tognon, G., Mammi, M. and Zanotti, G. (1981) In *Invertebrate Oxygen Binding Proteins: Structure, Active Site, and Function*, eds. J. Lamy and J. Lamy, J, 393–404. New York: Marcel Dekker.
11. Terwilliger, N.B., Terwilliger, R.C., Meyhofer, E. and Morse, M.P. (1988) *Comp. Biochem. Physiol.* **89**B: 189–195.
12. Wichertjes, T., Gielens, C., Schutter, W.G., Préaux, G., Lontie, R. and Van Bruggen, E.F.J. (1986) *Biochim. Biophys. Acta* **872**: 183–194.
13. Herskovits, T.T., Blake, P.A. and Hamilton, M.G. (1988) *Comp. Biochem. Physiol.* **90**B: 869–874.
14. Makino, N. (1971) *J. Biochem.* **70**: 149–155.
15. Senozan, N.M., Landrum, J., Bonaventura, J. and Bonaventura, C. (1981) In *Invertebrate Oxygen Binding Proteins: Structure, Active Site, and Function*, eds. J. Lamy and J. Lamy, 703–717. New York: Marcel Dekker.
16. Siezen, R.J. and Van Bruggen, E.F.J. (1974) *J. Mol. Biol.* **90**: 77–89.
17. Hamilton, M.G., Herskovits, T.T. and Wall, J.S. (1990) *Proc. XIIth Int. Congr. for Electron Microscopy*, 810–811.
18. Herskovits, T.T. (1988) *Comp. Biochem. Physiol.* **91**B: 597–611.
19. Condie, R.M. and Langer, R.B. (1964) *Science* **144**: 1138–1140.
20. Van Breemen, J.F.L., Wichertjes, T., Müller, M.F.J., Van Driel, R. and Van Bruggen, E.F.J. (1975) *Eur. J. Biochem.* **60**: 129–135.
21. Wall, J.S. and Hainfeld, J.F. (1986) *Annu. Rev. Biophys. Biophys. Chem.* **15**: 355–376.
22. Hamilton, M.G., Rodriguez, R.R., Herskovits, T.T. and Wall, J.S. (1989) In *Proc. 47th Ann. Mtg. Electron Microscopy Society of America*, ed. G.W. Bailey, 248–249. San Francisco: San Francisco Press.

5
Spectroscopic Analysis of Temperature-Induced Unfolding of the Hemocyanin from the Tarantula *Eurypelma californicum*

Franz Penz and Heinz Decker

Zoological Institute, University of Munich, Luisenstrasse 14, D-8000 Munich, Federal Republic of Germany

Introduction

Arthropod Hcs are multisubunit proteins (1). The smallest arthropod found in nature consists of six subunits. Each subunit (M_r = 75 kDa) is composed of three domains (2). The active site is located in the second domain, where O_2 is reversibly bound in a complex bridged between two copper atoms (3). Each copper atom is liganded to side-chain groups of the protein. Little data regarding the thermal stability of Hcs are available (4). This study presents the influence of the intact active site and of the quaternary state on the stability of an arthropod Hc with respect to temperature change.

Materials and Methods

Oligomeric and monomeric Hcs from the tarantula *Eurypelma californicum* were prepared according to the method of Schneider *et al.* (5). Apo-subunit *e* was obtained from dialysis against a 50 mM solution of KCN. The buffer systems used were glycine/OH⁻, 0.05M at pH 9.6 for the monomeric subunits, and Tris/HCl, 0.05M at pH 7.6 for the whole 24-meric Hc. The temperature-induced changes in the absorbances at 250, 292, and 340 nm were discontinuously determined with a Cary spectrophotometer, model 118, employing two thermostats. The reference cuvette was maintained at $20 \pm 0.1°C$, and the sample cuvette was set at the desired temperature, ranging from 10 to 75°C. Thermal equilibrium was reached after three minutes. Subsequently, the temperature was increased incrementally and the absorbance measured with a precision of ± 0.001. The protein concentration was about 1 mg/ml.

Results and Discussion

Unfolding of the 24-Meric Hc. Difference spectra of the 24-meric Hc were obtained after equilibration at different temperatures, and revealed distinct changes in absorption (Figure 1*C*). According to Herskovits (5) and Donovan (6), the changes in absorbance at 292 and 250 nm are mainly due to changes in the microenvironment of the amino acids Trp and Phe. The absorbance at 340 nm is due to the copper-oxygen complex (7). Between 10 and 75°C, the temperature dependence of the absorption was recorded at 250, 292 and 340 nm. The absolute changes in absorption (optical density units) were calculated to 0.2 at 250 and 340 nm; and to 0.02 at 292 nm. The data seem to produce two straight lines. The point of intersection was interpreted as the beginning of the unfolding process induced by temperature. The critical temperatures were different for all recorded wavelengths. For the absorbances at 250, 292 and 340 nm, the onsets of the unfolding processes were determined to occur at temperatures of 63, 60 and 51°C, respectively (Figures 1*A, B*). Nevertheless, heating the sample to temperatures below 75°C was fully reversible with respect to the absorbance measured after cooling the sample to room temperature.

Unfolding of the Native and Copper-Free Apo-Subunit e. The oligomeric Hc was dissociated and the different subunits were separated. Subunit *e* was chosen as a representative chain since much knowledge about it is available (8–11). It was prepared in native and copper-free forms, and three different points of intersection at different wavelengths were also found. Firstly, the absorbance at 340 nm was affected at a temperature of 46°C. The melting point was calculated to be 56°C for the absorbance at 250 nm, and that for 292 nm was found to be 61°C. After cooling the sample to room temperature, the absorbance at these wavelengths was no longer reversible. Obviously, no change in absorption should have been detected for copper-free Hc at the wavelength of 340 nm. The change in absorbance at the wavelengths of 250 and 292 nm revealed differences in the critical temperatures. It was found that transition at the wavelength of 250 nm took place at a temperature less than 6°C, compared with the unmodified subunit *e*. For the wavelength of 292 nm, the critical temperature was lowered by 5°C. The absolute changes in absorption were 0.12 at 250 and 340 nm, and 0.02 at 292 nm.

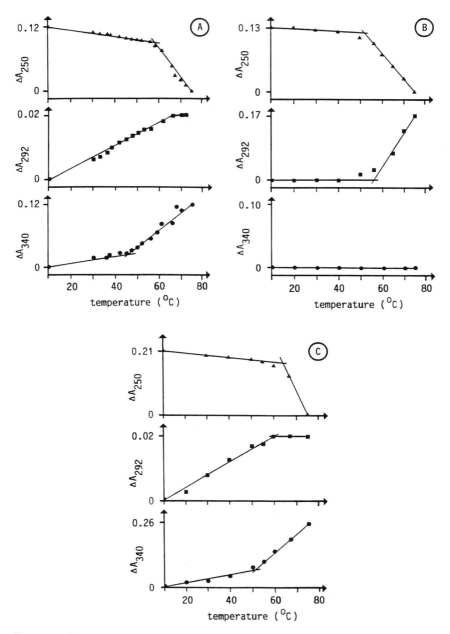

Figure 1. Changes in absorbance at wavelengths of 250, 292 and 340 nm of the Hc from the tarantula *E. californicum* over the temperature range 10 to 75°C: (*A*) monomeric Hc, (*B*) copper-free apo-subunit *e* and (*C*), 24-meric Hc. Conditions for (*A*) and (*B*): Glycine/OH⁻, 0.05M, pH 9.6. Conditions for (*C*): Tris/HCl, 0.05M, pH 7.6. The protein concentration was 1 mg/ml.

Conclusions

Analysis of the melting curves of the oligomeric and monomeric Hcs showed that the thermal unfolding may be roughly interpreted as a "two-state" transition. For the observed changes in absorbance at 250, 292 and 340 nm, only one turn-over point was observed, but it occurred sequentially at different temperatures. Within the experimental error, the critical temperatures of the two native Hcs were identical. A crucial difference, however, is the reversibility of the thermal unfolding of the oligomeric Hc below 75°C, which is probably due to the association state. In comparison to the native Hcs, the copper-free species are characterized by lower critical temperatures. It seems that an intact active center is necessary for the stability of Hc. This confirms the observation made after the incubation of subunit *e* and apo-subunit *e* with different concentrations of guanidine hydrochloride (12).

Acknowledgements

Financial support was provided by the Deutsche Forschungsgesellschaft (De 414/1-10).

References

1. Van Holde, K.E. and Miller, K.J. (1982) *Q. Rev. Biophys.* **15**: 1–129.
2. Volbeda, A. and Hol, W.G.J. (1989) *J. Mol. Biol.* **209**: 249–279.
3. Guzman-Casado, M., Parody-Morreale, A., Mateo, P.L. and Sanchez-Ruiz, J.M. (1990) *Eur. J. Biochem.* **188**: 181–185.
4. Schneider, H.-J., Markl, J., Schartau, W. and Linzen, B. (1977) *Hoppe-Seyler's Z. Physiol. Chem.* **358**: 1133–1141.
5. Herskovits, T.T. (1967) *Methods Enzymol.* **11**: 748–775.
6. Donovan, J.W. (1973) *Methods Enzymol.* **27**: 497–503.
7. Himmelwright, R.S., Eickmann, N.C., LuBien, C.D. and Solomon, E.I. (1980) *J. Am. Chem. Soc.* **102**: 5378–5388.
8. Markl, J., Decker, H., Linzen, B., Schutter, W.G. and Van Bruggen, E.F. (1981) *Hoppe-Seyler's Z. Physiol. Chem.* **363**: 73–82.
9. Schneider, H.-J., Drexel, R., Feldmaier, G. and Linzen, B. (1983) *Hoppe-Seyler's Z. Physiol. Chem.* **364**: 1357–1381.
10. Linzen, B., Soeter, N.M., Riggs, A., Schneider, H.-J., Schartau, W., Moore, M.D., Yokota, E., Behrens, P.Q., Nakashima, H., Takagi, T., Nemoto, T., Vereijken, J.M., Bak, H.J., Beintema, J.J., Volbeda, A., Gaykema, W.P. and Hol, W.G.J. (1985) *Science* **229**: 519–524.
11. Voit, R. and Schneider, H.-J. (1986) *Eur. J. Biochem.* **159**: 23–29.
12. Penz, F., Nahke, P. and Decker, H. (1991) In *Invertebrate Dioxygen Carriers*, ed. G. Préaux. Leuven: Leuven University Press. In press.

6
Subunit Structure of the Hemocyanin from the Arthropod *Ovalipes catharus*

H. David Ellerton[a] and Nik Rahimah Husain[b]

[a]School of Science and Technology, Charles Sturt University, P.O. Box 588, Wagga Wagga NSW 2650, Australia
[b]Department of Biochemistry, University of Kebangsaan, P.O. Box 12418, Kuala Lumpur 50778, Malaysia

Introduction

In spite of the considerable progress made in Hc research in recent years, comparatively little information on southern hemisphere crustacean Hc has been made available. Earlier studies have included those on the Hcs of *Jasus novaehollandiae* (1, 2), *Cherax destructor* (3) and studies from our group on *Jasus edwardsii* (4, 5) and *Ovalipes catharus* (6).

In this paper, we report the results of further studies on the Hc from the crab *Ovalipes*. We have re-investigated the subunit structure of the protein, and report the presence of seven different subunits, as shown by electrophoresis and chromatography. Reassociation of these subunits is described. The influence of the heavy metals Hg^{2+}, Cd^{2+}, Cu^{2+}, Mn^{2+} and Zn^{2+} on the O_2 binding and spectral properties of the Hc is reported.

Materials and Methods

The swimming crab *Ovalipes* was either captured from the sandy bars of Days Bay, near Wellington, New Zealand, or obtained by fishing boats from the deep waters of Cook Strait. Hemolymph was extracted from live animals and purified as described previously (6). Concentrations of purified Hc were measured spectrophotometrically using $A^{1\%} = 14.1$ as determined previously (5). Gel filtration media were Biogel A-5m (Bio-Gel) and Sephacryl S-200 (Pharmacia).

Polyacrylamide gel electrophoresis of native Hc was performed by modifications on the methods of Loehr and Mason (7) and Murray and Jeffrey (3). SDS-PAGE utilized the system of Laemmli (8, 9). Ion exchange chromatography was performed with DEAE-Sepharose $C1$–$6B$ and DEAE-Sephacel on 40 cm x 2.5 cm or 40 x 1.6 cm columns, respectively. Sedimentation equilibrium experiments were performed with a Beckman Model E Analytical Ultracentrifuge using Rayleigh optics, an An-H rotor and a six-channel centerpiece at 8,000, 10,000 and 26,000 rpm for the 25 S, 16 S and 5 S particles, respectively.

Oxygen-binding studies were performed by monitoring absorbance at 338 nm in a tonometer consisting of a glass bulb attached to a fused quartz cuvette (5) which fit into the sample compartment of the Cary 210 spectrophotometer. Difference spectra were obtained at 20°C according to the method of Brouwer *et al.* (10). Electron microscopic observations were performed in a Siemens Model 102 electron microscope at 40,000X magnification.

Results and Discussion

Electron Microscopy. The electron micrograph of *Ovalipes* Hc showed hexagonal, square and rectangular profiles of the hexameric 16 S protein. There were also profiles of a hexagon linked to a square or rectangle, which may represent the 25 S dodecamers. These profiles are consistent with the electron microscope images of other crustacean Hcs (11, 12).

Ultracentrifugation Studies. It has been shown that the native Hc has two components, sedimenting as 25 S and 16 S, the former being the predominant species (12, 13). As the pH was raised, dissociation took place, so that by pH 8.8, 25 S, 16 S and 5 S particles were present in roughly equal proportions; and by pH 9.8–10.0, only the 5 S particles were observed (12). Sedimentation velocity experiments on the separated components at various protein concentrations yielded $S_{20°,w}$ = 25.5 S, 17.1 S and 5.4 S, respectively (13).

In this study, we performed sedimentation equilibrium experiments on purified 25 S, 16 S and 5 S particles. The weight-average M_m of the 25 S particles, averaged over all the data points, was 928 kDa (averaged over three experiments). Similar data for the 16 S and 5 S particles yielded values of 476 kDa and 72.3 kDa, respectively. Point average M_w values obtained by computer averaging of the data were in agreement with these values. Thus, the data confirm that the 16 S and 25 S particles represent hexamers and dodecamers of the monomeric 5 S subunits.

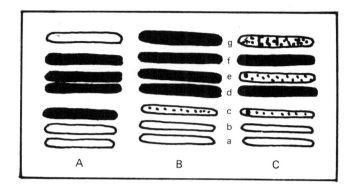

Figure 1. SDS-PAGE of *Ovalipes* Hc, using the method of Laemmli (8, 9). (*A*) 16 S component; (*B*) 25 S component; (*C*) whole Hc.

The M_m of the Subunits by Electrophoresis. SDS-PAGE of whole and fractionated 25 S and 16 S components using a modified Weber and Osborn method (14) showed the presence of two bands of 65 kDa and 83 kDa (the main band). A third band of 53 kDa was sometimes observed, especially on samples that had undergone extended dialysis.

Using the method of Laemmli (8, 9), seven, and sometimes eight, bands were observed. These are illustrated in Figure 1, and the data are summarized in Tables 1 and 2. The principal bands *a* to *g* have masses in the range of 70 to 83 kDa, with the main band at 75 kDa. In some experiments, bands *a* and *b* were observed to have the same mobility. Two additional bands, *h* and *i* (not shown in Figure 1), were seen in samples that had been dialyzed for one day or more at pH 10.6, and that had masses of 66 and 56 kDa, respectively. The SDS-PAGE data support the sedimentation equilibrium determination of the mass of the subunits: the average mass of the multiple subunit bands is in good agreement with the value observed by sedimentation.

This study confirms the presence of heterogeneous subunits in *Ovalipes* Hc. In an earlier study (6), Robinson and Ellerton found the presence of three subunits (masses 69, 101 and 112 kDa, respectively) by SDS-PAGE, using the electrophoretic method of Loehr and Mason (7). The Laemmli (8, 9) method used in this study resolved seven subunits (*a–g*) with M_ms in the range of 70 to 83 kDa. A similar difference between the two techniques has also been described by Larson *et al.* (15), who reported an improvement in resolution from two to six bands in a study on the Hc from *Cancer magister*.

Table 1. Proportion of subunits of the 25 S component of *Ovalipes* Hc.

Subunits	M_m	Percentage[a]	No. of Subunits per Dodecamer
a	70,000	6 ± 1	0.7 ± 0.1
b	72,000	4 ± 2	0.5 ± 0.2
c	73,000	8 ± 2	0.9 ± 0.2
d	75,000	36 ± 8	4.3 ± 0.9
e	77,000	13 ± 4	1.5 ± 0.5
f	79,000	18 ± 3	2.2 ± 0.4
g	83,000	14 ± 2	1.7 ± 0.2

[a]Values are the average of nine determinations.

Table 2.   Proportion of subunits of the 16 S component of *Ovalipes* H c (reelectrophoresed bands 1 and 2) from polyacrylamide gel electrophoresis at pH 7.8.

Band 1			Band 2		
	Percentage of Subunits	No. of Subunits per Hexamer		Percentage of Subunits	No. of Subunits per Hexamer
d	83 ± 3	5.0 ± 0.2	f	56 ± 3	3.4 ± 0.2
c	10 ± 2	0.6 ± 0.1	e	44 ± 3	2.6 ± 0.2
b	9 ± 2	0.5 ± 0.1			

An explanation has been given by Klarman and Daniel (16) on the necessity of subunit heterogeneity in arthropod Hc for the existence of 24 S, 37 S and 62 S particles *in vivo*, where not all positions in these structures are equivalent. Thus, specific subunits may be required for the formation of dodecamers: sometimes these may be specific dimers, as for the Hc of the crayfish *C. destructor* (17), or just monomers, as for the crab *Cancer pagurus* and the lobster *Homarus americanus* (18). The immunological relationship of *Cancer* and *Homarus* subunits has also been discussed, and an arrangement of subunits in the dodecamer suggested (18).

Ion Exchange Chromatography. Ion exchange chromatography of 5 S Hc on DEAE-Sephacel at pH 9.8 eluted with a gradient of 0.1 to 0.8 M NaCl, gave three peaks (Figure 2), designated I, II and III, with relative percentages of 17, 57 and 26%, respectively. SDS-PAGE of samples from the separation revealed that

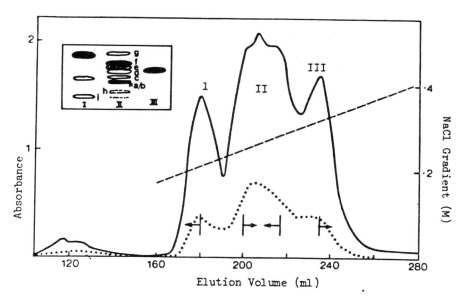

Figure 2. Ion exchange chromatography of 5 S monomers of *Ovalipes* Hc obtained by dialysis of the 25 S component (with traces of the 16 S component) for 24 hours in 0.05 M glycine-0.005 M EDTA, pH 9.8, on DEAE-Sephacel. The protein was eluted with a 0.1–0.8 M NaCl gradient in the same buffer. Hc was pooled in the direction of the arrows. (––––), (····), Absorbance at 280 nm and 338 nm, respectively. *Insert:* SDS-PAGE of the protein from peaks I, II and III.

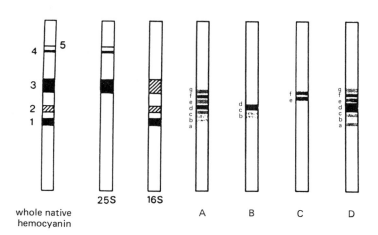

Figure 3. Polyacrylamide gel electrophoresis of whole native *Ovalipes* Hc and its 25 S and 16 S components purified on a Bio-Gel *A*-5 column at pH 7.8. (*A–D*) SDS-PAGE of the Hc. (*A*) whole Hc; (*B–D*) reelectrophoresis of bands 1, 2 and 3, respectively, isolated from whole native Hc.

the main component of peak I was band *g*, with lesser quantities of bands *a, b* and *i*. Bands *a, b, c, e* and *f* formed the main components of peak II, and peak III chromatographed as a single band *d*. Rechromatography of peak II showed that all subunits corresponding to band *g*, and most of bands *a, b* and *h*, eluted as peak I; peak II contained mainly bands *e* and *f*, and a trace of band *c*.

Search for Hc Dimers. In an attempt to detect and isolate subunit dimers, native Hc, and separated 25 S and 16 S components fractionated by chromatography on Bio-Gel A-5 were analyzed by SDS-PAGE according to the method of Laemmli (8, 9). Native Hc was also electrophoresed at pH 7.8, and the bands corresponding to the 25 S and 16 S components were excised and the protein reelectrophoresed.

PAGE of native Hc at pH 7.8 resulted in the observation of five bands (Figure 3). Bands 1 and 2 represent the major and minor 16 S forms respectively; band 3, the major 25 S component; and bands 4 and 5, the minor 25 S components. These observations are in agreement with an earlier study by Robinson and Ellerton (6).

Reelectrophoresis of the protein recovered from five bands in the presence of SDS is also illustrated in Figure 3. Band 1 yielded band *d* with traces of bands *b* and *c* in SDS, while band 2 electrophoresed as bands *e* and *f*. Band 3 yielded all the subunit bands seen in the native Hc, while bands 4 and 5, which were originally present in only very small amounts, did not yield any detectable bands.

Essentially, there was no difference in the patterns obtained in the presence or absence of 2-mercaptoethanol. Significantly, however, band *g* (M_m = 83,000, estimated at 1.7 subunits per dodecamer) was found only in protein derived from the 25 S particle, and may therefore be the polypeptide which promotes the association of 16 S species to the 25 S dodecamer.

Thus, no dimeric subunits were detected in this Hc, our observations instead suggesting that subunit *g* is involved in the hexamer-dodecamer association.

Reassociation. Hc monomers may reassociate under certain conditions to hexamers, dodecamers and higher aggregates, and Robinson (13) has reported some preliminary studies on the unfractionated monomers of *Ovalipes* and *Jasus* Hcs.

In this study, reassociation of the unfractionated 5 S subunit was carried out by first lowering the pH to 7, and then by the addition, through dialysis, of 0.01 M calcium ions. An attempt was also made to reassociate the protein fractions separated by ion exchange chromatography. Protein concentration was approximately 20 mg/ml. The addition of calcium ions during the reassociation resulted in the precipitation of about 40% of the protein.

Electrophoresis at pH 7.8 of the reassociated product yielded only about 10% of the original protein in the 25 S form (band 3, Figure 3) and 43% in the 16 S form. An additional 7% was observed in bands not found in the native protein.

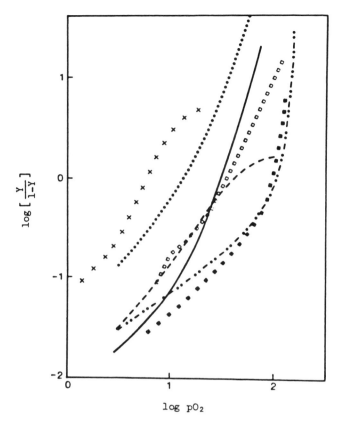

Figure 4. The effect of heavy metals on O_2-binding by *Ovalipes* Hc at 20°C. (———), in the absence of additional heavy metals; (X), 0.25 mM $CdCl_2$; (--·---), 0.24 mM $HgCl_2$; (□), 0.15 mM $CuCl_2$; (----), 0.20 mM $MnCl_2$; (····), 0.14 mM $ZnCl_2$; (■), 'aged' protein with 10 mM $CaCl_2$. The protein concentration was 3.5 mg ml^{-1} in 0.1 M Tris-HCl, pH 7.0, with 10 mM $CaCl_2$.

Reassociation of protein from peaks I, II and III of the ion exchange chromatography, mixed in the ratio of the areas under the peaks, lead to the formation of the main 16 S (45%) and 25 S (10%) components, together with small amounts of other high M_m aggregates. Reassociation of Hc from peaks II and III resulted in the formation of mainly 16 S (55% of original protein) and two 25 S forms (1% and 5%) not seen in the native Hc.

Reassociation experiments using lower concentrations of Hc (2 mg/ml) and calcium (0.002 M) were also performed. Electrophoresis at pH 7.8 revealed that most reaggregation was to the 16 S form (band 1, Figure 3) but a small amount of 25 S protein (up to approximately 2%) reformed when ion exchange bands I and III, or I, II and III were reassociated.

Effect of Heavy Metals. The binding of O_2 to arthropod Hc has been shown to be regulated by ions such as Ca^{2+}, Mg^{2+}, Cl^- and H^+ (12). Robinson (13) demonstrated positive Bohr effects and changes in cooperativity in the pH range of 6.2 to 10.6 for O_2 binding in *Ovalipes* Hc.

The influence of five heavy metals on the O_2-binding properties of the Hc is illustrated in Figure 4. The concentration of heavy metal was constrained by the fact that the self-association of Hcs was promoted above certain metal concentrations.

The effect of the metals on the O_2 affinity and cooperativity, as shown by the Hill coefficient, h_{50}, is summarized in Table 3. The presence of Cd^{2+}, Zn^{2+} and Cu^{2+} resulted in an increase in O_2 affinity, while a decrease was observed for Hg^{2+} and Mn^{2+}. A decrease in cooperativity was recorded under the influence of all five metals.

Oxygen saturation curves for the Hc in the presence of cadmium and zinc ions showed a hyperbolic curve similar to that of a binding curve of a single-site protein or that of a non-cooperative multi-site protein (19). Oxygen affinity increased in the presence of these two ions. Mercuric chloride and manganese chloride, however, caused a decrease in the O_2 affinity as well as in the cooperativity of the Hc. The effects of cadmium and mercuric ions on the O_2 binding are consistent with the results reported for *Callinectes sapidus* Hc (10), *Panulirus interruptus* Hc (for mercuric ions) (20) and *Panulirus elephas* (for mercuric ions) (21). For *P. interruptus* Hc, cadmium ions increased the O_2 affinity and shifted the lower asymptotes to higher O_2 affinities; the effects of mercuric ions were the opposite (20).

Ultraviolet difference spectra of the Hc when titrated with mercuric chloride are shown in Figure 5. The copper-oxygen absorption band at 338 nm increases (negatively) as the concentration of Hg^{2+} increases. Since Hg^{2+} decreases the O_2 affinity of the Hc (Table 3), this decrease in absorption is most likely due to a slight dissociation of the copper-oxygen complex upon the addition of Hg^{2+}. In *C. sapidus* Hc, a similar decrease in absorbance at 338 nm was reversed by the addition of EDTA, implying that only weak interactions between the mercuric ions and the amino acid side chains are involved here (10).

Positive difference spectra bands observed at 276, 285 and 297 nm may reflect perturbations of the Trp and Tyr chromophores. The large positive absorbance difference peak between 270 nm and 240 nm may be due partly to the formation of an ultraviolet-absorbing Tris-HCl-Hg complex and partly to the formation of a mercuric-mercaptide bond (10).

The ultraviolet difference spectra induced by the presence of cadmium, copper, manganese and zinc chlorides shown in Figure 6 are somewhat more difficult to interpret. The increase in absorbance at about 338 nm in the presence of cadmium, copper and zinc is probably the spectral change reflected by the increase in O_2 affinity seen in the O_2-binding studies of the Hc in the presence of these heavy metals. The influence of metals on the protein is also reflected in the absorption below 300 nm, and it seems likely that the same amino acids are involved in each case.

Table 3. O_2 affinity and cooperativity of *Ovalipes* Hc in the presence of heavy metals.

Heavy Metal	Molar Ratio[a] (Metal : Protein)	P_{50} (mm Hg)	h_{50}
—	—	32	3.2
$CdCl_2$ (0.25 mM)	5.4 : 1	5	2.2
$HgCl_2$ (0.24 mM)	5.2 : 1	101	2.6
$CuCl_2$ (0.15 mM)	3.2 : 1	26	1.9
$MnCl_2$ (0.20 mM)	4.3 : 1	41	1.0
$ZnCl_2$ (0.14 mM)	3.0 : 1	13	1.8

[a]A subunit M_m of 76,000 was used for the metal:protein molar ratio (since there is one O_2 binding site per subunit).

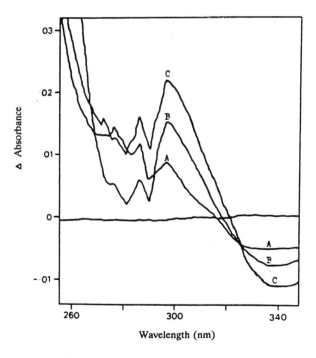

Figure 5.   Absorbance difference spectra of *Ovalipes* Hc induced by the presence of mercuric chloride. Three ml of 0.69 mg ml^{-1} Hc in 0.05 M Tris-HCl, 10 mM $CaCl_2$ at pH 7.0 was used in the sample and reference beams. Five, 10 and 30 μl of mercuric chloride (10 mM) dissolved in the same buffer was added to the Hc in the sample beam, and the absorption spectra scanned immediately from 350 to 250 nm: [A] 0.017 mM $HgCl_2$; [B] 0.033 mM $HgCl_2$; [C] 0.098 mM $HgCl_2$.

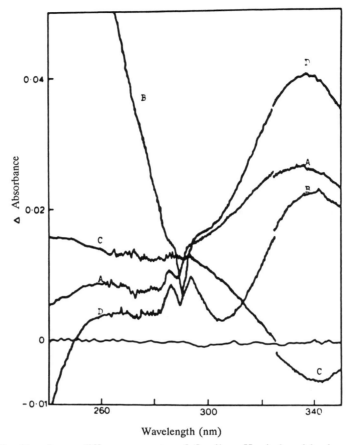

Figure 6. Absorbance difference spectra of *Ovalipes* Hc, induced by heavy metals. Three ml of 0.69 mg ml^{-1} Hc in 0.05 M Tris-HCl, 10 mM CaCl$_2$ at pH 7.0 was used in the sample and reference beams and 5–10 μl of the heavy metal (10 mM) dissolved in the same buffer was added to the sample beam. The difference spectra were recorded from 350 to 240 nm: [A] 0.033 mM CdCl$_2$; [B] 0.033 mM CuCl$_2$; [C] 0.017 mM MnCl$_2$ and [D] 0.033 mM ZnCl$_2$.

The binding of cadmium, (added) copper, manganese and zinc ions showed slight, positive peaks at 297 and 285 nm, indicative of Trp and/or Tyr bonds. Bere and Helene (23) have also shown that zinc ions bind to the Tyr residues of a synthetic polypeptide. Furthermore, copper ion-binding to this polypeptide was accompanied by a strong absorbance difference band around 245 nm, probably due to the involvement of carboxylic acid groups. Likewise, in this study, the large positive absorbance difference below 270 nm induced by the binding of the added copper ions to *Ovalipes* Hc may be due to the binding of these ions to carboxylic acids.

Acknowledgements

This study was performed in the laboratories of the Department of Biochemistry, Victoria University, Wellington, New Zealand. We acknowledge financial support for the project from the Internal Research Committee of Victoria University, and the New Zealand University Grants Committee. N.R.H. was the holder of a scholarship from the University of Kebangsaan.

References

1. Moore, C.H., Henderson, R.W. and Nichol, L.W. (1968) *Biochemistry* 7: 4075–4085.
2. Boas, J.F., Pilbrow, J.R., Troup, G.J., Moore, C. and Smith, T.D. (1969) *J. Chem. Soc.* A: 965–971.
3. Murray, A.C. and Jeffrey, P.D. (1974) *Biochemistry* 13: 3667–3671.
4. Ellerton, H.D., Blazey, N.D. and Robinson, H.A. (1977) *Biochim. Biophys. Acta* 495: 140–150.
5. Ellerton, H.D., Collins, L.B., Gale, J.S. and Yung, A.Y.P. (1977) *Biophys. Chem.* 6: 47–57.
6. Robinson, H.A. and Ellerton, H.D. (1977) In *The Structure and Function of Haemocyanin*, ed. J.V. Bannister, 55–70. Heidelberg: Springer-Verlag.
7. Loehr, J.S. and Mason, H.S. (1973) *Biochem. Biophys. Res. Commun.* 51: 741–745.
8. Laemmli, U.K. (1970) *Nature* 227: 680–685.
9. Laemmli, U.K. and Favre, M. (1973) *J. Mol. Biol.* 80: 575–599.
10. Brouwer, M., Bonaventura, C. and Bonaventura, J. (1982) *Biochemistry* 21: 2529–2538.
11. Van Bruggen, E.F.J., Schutter, W.G., Van Breemen, J.F.L., Bijlholt, M.M.C. and Wichertjes, J. (1981) In *Electron Microscopy of Proteins*, ed. J.R. Harris, vol. 1, 1–38.
12. Ellerton, H.D., Ellerton, N.F. and Robinson, H.A. (1983) *Prog. Biophys. Mol. Biol.* 41: 143–248.
13. Robinson, H.A. (1978) Ph.D. Thesis. Victoria University.
14. Weber, K. and Osborn, M. (1975) In *The Proteins*, eds. H. Neurath and R.L. Hill, vol. 1, 179–223. New York: Academic Press.
15. Larson, B.A., Terwilliger, N.B. and Terwilliger, R.C. (1981) *Biochim. Biophys. Acta* 667: 294–302.
16. Klarman, A. and Daniel, E. (1981) *Comp. Biochem. Physiol.* 70B: 115–123.
17. Jeffrey, P.D., Shaw, D.C. and Treacy, G.B. (1978) *Biochemistry* 17: 3078–3084.
18. Markl, J., Hofer, A., Bauer, G., Markl, A., Kempter, B., Brenzinger, M. and Linzen, B. (1979) *J. Comp. Physiol.* 133B: 167–175.
19. Van Holde, K.E. and Miller, K.I. (1982) *Q. Rev. Biophys.* 15: 1–129.
20. Kuiper, H.A., Zolla, L., Calabrese, L., Vecchini, P., Constantini, S. and Brunori, M. (1981) *Comp. Biochem. Physiol.* 69C: 253–259.

21. Zolla, L., Bellelli, A., Brunori, M., Cau, A. and Giardina, B. (1983) *Life Chem. Rep. Supp. Ser.* **1**: 269–272.
22. Herskovits, T.T. (1967) *Methods Enzymol.* **11**: 748–775.
23. Bere, A. and Helene, C. (1979) *Int. J. Biol. Macromol.* **1**: 227–232.

7

Crosslinking with Bifunctional Reagents and its Application to the Determination of the Quaternary Structures of Invertebrate Extracellular Hemoglobins

Abdussalam Azem, Amnon Pinhasy and Ezra Daniel

Department of Biochemistry, Tel Aviv University, Tel Aviv 69978, Israel

Introduction

Crosslinking with bifunctional reagents is a potentially powerful method for the investigation of the structure of multisubunit proteins. The method can yield information on the number of subunits and their arrangement within the molecule. In the case of a protein composed of identical subunits, this leads to a determination of the symmetry underlying the molecular structure. Theoretical treatments directed specifically at the case of four subunit assemblies have been carried out and successfully tested for a number of tetrameric proteins (1, 2). Recently, the theoretical framework for analyzing crosslinking has been extended to proteins composed of more than four subunits (3). In the following discussion, we briefly review the theory and practice of the crosslinking method.

We shall be concerned with problems encountered in the application of the method to large assemblies of identical subunits ($n \geq 6$, where n is the number of subunits). Examples for such systems are provided by extracellular Hbs of arthropod, mollusc and nematode origin.

According to theory (4, 5), allowed structures for an oligomeric protein composed of identical subunits must satisfy two criteria: 1) the subunits must occupy environmentally equivalent positions and 2), the arrangement must lead to a closed structure. These requirements lead to structures with three types of symmetry: cyclic, dihedral, and cubic. Cyclic symmetry is possible with any number of subunits. Dihedral symmetry is restricted to an even number of subunits. Cubic symmetry, which will not be further considered here, requires fixed numbers of subunits, 12, 24 or 60.

The symmetry of a structure is reflected in the spatial arrangement of the subunits. Cyclic symmetry is associated with a ring structure. Dihedral symmetry is consistent with a ring arrangement of subunits with alternating interactions, a so-called alternating ring structure (3), and with two-layered subunit arrangements going from eclipsed to staggered orientations. Direct information about the spatial arrangement of subunits can be obtained from electron microscopy.

The symmetry of the molecule also finds its expression in the nature of the bonds linking the subunits together. Monod *et al.* (6) introduced a classification according to which one may distinguish between isologous bonding where each subunit contributes an identical bonding site, and heterologous bonding where each subunit contributes a different bonding site. In the cyclic structure, the

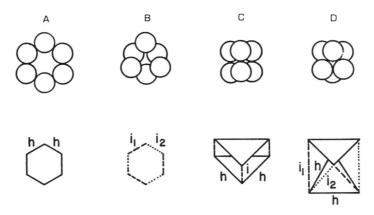

Figure 1. Arrangements of subunits and bonding patterns for an assembly of six identical subunits. Upper panel shows arrangements for (A) cyclic, (B) alternating ring, and two-layered (C) eclipsed and (D) staggered structures. Lower panel shows corresponding bonding patterns. Heterologous and isologous bonds are indicated respectively by *h* and *i*.

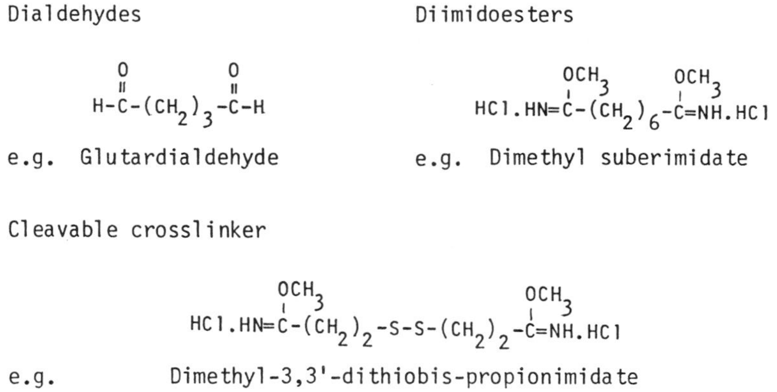

Dialdehydes

Diimidoesters

$$O \qquad O$$
$$\parallel \qquad \parallel$$
$$H-C-(CH_2)_3-C-H$$

e.g. Glutardialdehyde

$$OCH_3 \qquad OCH_3$$
$$| \qquad\quad |$$
$$HCl.HN=C-(CH_2)_6-C=NH.HCl$$

e.g. Dimethyl suberimidate

Cleavable crosslinker

$$OCH_3 \qquad\qquad\qquad OCH_3$$
$$| \qquad\qquad\qquad\qquad |$$
$$HCl.HN=C-(CH_2)_2-S-S-(CH_2)_2-C=NH.HCl$$

e.g.　　　Dimethyl-3,3'-dithiobis-propionimidate

Figure 2. Bifunctional reagents commonly used in the crosslinking of oligomeric proteins.

intersubunit bonds are heterologous and all of one type. In dihedral structures, the bonds linking the subunits cannot all be of the same type. Spatial arrangements of subunits and corresponding bonding patterns for oligomers containing six identical subunits are presented in Figure 1.

Materials and Methods

Crosslinking is usually carried out with diimidoesters or with glutaraldehyde (Figure 2). These bifunctional reagents react with primary amines, and ε-amino groups of lysyl residues of the protein constitute sites for reaction with the crosslinker. In general, the two reacting groups on the protein may be situated on the same subunit (intrasubunit crosslinks), on two subunits within the same oligomer (intersubunit), or on two subunits from two different oligomeric molecules (intermolecular). For the purpose of analyzing the subunit structure, intrasubunit crosslinks are irrelevant and intermolecular crosslinks create an additional complication. Reaction with the bifunctional reagent should therefore be carried out under conditions in which intermolecular crosslinking is negligible. To begin this procedure, the oligomer must first be exposed to the crosslinker in solution for a set period of time. Then, SDS-PAGE is carried out, which terminates the crosslinking reaction and disrupts the non-covalent

Table 1. Crosslinking rate constants associated with models of different symmetries.

Model[a]	Expected number of rate constants
Cyclic ring	1
Alternating ring	2
Two-layered eclipsed	2
Two-layered staggered	3

[a]See Figure 1

interactions in the original structure. Finally, the various electrophoretic bands are identified in terms of the number of polypeptide chains crosslinked together, and the area corresponding to each band is measured in a densitometric scan of the stained gel. This results in the determination of the distribution of protein among crosslinked species.

Results and Discussion

Analysis. The principle underlying the interpretation is that intersubunit crosslinks, like the bonds linking the subunits in the native structure, must be consistent with the symmetry of the molecule. This means that different types of crosslinks, if formed, must reflect different types of bonds. Models for subunit arrangement will generally differ in both the crosslink pattern they generate and in the number of rate constants involved in the crosslinking reaction (Table 1). Analysis of the crosslinking and its progress with time will therefore provide information concerning the original bonding pattern.

In the derivation of a theory for crosslinking, a number of assumptions are usually made: (1) association or dissociation of protein oligomers during crosslinking is negligible; (2) formation of a crosslink at one site does not affect subsequent reaction with crosslinking reagent at other sites; (3) no intermolecular crosslinks are formed; and (4), the second function of the reagent reacts considerably faster than does the first so that crosslinking can be described solely by the rate constant for the first binding reaction.

For the prediction of crosslinking patterns, two approaches have been taken. The first is the approach of Hajdu *et al.* (2), in which differential equations describing the rates of formation of crosslinked species are written and solved sequentially. An illustration is provided by the case of an oligomer composed of three identical subunits that has been exposed to crosslinker for a time interval t. One finds (Figure 3):

$$x_i = Mx_0 \, (e^{k_p\tau} - 1)^i \quad , \tag{1}$$

where x_i is the fraction of protein with i crosslinks ($0 \leq i \leq 3$), M is a numerical factor that depends upon the number of possible arrangements of crosslinks, k_p is the rate constant of the crosslinking reaction at an intersubunit site of type p, and τ is the extent of reaction. The relation between τ and t is given by

$$\tau = \int_0^t |B| \, \partial t \, , \tag{2}$$

where $|B|$ is the concentration of crosslinker at time t. This approach was originally applied to a tetrameric assembly.

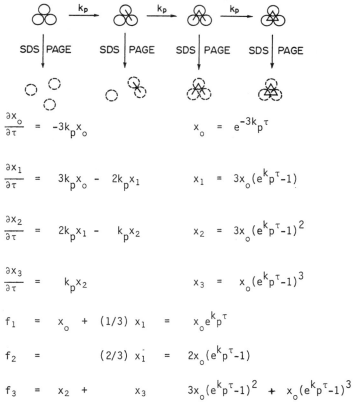

$$\frac{\partial x_0}{\partial \tau} = -3k_p x_0 \qquad\qquad x_0 = e^{-3k_p\tau}$$

$$\frac{\partial x_1}{\partial \tau} = 3k_p x_0 - 2k_p x_1 \qquad\qquad x_1 = 3x_0(e^{k_p\tau} - 1)$$

$$\frac{\partial x_2}{\partial \tau} = 2k_p x_1 - k_p x_2 \qquad\qquad x_2 = 3x_0(e^{k_p\tau} - 1)^2$$

$$\frac{\partial x_3}{\partial \tau} = k_p x_2 \qquad\qquad x_3 = x_0(e^{k_p\tau} - 1)^3$$

$$f_1 = x_0 + (1/3) \, x_1 = x_0 e^{k_p\tau}$$

$$f_2 = (2/3) \, x_1 = 2x_0(e^{k_p\tau} - 1)$$

$$f_3 = x_2 + x_3 \qquad 3x_0(e^{k_p\tau} - 1)^2 + x_0(e^{k_p\tau} - 1)^3$$

Figure 3. Analysis of crosslinking for an assembly composed of three identical subunits. The fraction of protein resolving under SDS-PAGE as i crosslinked polypeptide chain species ($1 \leq i \leq 3$) is denoted by f_i. For the definition of other symbols, see text.

In the approach of Sculley *et al.* (3), we define p, the probability for a given site to be occupied (for a crosslink to be formed), and u, the probability for it to be unoccupied. For a set of identical crosslinking sites governed by a single rate constant k_p,

$$u = e^{-k_p \tau} \tag{3}$$

$$p = 1 - e^{-k_p \tau} . \tag{4}$$

This approach was originally applied to a hexameric assembly.

The way to obtain a model that describes the symmetry and arrangement of subunits in an oligomer should now be clear. Theory permits calculation of predicted distributions of protein among crosslinked species, assuming a given model for subunit assembly. These distributions are then compared with the observed distribution at the same extent of reaction τ. The model which best describes the crosslinking behavior is the one which carries information for determining the subunit structure.

Application to Extracellular Hbs.   The crosslinking method has been successfully applied to extracellular Hbs from a parasitic nematode, *Ascaris suum*, and from an arthropod crustacean, *Caenestheria inopinata*, composed of eight and ten subunits, respectively. Dihedral symmetry was found for both Hbs (7, 8). In the case of *Caenestheria* Hb, the crosslinking data indicate a two-layered arrangement of subunits and strongly favor an eclipsed orientation of the two layers. This arrangement of subunits is in agreement with the one previously proposed on the basis of results from electron microscopy (9).

Choice of a Model.   To derive a model from crosslinking data, Sculley *et al.* compared observed fractions of crosslinked species plotted versus t, the duration of exposure to crosslinker with fractions predicted from the model plotted versus the extent of reaction τ (3). As seen from equation (2), τ is directly proportional to t, and the comparison is legitimate provided the concentration of crosslinker |B| is constant throughout the duration of the crosslinking reaction; for example, when the crosslinker is in a large excess relative to the amount of protein. Often however, the condition of excess reagent cannot be realized. To circumvent this difficulty, the fraction of remaining monomer may be used as a parameter for describing the extent of reaction. The usefulness of such a "polymer versus monomer" plot, which is independent of the relative concentrations of protein and crosslinker, is illustrated in Figure 4.

Consecutive Crosslinking. In the study of Hb from *Caenestheria*, equally good descriptions of the progress of crosslinking were provided by an alternating ring or by a two-layered eclipsed model. The ambiguity was resolved by carrying out consecutive crosslinking with two types of crosslinkers, a regular crosslinker and one that may be cleaved by sulfhydryl reagents like 2-mercaptoethanol (Figure 2) (8). The potential use of consecutive crosslinking in solving other types of ambiguities, for example between an eclipsed and a staggered orientation, is being explored.

Insufficient Resolution of the Electrophoretic Pattern. In the crosslinking of *Ascaris* (eight subunits) and *Caenestheria* (ten subunits) Hbs, all bands in the electrophoretic pattern were well-resolved. This was not the situation, however, with *Daphnia magna* Hb. Here, due to the larger number of subunits in the molecule, difficulties were encountered in resolving the low-mobility bands. The electrophoretic pattern (not shown) reveals, besides the monomeric band, 10 bands that may be identified as 2 to 11 crosslinked polypeptide chains and, in addition, a region of unresolved protein stain. Efforts are currently being made to overcome this problem.

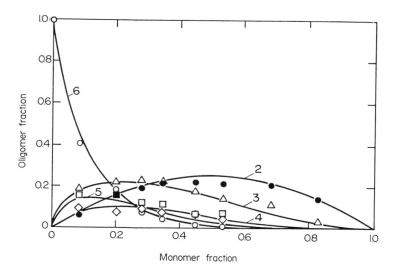

Figure 4. Distribution of protein among crosslinked species in urease exposed to glutaraldehyde. The observed fractions of protein in (●) dimer, (Δ) trimer, (□) tetramer, (◊) pentamer and (O) hexamer are plotted versus the fraction of remaining monomer: (—) theoretical curves, designated by numbers from 2 (dimers) to 6 (hexamers), calculated for an assembly of six identical subunits assuming a two-layered eclipsed model with $k_q/k_p = 2$, where k_q and k_p represent crosslinking rate constants for subunits in different and same layers, respectively.

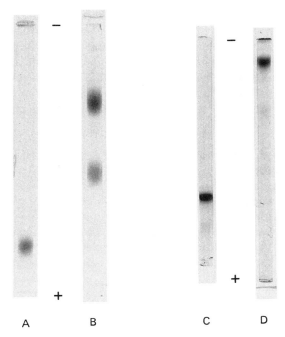

Figure 5. SDS-PAGE of oligomeric proteins following long exposures to crosslinker. Left: Human HbO_2 ($\approx$ 0.07 mg/ml): (A) not exposed and (B) exposed to glutaraldehyde for 22 h. The bands were identified, in decreasing order of mobilities, as monomer, dimer and tetramer. Right: Urease ($\approx$ 0.1 mg/ml): (C) not exposed and (D) exposed to glutaraldehyde for 24 h. The bands were identified, in decreasing order of mobilities, as monomer and hexamer.

Oligomer Dissociation. Current theories for crosslinking may not be applied to proteins that exist in dissociation-association equilibrium. It is important, therefore, to be able to distinguish between an oligomer that dissociates under the crosslinking conditions and one that does not. Crosslinking of an oligomeric protein undergoing dissociation is not expected to yield, even after long exposure to crosslinker, the fully crosslinked molecule as an exclusive product. This is illustrated in the crosslinking of human HbO_2 which is known to undergo a tetramer-dimer equilibrium with $K_d = 2 \times 10^{-6}$M. Here, appreciable amounts of dimeric crosslinked species are always present along with the tetrameric species in the electrophoretic pattern (Figure 5). The case of an oligomer that does not dissociate under the crosslinking conditions is illustrated by urease. Here, the electrophoretic pattern at long exposures to crosslinker shows the presence of virtually one band corresponding to the fully crosslinked hexamer (Figure 5). It is clear that a situation in which crosslinking of the whole subunit assembly may be attained opens the way for the application of theoretical treatments to derive a model for the crosslinking.

References

1. Hucho, F., Mullner, H. and Sund, H. (1975) *Eur J. Biochem.* **59**: 79–87.
2. Hajdu, J., Bartha, F. and Friedrich, P. (1976) *Eur. J. Biochem.* **68**: 373–383.
3. Sculley, M.J., Treacy, G.B. and Jeffrey, P.D. (1984) *Biophys. Chem.* **19**: 39–47.
4. Klotz, I.M., Darnall, D.M. and Langerman, N.R. (1975) *The Proteins*, eds. H. Neurath and R.L. Hill, 293–411. New York: Academic Press.
5. Matthews, B.W. and Bernhard, S.A. (1973) *Annu. Rev. Biophys. Bioeng.* **2**: 257–312.
6. Monod, J., Wyman, J. and Changeux, J.-P. (1965) *J. Mol. Biol.* **12**: 88–118.
7. Darawshe, S. and Daniel, E. (1991) In *Invertebrate Dioxygen Carriers,* ed. G. Préaux. Leuven: Leuven University Press. In press.
8. Tsfadia, Y., Shaked, I. and Daniel, E. (1991) In *Invertebrate Dioxygen Carriers,* ed. G. Préaux. Leuven: Leuven University Press. In press.
9. Ilan, E., David, M.M. and Daniel, E. (1981) *Biochemistry* **20**: 6190–6194.

8
Arthropod (*Cyamus scammoni*, Amphipoda) Hemoglobin Structure and Function

Nora B. Terwilliger

Oregon Institute of Marine Biology, Charleston, OR 97420 and Department of Biology, University of Oregon, Eugene, OR 97403, USA

Introduction

Respiratory protein expression in the Arthropoda generally occurs along phylogenetic lines. Hemocyanin, the blue, copper-containing respiratory protein, is found in the Chelicerata, in at least one Uniramia and in most of the Crustacea (1, 2). Hemoglobin, despite its ubiquitous appearance in both plant and animal kingdoms, is more limited in distribution amongst the Arthropoda than is Hc. The red, iron-containing Hb molecule has been described for only a few insects and four classes of Crustacea: the Branchiopoda, Ostracoda, Copepoda and Cirrepedia (3–7). Thus, the occurrence of Hb in an amphipod, a more advanced crustacean belonging to the class Malacostraca, is an unexpected finding. This paper describes preliminary studies on the structure and function of Hb from *Cyamus scammoni*, a cyamid amphipod that is an obligate ectosymbiont on the gray whale, *Eschrictius robustus*.

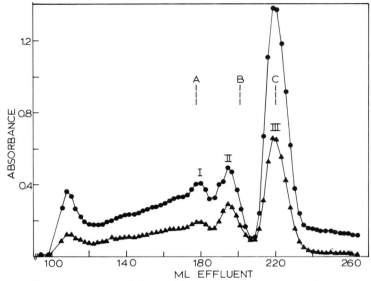

Figure 1.   Chromatography of *Cyamus* hemolymph on 1.8 x 109 cm column of BioGel A-5m.   Buffer, 0.1 ionic strength Tris-HCl, 0.1 M NaCl, 16 mM CaCl$_2$, 8 mM MgCl$_2$, pH 7.5.   Absorbance at 280 nm (●) and 417 nm (▲).   Calibrants: (A) *Helisoma trivolvis* Hb, $M_w = 1.7 \times 10^6$.   (B) *Cancer magister* 25 S Hc, $M_w =$ 940,000.   (C) *C. magister* 16 S Hc, $M_w = 450,000$.

Materials and Methods

Live *Cyamus* were removed from the surface of several gray whales, *E. robustus*, that had just died.   Hemolymph samples were obtained by microcapillary pipette from the base of the legs, the spiral gill structures and the oostegites of the brood pouch.   In order to prevent gelation and proteolysis, the red hemolymph was diluted immediately in an equal volume of ice cold Tris-HCl buffer, 0.1 M NaCl, 16 mM CaCl$_2$, 8 mM MgCl$_2$, pH 7.5, containing 1 mM phenylmethyl sulfonyl fluoride, a protease inhibitor, and centrifuged at 2000X g for 5 min (4°C).   The red color remained in the supernatant, which was then centrifuged at 10,000 g for 15 min (4°C) and used for further analysis.

Results and Discussion

Spectral analysis of the centrifuged hemolymph showed absorbance maxima at 417, 544 and 577 nm, corresponding to an oxyHb spectrum.   Flushing the sample with CO shifted the maxima to 420, 542 and 577 nm, typical carboxyHb values.   Thus, these amphipods contain Hb, and constitute the first example of Hb expressed in a malacostrachan.

Extracellular Hbs range in size from 15 kDa in the insect larva, *Chironomus*, to 12,000 kDa in the bivalve *Cardita* (8, 9). In order to determine the size of this amphipod Hb molecule, the hemolymph was purified on a column of BioGel A-5m (Figure 1). The hemolymph fractionated into three peaks. Peak I has an apparent M_w of 1,800 kDa, peak II has an M_w of 1,100 kDa, and peak III, an M_w of about 500 kDa. When peak I was rechromatographed on the same column, it partially dissociated into the same three peaks. This result suggests that the peaks are in a state of association-dissociation with the 500 kDa peak III, associating into perhaps a dimeric and then a tetrameric polymer. All three peaks absorb at 417 nm, indicating the presence of heme groups; the absorbance spectrum is indistinguishable for all three peaks and resembles that of the crude hemolymph. Furthermore, the 280/417 nm absorbance ratio, the proportion of protein to heme, is similar for all three peaks.

When *Cyamus* hemolymph, negatively stained with 0.5% uranyl acetate, was examined by TEM, almost all of the molecules appeared to be about the same size and various orientations of the same kind of molecule. The predominant structure is a chevron-shaped aggregate. The apparent molecular homogeneity of the hemolymph is in marked contrast to the variety of molecules, including Hb, present in the hemolymph of the barnacle, *Briarosaccus callosus*, parasitic on the king crab (7). TEM of purified *Cyamus* Hb showed the same chevron-shaped molecules as did the hemolymph, an architecture unique among the invertebrate Hbs (10).

Hemolymph was removed directly from an active *Cyamus*, mixed with boiling SDS incubation buffer containing reducing agents and run on SDS-PAGE along with M_w standards and an aliquot of gray whale blood (11). The protein electrophoresed as a major component with an M_w of 175 kDa, regardless of whether the hemolymph was obtained from leg, gill or oostegite. No 15–16 kDa component was observed. These results, in agreement with the M_w of the native molecule and the TEM results, indicate first that the red fluid in the gills of *Cyamus* is not ingested whale Hb. Secondly, the large size of this Hb subunit resembles the 125 kDa M_w polypeptide chain made up of covalently linked heme-containing domains that is found in the extracellular Hb of the branchiopod crustacean, *Artemia* (12). Preliminary heme analysis and proteolysis experiments suggest that the *Cyamus* Hb subunit is made up of a series of low M_w domains; unlike *Artemia* Hb, however, not all of the domains contain a heme. Other crustacean extracellular Hbs have subunits ranging in size from 16 kDa M_w monomers containing one heme group in the barnacle *Briarosaccus* (7), to 34 kDa M_w subunits with two heme O_2-binding sites per chain in some branchiopods, such as *Lepidurus* and *Daphnia* (4, 13). There is clearly a great deal of variability amongst the crustacean Hbs with respect to correspondence between the M_w of the intact molecule and length of the polypeptide chain.

The extracellular Hb of *Cyamus* binds O_2 reversibly. Figure 2 illustrates the change in spectra around 417 nm from deoxy to oxyHb. Compared to many other Hbs (14), *Cyamus* Hb has a relatively low O_2 affinity with P_{50} values of

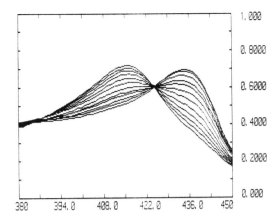

Figure 2. Change in spectra from deoxy to oxyHb of purified *Cyamus* Hb. Buffer, 0.1 ionic strength Tris-HCl, 0.1 M NaCl, 16 mM CaCl$_2$, 8 mM MgCl$_2$, pH 7.3. Ordinate, absorbance; abscissa, wavelength in nm.

14–24 torr between pH 7.0 and 7.9 and a cooperativity of about 1.5. While this cooperativity is lower than that of many annelid extracellular Hbs, for example, there are other large invertebrate Hbs which show minimal or no cooperativity (14). The moderately low affinity Hb should be well-suited to loading O$_2$ at the gills of *Cyamus* as its host, the gray whale, swims through the well-oxygenated waters of the Pacific Ocean.

The presence of a high M_w extracellular Hb in this amphipod is the first report of Hb expression in the more advanced crustaceans, the peracaridans and eucaridans. It is particularly intriguing since other peracaridans previously studied, the isopods, amphipods and mysids, contain Hc as their respiratory protein. One wonders whether both the Hb and Hc genomes might be present throughout the crustaceans, though only one is expressed within a particular genus. If this is so, what are the factors that determine the expression of one respiratory protein or the other? Future investigations on crustacean Hb and Hc genomes and their regulation may help to answer these questions.

Acknowledgements

This work is dedicated to the memory of Robert C. Terwilliger, husband and colleague, whose enthusiasm for the beauty and diversity of invertebrates and their respiratory proteins has inspired all who worked with him. I am grateful to Steven Jeffries, of the Washington Department of Wildlife, Marine Mammal Investigations, for providing some of the amphipods. The study was supported by National Science Foundation grants DMB 85-11150 and DCB 89-08362.

References

1. Van Holde, K.E. and Miller, K. (1982) *Q. Rev. Biophys.* **15**: 1–129.
2. Mangum, C.P., Scott, J.L., Black, R.E.I., Miller, K.I. and Van Holde, K.E. (1985) *Proc. Natl. Acad. Sci. U.S.A.* **82**: 3721–3725.
3. Fox, H.M. (1957) *Nature* **179**: 148–150.
4. Ilan, E. and Daniel, E. (1979) *Comp. Biochem. Physiol.* **63**B: 303–308.
5. Dangott, L.J. and Terwilliger, R.C. (1981) *Comp. Biochem. Physiol.* **70**B: 549–537.
6. Wolf, G., Van Pachtenbeke, M., Moens, L. and Van Hauwaert, M. (1983) *Comp. Biochem. Physiol.* **76**B: 731–736.
7. Terwilliger, R.C., Terwilliger, N.B. and Schabtach, E. (1986) In *Invertebrate Oxygen Carriers*, ed. B. Linzen, 125–127. Berlin: Springer-Verlag.
8. Thompson, P., Bleeker, W. and English, D.S. (1968) *J. Biol. Chem.* **243**: 4463–4467.
9. Terwilliger, N.B. and Terwilliger, R.C. (1978) *Biochim. Biophys. Acta* **537**: 77–85.
10. Terwilliger, R.C. (1980) *Am. Zool.* **20**: 53–67.
11. Terwilliger, N.B. and Terwilliger, R.C. (1982) *J. Exp. Zool.* **221**: 181–191.
12. Moens, L. and Kondo, M. (1978) *Eur. J. Biochem.* **82**: 65–72.
13. Dangott, L.J. and Terwilliger, R.C. (1979) *Biochim. Biophys. Acta* **579**: 452–461.
14. Weber, R.E. (1980) *Am. Zool.* **20**: 79–101.

9
The Principal Subunit of Earthworm Hemoglobin is a Dodecamer of Heme-Containing Chains

Pawan K. Sharma,[a] Aziz N. Qabar,[a]
Oscar H. Kapp,[b] Joseph S. Wall[c] and
Serge N. Vinogradov[a]

[a]Department of Biochemistry, Wayne State University School of Medicine, Detroit, MI 48201, USA
[b]Enrico Fermi Institute, University of Chicago, Chicago, IL 60637, USA
[c]Department of Biology, Brookhaven National Laboratory, Upton, NY 11973, USA

Introduction

The extracellular Hbs of the annelids are giant molecules with an acidic isoelectric point, a sedimentation coefficient of 60 S and a low iron content of 0.23 ± 0.03 wt.% (1–6). In addition, these Hbs have a symmetrical, hexagonal bilayer appearance in electron micrographs (7–9). The extracellular Hb of *Lumbricus terrestris* is the most well-studied Hb of the annelid group with an estimated M_w of about 3,600 kDa (10). SDS-PAGE shows that it consists of four subunits: monomer, M (17 kDa), linkers, $D1$ and $D2$ (*ca.* 30 kDa) and trimer, T (50 kDa). The subunits M and T contain heme and consist of globin chains I and disulfide-bonded chains II, III and IV, respectively. The complete amino acid sequences of these chains are known (11–13). The $D1$ and $D2$ subunits contain very little heme, and consist of two chains, VA and VB, and of chain VI, respectively (14). Studies of the dissociation of *Lumbricus* Hb at alkaline pH (15), at acid pH (14) and at neutral pH in the presence of various dissociating agents (16) have shown that, although dissociation products having an estimated M_m of *ca.* 200 kDa were obtained under equilibrium conditions, they never had the same subunit composition as does the native Hb. To explain the results, a heteromultimeric model was proposed (16, 17) in which a "bracelet," or scaffolding, of 30 kDa "structural" subunits having little or no heme is decorated by twelve complexes (each of *ca.* 200 kDa) of heme-containing subunits, each complex consisting of three copies of the monomer (17 kDa) and three copies of the disulfide-bonded trimer (51 kDa). In this work, we report the isolation and characterization of the *ca.* 200 kDa complex by mild dissociation of *Lumbricus* Hb at neutral pH.

Materials and Methods

Live *Lumbricus* were obtained from Forest Bait Farm (London, Ontario, Canada). The Hb was prepared as described by Shlom and Vinogradov (18), and was kept in CO-saturated 0.1 M Tris HCl buffer pH 7.2, 1 mm EDTA. Repeated dissociation of *Lumbricus* Hb was carried out according to the method

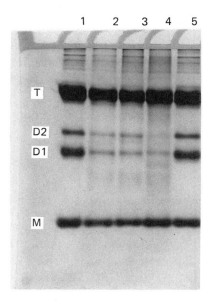

Figure 1.   Unreduced SDS-PAGE pattern of *Lumbricus* Hb (lanes 1 and 5), first generation peak I*b* (lane 2), second generation peak I*b'* (lane 3) and third generation peak I*b"* (lane 4) obtained by repeated dissociation in 4 M urea and gel filtration on a 6.5 x 50 cm Sephadex G-200 column (first generation peak) and a 5.5 x 105 cm Sephadex G-100 column (second and third generation peaks) at neutral pH.

Table 1. Relative proportions (% of total area) of subunits T, $D1$, $D2$ and M in the three generation peaks Ib, Ib' and Ib'' obtained by repeated dissociation in 4 M urea and gel filtration at neutral pH, determined by densitometry of unreduced SDS-PAGE patterns.

Subunit	Control ($n = 10$)	Peak Ib ($n = 6$)	Peak Ib' ($n = 6$)	Peak Ib'' ($n = 6$)	Peak R
M	20 ± 1.6	24 ± 2.5	24 ± 3.2	26 ± 1.9	20
$D1$ and $D2$	24 ± 3.3	6.0 ± 1.9	5.3 ± 2.0	2.3 ± 1.4	22
T	56 ± 3.7	70 ± 3.3	70 ± 4.2	71 ± 2.9	58

of Vinogradov *et al.* (16) in the presence of urea at neutral pH using columns of Sephadex G-200 and G-100 for gel filtration. Analytical gel filtration was performed on 1 x 30 cm columns of Superose S-6 (Pharmacia, Piscataway, NJ, USA) using a low pressure FPLC system (Pharmacia). SDS-PAGE was carried out in 0.1% SDS using 1 mm x 10 cm x 20 cm slab gels and the buffer system of Laemmli (19). The gels were scanned on a Zeineh densitometer and the areas under the peaks determined by weighing. Unstained, freeze-dried specimens were used for mass measurements by the Brookhaven STEM using the method of Wall and Hainfeld (20). The negatively stained samples were examined as described by Kapp *et al.* (21) in the dark field mode with a high-resolution field emission STEM at the University of Chicago.

Results and Discussion

The first generation peak Ib obtained by gel filtration on a Sephadex G-200 column after the dissociation of *Lumbricus* Hb in 4M urea was subjected to dialysis against 4 M urea and gel filtration on a Sephadex G-100 column to obtain the second generation peak Ib'. The latter was treated similarly to produce the third generation peak Ib''.

Figure 1 depicts the SDS-PAGE pattern of unreduced *Lumbricus* Hb (lanes 1 and 5), the first generation peak Ib (lane 2), the second generation peak Ib' (lane 3) and the third generation peak Ib'' (lane 4), each obtained by repeated dissociation in 4 M urea. The results of the densitometry of the unreduced SDS-PAGE patterns are shown in Table 1. Peak Ib is the largest fragment of *Lumbricus* Hb obtained by its dissociation in 4 M urea at neutral pH. Yet, its content of subunits $D1$ and $D2$ is only about 25% of the native molecule, demonstrating clearly that Ib is not a protomer and that the Hb is heteromultimeric, which is in agreement with previous results (16). Further dissociation results in a decrease in the relative proportions of subunits $D1$ and

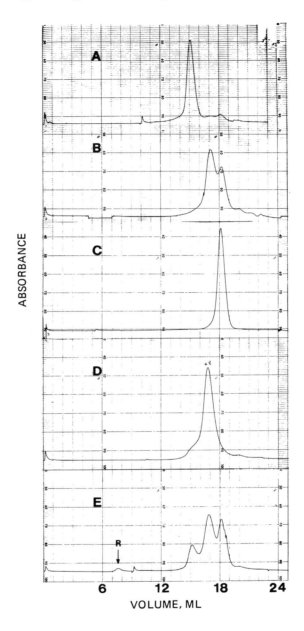

Figure 2. FPLC elution profiles on a 1 x 30 cm column of Superose 6: (*A*) peak I*b*′;
(*B*) peak I*b*′ in 4 M Gdn·HCl; (*C*) peak II′; (*D*) peak III′ and (*E*) an equimolar mixture of
peaks II′ and III′.

$D2$ from 6.0 ± 1.9 in peak Ib to 5.3 ± 2.0 in peak Ib', and 2.3 ± 1.4 in peak Ib''. Analytical gel filtration on Superose S6 column indicates an increase in the elution volumes by 0.05 ml from peak Ib to peak Ib', and by 0.1 ml from peak Ib' to peak Ib'', with the relative changes in mass proportional to the changes in the percent content of $D1$ and $D2$ (Table 1).

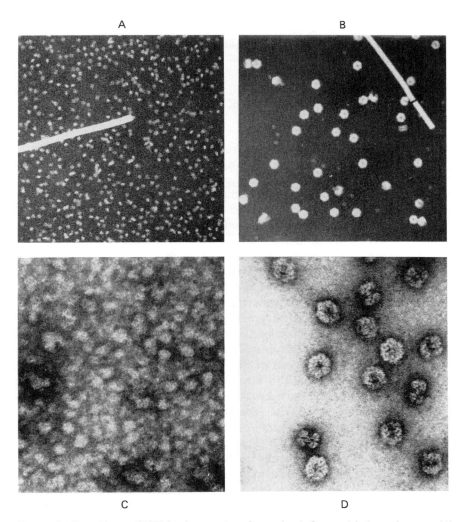

Figure 3. Brookhaven STEM micrographs of unstained, freeze-dried specimens: (*A*) peak Ib' and (*B*) *Lumbricus* Hb. The dimensions of the field of view are 1525 x 1525 Å. University of Chicago STEM micrographs of negatively stained specimens: (*C*) peak Ib' and (*D*) peak R in Figure 2 (magnification x 149,000).

Table 2. Results of mass determinations by gel filtration and of STEM mass measurements.

Material	Gel Filtration (kDa)	STEM (kDa)	Calculated[a] (kDa)
Peak I*b'*	191	197 ± 19 ($N = 91$)	209
Peak II (subunit *T*)	50	–	52.9
Peak III (subunit *M*)	17.2	–	16.75
Peak II+III	182	201 ± 39 ($N = 65$)	–
Lumbricus Hb[b]	–	188 ± 21 ($N = 18$)	–

[a]From amino acid sequences (12, 13).
[b]Calculated for the particles present in Figure 3B due to the partial dissociation of *Lumbricus* Hb during specimen preparation.

The elution profile obtained by analytical gel filtration of peak I*b'* at neutral pH on Superose S6 column is shown in Figure 2A. Only one peak was observed and its mass, estimated from plots of log M versus elution volume, was 191 kDa (Table 2). This particle could be dissociated completely at neutral pH in 4 M guanidine HCl into the constituent subunits M and T (Figure 2B). Figures 2C and D show the FPLC elution profiles of the subunits M and T, isolated concurrently with peak I*b'*. As shown in Figure 2E, mixing the subunits M and T resulted in the partial reassociation into particles the elution volume of which was the same as that of peak I*b'* in Figure 2A, and the mass of which was determined to be *ca*. 182 kDa.

The mean mass of the unstained, freeze-dried specimen of peak I*b'* obtained with the Brookhaven STEM (Figure 3A) was 197 ± 19 kDa (Table 2). Figure 3B shows a micrograph of unstained, freeze-dried *Lumbricus* Hb. It is apparent that some of the hexagonal bilayer molecules have dissociated into particles of a size which corresponds to that of the I*b'* particles. The mean mass of the particles in Figure 3B is 188 kDa (Table 2). Figures 3C and D show micrographs of negatively stained peak I*b'* and peak R (Figure 2E) obtained with the University of Chicago STEM. Due to the presence of traces of subunits $D1$ and $D2$, it appears that some reassociation into complete hexagonal bilayer molecules has occurred: peak R (Figure 2E) has an SDS-PAGE composition similar to that of the native *Lumbricus* Hb (Table 1).

The experimental results clearly show that a particle of *ca*. 200 kDa, consisting of subunits T and M, exists within the structure of *Lumbricus* Hb. The experimental estimates of its mass are in very good agreement with the theoretical value of 209 kDa for a trimer of the T and M complex, calculated using the masses from the known amino-acid sequences of the four chains I, II, III and IV (Table 2). Hence, a 209 kDa dodecameric complex, 3(I + II + III + IV), represents the largest subunit of *Lumbricus* Hb. This subunit has been shown recently to have the same oxygen affinity as does the native Hb but only two thirds of its cooperativity (22). The demonstration of the existence of a stable and functional dodecameric complex of globin chains within the quaternary structure of *Lumbricus* Hb, represents an important step in the understanding of the architecture of annelid Hbs and Chls and provides very convincing support for a model wherein twelve dodecameric subunits arranged in two hexagonal layers are held together by heme-deficient, *ca.* 30 kDa chains (16, 17).

Acknowledgements

This research was partially supported by National Institutes of Health grant DK 38674 (S.N.V.) and by Department of Energy grant DE-FG02-86-ER60437 (O.H.K.). The Brookhaven STEM facility is supported by the U.S. Department of Energy and by the National Institutes of Health Biotechnology Resources Branch grant number RR 01777 (J.S.W.).

References

1. Mangum, C. (1976) In *Adaptations to Environment: Physiology of Marine Animals*, ed. P.C. Newell, 191–278. London: Butterworth's.
2. Weber, R.E. (1978) In *Physiology of Annelids*, ed. P. J. Mill, 393–445. London: Academic Press.
3. Chung, M.C.C. and Ellerton, H.D. (1979) *Prog. Biophys. Mol. Biol.* **35**: 51–102.
4. Vinogradov, S.N., Shlom, J.M., Kapp, O.H. and Frossard, P. (1980) *Comp. Biochem. Physiol.* **67**B: 1–16.
5. Vinogradov, S.N. (1985) *Comp. Biochem. Physiol.* **82**B: 1–15.
6. Vinogradov, S.N. (1985) In *Respiratory Pigments in Animals*, eds. J. Lamy, J.P. Truchot and R. Gilles, 9–20. Berlin: Springer-Verlag.
7. Levin, O. (1963) *J. Mol. Biol.* **6**: 93-99.
8. Kapp, O.H., Vinogradov, S.N., Ohtsuki, M. and Crewe, A.V. (1982) *Biochim. Biophys. Acta* **704**: 546–548.
9. Vinogradov, S.N., Kapp, O.H. and Ohtsuki, M. (1982) In *Electron Microscopy of Proteins*, ed. J. Harris, vol. 3, 135–163.
10. Vinogradov, S.N. and Kolodzej, P. (1988) *Comp. Biochem. Physiol.* **91**B: 577–579.
11. Garlick, R.L. and Riggs, A. (1982) *J. Biol. Chem.* **257**: 9005–9015.
12. Shishikura, F., Snow, J.W., Gotoh, T., Vinogradov, S. N. and Walz, D. A. (1987) *J. Biol. Chem.* **262**: 3123–3131.
13. Fushitani, K., Matsuura, M.S.A. and Riggs, A.F. (1988) *J. Biol. Chem.* **263**: 6502–6517.
14. Mainwaring, M.G., Lugo, S.D., Fingal, R.A., Kapp, O.H. and Vinogradov, S.N. (1986) *J. Biol. Chem.* **261**: 10899–10908.
15. Kapp, O.H., Polidori, G., Mainwaring, M.G., Crewe, A.V. and Vinogradov, S.N. (1984) *J. Biol. Chem.* **259**: 628–639.

16. Vinogradov, S.N., Lugo, S.L., Mainwaring, M.G., Kapp, O.H. and Crewe, A.V. (1986) *Proc. Natl. Acad. Sci. U.S.A.* **83**: 8034–8038.
17. Vinogradov, S.N. (1986) In *Invertebrate Oxygen Carriers*, ed. B. Linzen, 25–36. Berlin: Springer-Verlag.
18. Shlom, J. and Vinogradov, S.N. (1973) *J. Biol. Chem.* **248**: 7904–7912.
19. Laemmli, U. (1970) *Nature* **227**: 680–685.
20. Wall, J.S. and Hainfeld, J.F. (1986) *Annu. Rev. Biophys.* **15**: 355–376.
21. Kapp, O.H., Qabar, A.N., Bonner, M.C., Stern, M.S., Walz, D.A., Schmuck, M., Pilz, I., Wall, J.S. and Vinogradov, S.N. (1990) *J. Mol. Biol.* **213**: 141–158.
22. Vinogradov, S.N., Sharma, P.K., Qabar, A.N., Wall, J.S., Westrick, J.A., Simmons, J.H. and Gill, S.J. (1991) *J. Biol. Chem.* In press.

10
Heterogeneity of the Products of Dissociation of *Lumbricus terrestris* Hemoglobin at Alkaline pH

Pawan K. Sharma,[a] Serge N. Vinogradov[a] and Daniel A. Walz[b]

Departments of Biochemistry[a] and Physiology,[b] Wayne State University School of Medicine, Detroit, MI 48201, USA

Introduction

The extracellular Hb of the common North American earthworm *Lumbricus terrestris* is a symmetrical, hexagonal bilayer molecule with an acidic isoelectric point and an iron content of about 2/3 the normal value of 0.33 wt.% (1–5). It has an M_r of *ca.* 3,600 kDa, contains 140–150 heme groups and consists of *ca.* 180 polypeptide chains divided into two classes: four different *ca.* 17 kDa, heme-containing chains and at least three chains of *ca.* 30 kDa (6) which contain little or no heme (7).

The separation of the dissociation products of *Lumbricus* Hb at alkaline pH by gel filtration was first reported by Shlom and Vinogradov (8). The basic profile, in addition to the undissociated Hb, was comprised of two peaks: the first peak to elute consisted of the disulfide-bonded trimer (9) of globin chains and of the nonreducible "dimer" chains of *ca.* 30 kDa (6), and the last peak to elute consisted of the monomer subunit *M*, one of the four heme-containing chains (10). The amino acid sequences of the three chains of the trimer have been reported by Fushitani *et al.* (11), and those of the monomer by Shishikura *et al.* (12). The amino acid sequences of the "dimer" chains are not known. Similar patterns of dissociation at alkaline pH have been demonstrated for many other hexagonal bilayer Hbs such as those of *Arenicola cristata* (13, 14), *Thelepus crispus* (15), *Tylorrhynchus heterochaetus* (16), *Abarenicola affinis affinis* (17), *Tubifex tubifex* (18), *Lumbricus rubellus* (19) and *Maodrilus montanus* (20).

In the present study, we report the proportions of the three subunits of *Lumbricus* Hb, *T*, *D*1 and *D*2 and *M*, obtained by the SDS-PAGE of fractions across the elution profile of the Hb subjected to gel filtration at alkaline pH on different column media.

Materials and Methods

Lumbricus Hb was prepared from live worms as described by Shlom and Vinogradov (9). Gel filtration was carried out on a 2.2 x 65 cm column of Ultragel AcA44, a 5.5 x 65 cm column of Sephadex G-200 and a 5.5 x 105 cm column of Sephadex G-100, using 0.1 M sodium borate buffer pH 9.8. The loads ranged from 80 mg on the smaller columns to 300 mg on the larger one.

SDS-PAGE was carried out using minigels, 1.5 mm x 8 cm x 6 cm, of 15% polyacrylamide and the Laemmli buffer system (21). Protein loads varied from 12 to 60 mg. The gels were stained for periods ranging from 0.5 hr to 18 hrs with 0.1% Coomassie Brilliant Blue R 250. The destained gels were scanned on a Zeineh Soft Laser Scanning Densitometer. The areas under the peaks in the tracings were determined by their weight.

Results and Discussion

Figure 1 shows the elution profile of *Lumbricus* Hb at pH 9.8 on a column of Ultragel AcA44. At this pH, the molecule is completely dissociated into two peaks, *A* and *B* (22). Figure 2 shows the SDS-PAGE patterns of several fractions taken throughout the elution profile; the areas under the peaks of the corresponding densitometric scans are presented in Table 1 as percent of the total stained area. Peak *A* consists of subunit *T*, the disulfide-bonded trimer and subunits *D*1 and *D*2 in varying proportions. The trough between peaks *A* and *B* and the left-hand moiety of peak *B* consists of a mixture of the subunits, with *D*1 and *D*2 predominant. The proportions of the three subunits vary along the elution profile of peak *A* and no one fraction consists of a pure subunit. Similar results were obtained with Sephadex G-200 and Sephadex G-100 columns. This is not in agreement with the reported separation of *Lumbricus* Hb into trimer, *T*, and monomer, *M*, subunits by gel filtration at alkaline pH on a column of Ultragel AcA44 (23). The right-hand side of peak *B* consists of pure monomer, as reported earlier (8, 10).

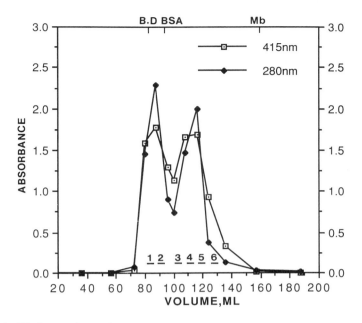

Figure 1. Elution profile of *Lumbricus* Hb on a 2.2 x 65 cm column of Ultragel AcA44 in 0.1 M sodium borate buffer pH 9.8, 1 mM EDTA.

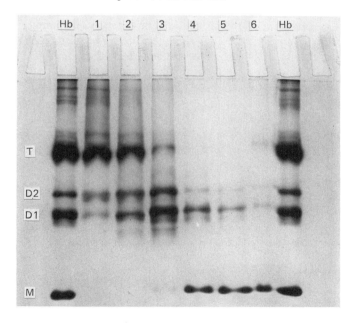

Figure 2. Unreduced SDS-PAGE of the fractions indicated across the elution profile in Figure 1. First and last lanes: *Lumbricus* Hb.

Table 1. Relative proportions of subunits $T, D1, D2$ and M across the elution profile of *Lumbricus* Hb on an Ultragel AcA 44 column at pH 9.8 (Figure 1), determined by densitometry of unreduced SDS-PAGE patterns (Figure 2).

Subunit	Control	Fractions					
		1	2	3	4	5	6
M	23 ± 3	0	0	0	48	69	94
$D1$ and $D2$	22 ± 5	10	39	88	52	31	6
T	55 ± 5	90	61	12	0	0	0

We considered the possibility that the discrepancy between our results, which show that the first peak consists of variable proportions of $D1$ and $D2$ and T, and the results reported by Fushitani *et al.* (23), which indicate that the first peak consists only of the T subunit, could be due to differences in SDS-PAGE. Hence, we investigated whether the relative proportions of subunits $T, D1$ and $D2$ were affected either by protein load or by the duration of staining. SDS-PAGE of individual fractions from peak A was performed with protein loads of 12 to 60 mg and with staining periods of 0.5 to 18 hrs. The densitometric results did not reveal any systematic variation in the proportion of T to $D1$ and $D2$ subunits as a function of the protein load or duration of staining. It is clear from our results that gel filtration at alkaline pH may be used to obtain a pure subunit M, but that subunit T cannot be obtained free of subunits $D1$ and $D2$.

Acknowledgements

This research was supported in part by grants from the National Institutes of Health, DK 38674 (S.N.V.) and DK 30382 (D.A.W.).

References

1. Mangum, C. (1976) In *Adaptations to Environment: Physiology of Marine Animals*, ed. P.C. Newell, 191–278. London: Butterworth's.
2. Weber, R.E. (1978) In *Physiology of Annelids*, ed. P. J. Mill, 393–445. London: Academic Press.
3. Chung, M.C.C. and Ellerton, H.D. (1979) *Prog. Biophys. Mol. Biol.* **35**: 51–102.
4. Terwilliger, R.C. (1980) *Am. Zool.* **20**: 53–67.
5. Vinogradov, S.N., Shlom, J.M., Kapp, O.H. and Frossard, P. (1980) *Comp. Biochem. Physiol.* **67**B: 1–16.

6. Walz, D.A., Snow, J.S., Mainwaring, M.G. and Vinogradov, S.N. (1987) *Fed. Pro.* **46**: 2266.

7. Mainwaring, M.G., Lugo, S.D., Fingal, R.A., Kapp, O.H. and Vinogradov, S.N. (1986) *J. Biol. Chem.* **261**: 10899–10908.

8. Shlom, J. and Vinogradov, S.N. (1973) *J. Biol. Chem.* **248**: 7904–7912.

9. Shishikura, F., Mainwaring, M.G., Yurewicz, E., Lightbody, J.L., Walz, D.A. and Vinogradov, S.N. (1986) *Biochim. Biophys. Acta* **869**: 314–321.

10. Vinogradov, S.N., Shlom, J.M., Hall, B.C., Kapp, O.H. and Mizukami, H. (1977) *Biochim. Biophys. Acta* **492**: 136–155.

11. Fushitani, K., Matsuura, M.S.A. and Riggs, A.F. (1988) *J. Biol. Chem.* **263**: 6502–6517.

12. Shishikura, F., Snow, J.W., Gotoh, T., Vinogradov, S.N. and Walz, D.A. (1987) *J. Biol. Chem.* **262**: 3123–3131.

13. Waxman, L. (1971) *J. Biol. Chem.* **216**: 7318–7327.

14. Vinogradov, S.N., Shlom, J.M. and Doyle, M. (1979) *Comp. Biochem. Physiol.* **65**B: 145–150.

15. Garlick, R.L. and Terwilliger, R. (1975) *Comp. Biochem. Physiol.* **51**A: 849–857.

16. Gotoh, T. (1980) *J. Sci. Univ. Tokushima* **13**: 1–7.

17. Chung, M.C.C. and Ellerton, H.D. (1982) *Biochim. Biophys. Acta* **702**: 17–22.

18. Polidori, G., Mainwaring, M.G., Kosinski, T., Schwarz, C., Fingal, R. and Vinogradov, S.N. (1984) *Arch. Biochem. Biophys.* **233**: 800–814.

19. Ellerton, H.D., Chen, C.T. and Lim, A.K. (1987) *Comp. Biochem. Physiol.* **87**B: 1011–1016.

20. Ellerton, H.D., Bearman, C.H. and Loong, P.C. (1987) *Comp. Biochem. Physiol.* **87**B: 1017–1023.

21. King, J. and Laemmli, U.K. (1971) *J. Mol. Biol.* **62**: 465–477.

22. Kapp, O.H., Polidori, G., Mainwaring, M.G., Crewe, A.V. and Vinogradov, S.N. (1984) *J. Biol. Chem.* **259**: 628–239.

23. Fushitani, K., Imai, K. and Riggs, A.F. (1986) In *Invertebrate Oxygen Carriers*, ed. B. Linzen, 77–79. Berlin: Springer-Verlag.

11
Studies on the Dissociation of *Eudistylia vancouverii* Chlorocruorin

Aziz N. Qabar,[a] Oscar H. Kapp,[b]
Joseph S. Wall[c] and Serge N. Vinogradov[a]

[a]Department of Biochemistry, Wayne State University School of Medicine, Detroit, MI 48201, USA
[b]Enrico Fermi Institute, University of Chicago, Chicago, IL 60637, USA
[c]Brookhaven National Laboratory, Upton, NY 11973, USA

Introduction

Annelid extracellular Hbs and Chls form one of four groups which may be distinguished on the basis of their quaternary structure (1). These molecules have a characteristic hexagonal bilayer appearance in electron micrographs, a sedimentation coefficient of about 60 S, an acidic isoelectric point and a low iron content of $0.23 \pm 0.03\%$ (2–6).

Recent work on the dissociation of *Lumbricus terrestris* Hb, the best-studied of these molecules, at alkaline pH (7), at acid pH (8) and at neutral pH in the presence of a number of dissociating agents (9) has clearly demonstrated that this giant, hexagonal bilayer molecule is heteromultimeric. A working model of *Lumbricus* Hb was suggested by Vinogradov *et al.* (9), in which a "bracelet" of "structural," heme-free, *ca.* 30 kDa subunits $D1$ and $D2$ is decorated with twelve complexes of functional, heme-containing, *ca.* 200 kDa subunits consisting of three copies of the monomeric subunit M and three copies of the trimeric subunit T, the latter of which is a disulfide-bonded trimer of three different chains (10).

We have extended our studies on the dissociation of *Lumbricus* Hb to Chls that differ from the annelid Hbs only in having an altered heme group in which the vinyl group at position 2 is substituted by a formyl group (4). The extracellular Chl of the marine polychaete *Eudistylia vancouverii* appears as an hexagonal bilayer in electron micrographs (11, 12) and consists of two types of subunits: subunit 1 consisting of at least two different chains of 30 kDa; and subunit 2, a disulfide-bonded tetramer of at least two different 16 kDa chains (13). In this work we report some preliminary results on the dissociation of *Eudistylia* Chl in 4M urea at neutral pH.

Materials and Methods

The Chl was prepared from live worms collected from Sequim Bay (Sequim, WA, USA). The Chl was prepared as described earlier (1). Dissociation in urea was carried out as described by Vinogradov *et al.* (9). The separation of subunits was by gel filtration on a 5 x 105 cm Sephadex G-100 column at 4°C and CO-saturated, 0.1 M Tris.Cl buffer, pH 7.0, 1 mM EDTA.

Analytical gel filtration was carried out on a 1 x 30 cm column of Superose S12 using an FPLC system (Pharmacia, Piscataway, NJ, USA). Flow rates of 0.4 ml/min were used, and the eluate was monitored at 280 nm. Proteins ranging in mass from 665 kDa to 17 kDa were used as M_m standards.

SDS-PAGE was carried out in 0.1% SDS in the presence or absence of 2-mercaptoethanol, using 1.5 x 10 x 20 cm slab gels and the buffer system of Laemmli (14). The gels were electrophoresed for two hours at 15 mA/gel, stained with Coomassie Brilliant Blue R250 in 25% methanol and 7.5% acetic acid solution overnight, and destained in the same solvent.

Mass measurements of unstained, freeze-dried specimens were carried out with the STEM at Brookhaven National Laboratory as described by Wall and Hainfeld (15). Negatively stained specimens, prepared as described by Kapp *et al.* (16), were examined with the STEM at the University of Chicago.

Results and Discussion

The Chl dissociated in 4M urea and subjected to gel filtration on a Sephadex G-100 column produced three peaks. The profile was divided into five fractions, *A* through *E*. The pooled, concentrated fractions were examined by FPLC on Superose 12 column, and their subunit compositions were determined by SDS-PAGE. The results are shown in Figures 1 and 2.

Fraction *A* (Figure 1*A*) consisted of the undissociated material and peak *a* eluting at *ca.* 12 ml. Fraction *B* consisted of a single peak *a* also eluting at *ca.* 12 ml. Fraction *C* consisted of two overlapping peaks, *a* and *b*, eluting at *ca.* 12 ml and 13.4 ml, respectively. Fraction *D* showed a single peak *b* eluting at 13.4 ml and corresponding to an M_m of 66 kDa. Fraction *E* provided a single peak

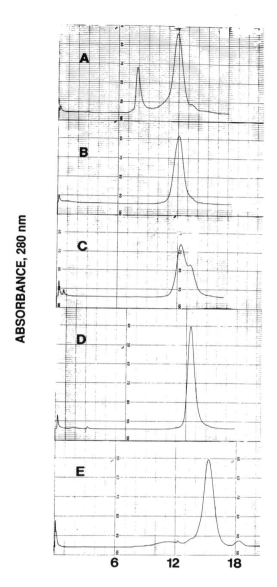

VOLUME, ML

Figure 1. Elution profiles at 280 nm of the five fractions *A* through *E* on a 1 x 30 cm Superose S-12 column in 0.1 M Tris.HCl buffer pH 7.0, 1 mM EDTA.

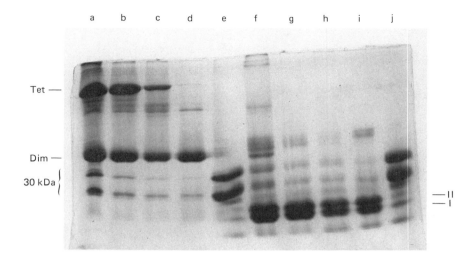

Figure 2. SDS-PAGE of the five fractions *A* through *E* produced by dissociation of *Eudistylia* Chl in 4 M urea. Lanes *a, b, c, d* and *e* are without 2-mercaptoethanol, lanes *f, g, h, i* and *j* are with 2-mercaptoethanol.

eluting at 15.3 ml, corresponding to an M_m of about 30 kDa, and consisted of the two chains of subunit 1 (Figure 2, lane *e*). The reduced Fractions *B, C* and *D* had the same SDS-PAGE pattern (Figure 2, lanes *g–i*), and consisted of the two *ca.* 16 kDa chains. The appearance of the two chains of the 30 kDa subunit resolved from the dimer subunit (Figure 2, lane *a*) was probably due to the high acrylamide concentration used (15%). The tetrameric subunits have been observed to dissociate into a disulfide-bonded dimeric subunit. However, when reduced (Figure 2, lanes *g–i*), both subunits produce at least two chains of *ca.* 16 kDa (chains I and II). The shift in the elution volume of peak *a* (Figures 1*A, B* and *C*) is consistent with the gradual loss of the 30 kDa chains (Figure 2, lanes *b–d*). The mean masses of particles in peaks *a* and *b*, determined by STEM, were found to be 197 ± 31 ($n = 233$) and 67 ± 21 ($n = 96$), respectively. All the experimental values for the masses of peaks *a* and *b* obtained in the present study are given in Table 1. A histogram of the peak *a* particle mass is shown in Figure 4.

In contrast to the Hb from *Lumbricus*, *Eudistylia* Chl has a simpler subunit structure consisting of 30 kDa nonreducible subunit(s), and a tetrameric subunit

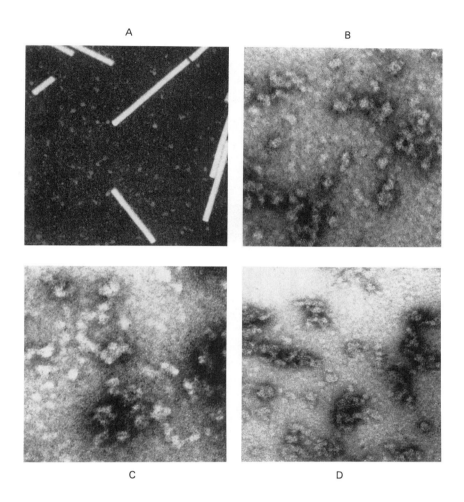

Figure 3. (*A*) STEM micrographs of unstained freeze-dried specimens of peak *a*, fraction *C* (1525 x 1525 Å). (*B*), (*C*) and (*D*) STEM micrographs of negatively stained fractions *B, C* and *D* (x 149,000). TMV was used as an internal standard for STEM mass measurements and appears as rods in panel (*A*).

of *ca*. 65 kDa. The presence of peak *a* in fractions *A, B* and *C* (Figure 1) with an M_m of *ca*. 197 kDa, and consisting of the tetrameric subunit, suggests that this is the principal subunit in the hexagonal bilayer structure of *Eudistylia* Chl. Moreover, peak *b* is probably a dissociation product of the tetrameric subunit. Its reduced SDS-PAGE pattern is the same as that of peak *a*. Peak *b* was found in fraction *C* along with peak *a* and in fraction *D* by itself with an M_m of 66 kDa. This leads us to believe that peak *a* represents the principal subunit

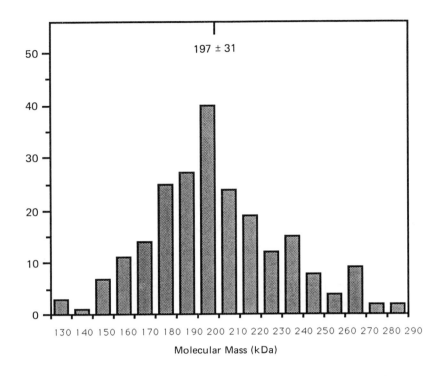

Figure 4. A histogram of the STEM M_m values for 233 particles from peak *a* of fraction *C*.

of *Eudistylia* Chl of *ca*. 200 kDa made up of a trimer of the tetrameric subunit (subunit 2). STEM micrograph images of stained, freeze-dried samples show some differences in size between fractions *B*, *C* and *D* (Figure 3). Fraction *C* particles are heterogenous in size, probably because it consists of dimers and monomers of the tetrameric subunit. Fraction *B* particles consist mainly of trimers and fraction *D* only of monomers of the same subunit. The above explanation is in agreement with the FPLC and SDS-PAGE results of fractions *B*, *C* and *D*. The M_m of peak *a*, determined by STEM to be 197 ± 31 kDa, is in good agreement with the expected M_m of a trimeric complex of subunit 2 (3 x 66 = 198 kDa).

The presence of a structurally stable dodecameric complex of globin chains within the quaternary structure of *Eudistylia* Chl appears to be entirely analogous

Table 1. *Masses* of peaks *a* and *b* of fraction *C* determined by STEM and by gel filtration on Superose S-12.

Method	Peak *a* (kDa)	Peak *b* (kDa)
STEM	197 ± 31	67 ± 21
FPLC	162 ± 15	66 ± 2

to *Lumbricus* Hb (17). It is likely that annelid Chls and Hbs have similar quaternary structures in which twelve dodecameric complexes of heme-containing chains are held together in a hexagonal bilayer by *ca*. 30 kDa chains deficient in heme.

Acknowledgements

This research was partially supported by National Institutes of Health grant DK 38674 (S.N.V.) and by a grant from the U.S. Department of Energy, number DE-FG02-86-ER60437 (O.H.K.). The Brookhaven STEM facility is supported by the U.S. Department of Energy and by the National Institutes of Health Biotechnology Resources Branch grant number RR 01777 (J.S.W.).

References

1. Vinogradov, S.N. (1985) *Comp. Biochem. Physiol.* **82**B: 1–15.
2. Mangum, C.P. (1976) In *Adaptation to the Environment*, ed. R.C. Newell, 191–278. London: Butterworth's.
3. Weber, R.E. (1978) In *Physiology of Annelids*, ed. P. J. Mill, 393–446. London: Academic Press.
4. Chung, M.C.C. and Ellerton, D. (1979) *Prog. Mol. Biol.* **35**: 51–102.
5. Vinogradov, S.N., Shlom, J.M., Kapp, O.H. and Frossard, P. (1980) *Comp. Biochem. Physiol.* **67**B: 1–16.
6. Terwilliger, R.C. (1980) *Am. Zool.* **20**: 53–67.
7. Kapp, O.H., Polidori, G., Mainwaring, M.G., Crewe, A.V. and Vinogradov, S.N. (1984) *J. Biol. Chem.* **259**: 628–639.
8. Mainwaring, M.G., Lugo, S.D., Fingal, R.A., Kapp, O.H., Crewe, A.V. and Vinogradov, S.N. (1986) *J. Biol. Chem.* **261**: 10899–10908.
9. Vinogradov, S.N., Lugo, S.L., Mainwaring, M.G., Kapp, O.H. and Crewe, A.V. (1986) *Proc. Natl. Acad. Sci. U.S.A.* **83**: 8034–8038.
10. Shishikura, F., Walz, D., Standley, P.G., Mainwaring, M.G. and Vinogradov, S.N. (1986) *Biochim. Biophys. Acta* **869**: 314–321.

11. Terwilliger, R.C., Garlick, R.L., Terwilliger, N.B. and Blair, D.P. (1975) *Biochim. Biophys. Acta* **400**: 1302–1309.
12. Terwilliger, R.C., Terwilliger, N.B. and Schabtach, E. (1976) *Comp. Biochem. Physiol.* **55A**: 51–55.
13. Qabar, A.N., Kapp, O.H., Stern, M.S., Walz, D.A., Wall, J.S. and Vinogradov, S.N. (1991) In preparation.
14. Laemmli, U.K. (1970) *Nature* **227**: 680–685.
15. Wall, J.S. and Hainfeld, J.F. (1986) *Annu. Rev. Biophys. Chem.* **15**: 555–376.
16. Kapp, O.H., Qabar, A.N., Bonner, M.C., Stern, M.S., Walz, D.A., Schmuck, M., Pilz, I., Wall, J.S. and Vinogradov, S.N. (1990) *J. Mol. Biol.* **213**: 141–158.
17. Sharma, P.K., Qabar, A.N., Kapp, O.H., Wall, J.S. and Vinogradov, S.N. (1991) This volume.

Part II
Structure and Function

12
Nested Allostery of Arthropod Hemocyanins

Heinz Decker

Zoological Institute, University of Munich, Luisenstrasse 14, 8000 Munich 2, Federal Republic of Germany

Introduction

Arthropod Hcs are multisubunit respiratory proteins (1, 2, 3). As extracellular proteins, Hcs are more frequently exposed to changing conditions than is human Hb, which is kept in an almost homeostatic environment within the erythrocytes. The O_2 binding of Hcs is highly cooperative. The cooperativity and O_2 affinity are sensitive to internal and environmental allosteric effectors. These effectors regulate the pO_2 gradient from the hemolymph to the tissue by affecting the O_2-binding behavior of the Hcs (Figure 1). It is of biophysical and physiological interest to understand the functional properties of arthropod Hcs on the basis of their structure.

Hcs found in nature consist of six subunits (Figure 2) or multiples thereof (1x6-, 2x6-, 4x6-, 6x6- and 8x6-mer). Each subunit is in the shape of a kidney. Single hexamers have a three-fold axis and gain their cubic structure by two rings of trimers positioned above each other (4, 5). The number of binding sites and quaternary structures depend upon the species (6). Each subunit binds with one molecule of O_2 bridging two copper atoms as peroxide. A change in the valency of the copper atoms from Cu^+ when in the deoxy state to Cu^{2+} when in the oxy state is observed (7).

Results and Discussion

O_2-Binding Behavior. Comparing the functional properties of morphologically similar Hcs reveals striking differences. Figure 3 shows the dependence of the O_2 affinities and cooperativities on pH for the two 4x6-meric cheliceratan Hcs

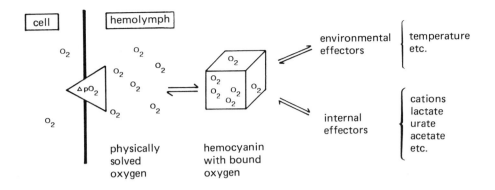

Figure 1. Scheme of the *in vivo* function of Hc.

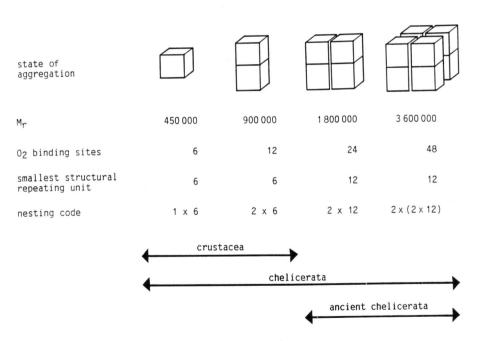

Figure 2. Scheme of quaternary structures of arthropod Hcs.

from the tarantula *Eurypelma californicum* and from the scorpion *Pandinus imperator*, and for the 2x6-meric crustacean Hc from the lobster *Homarus americanus*. Because the functional properties of these Hcs and of other arthropod Hcs that consist of more than six subunits could not be described by the classical two state model (8, 9), the "nesting model" was developed (10, 11).

Until now the nesting model had been successfully applied to several Hcs (11–14), considering the half-molecules to be "allosteric units" in all cases (allosteric units include all subunits that synchronously take part in the conformational transition (1)). A comparison of the results reveals some interesting features (Figure 4): 1) the affinities that characterize the assumed conformations are almost independent of pH; 2) protons, as allosteric effectors, influence only the allosteric equilibrium constants; and 3), the affinities of the assumed conformations of chelicerate Hcs are very similar. This feature indicates that aggregated subunits of oligomeric Hcs from different species may adopt virtually identical conformations. Similar results were obtained when the two 2x6-meric crustacean Hcs from the lobster *H. americanus* and from the crab *Carcinus maenas* were compared (15).

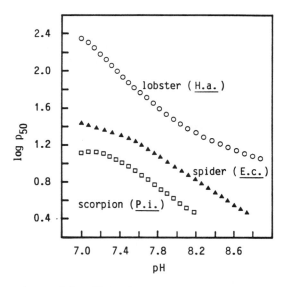

Figure 3*A*. Dependence of O$_2$ affinity P_{50} on pH; the results are for Hcs from three different species (spider *E. californicum*, scorpion *P. imperator* and lobster *H. americanus*). The data are taken from references 13 and 14.

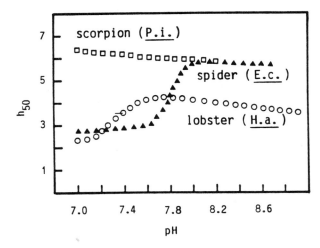

Figure 3*B*. Dependence of O_2 binding cooperativity h_{50} on pH; the results are for Hcs from three different species (spider *E. californicum*, scorpion *P. imperator* and lobster *H. americanus*). The data are taken from references 13 and 14.

The presence of respiratory proteins with nested allosteric units could possibly provide the animals with some advantages in the maintenance and regulation of O_2 transport. Depending on the level of structural hierarchy, different allosteric effectors like protons or lactate might regulate both the physiologically important pO_2 gradient between the hemolymph and the tissue, and the O_2-binding capacity (16).   Experimental evidence for the nesting model is summarized elsewhere (15).

Conformational Transitions Mechanism. Here several issues are of interest: the existence of different conformations, the number of such conformations and the occurrence of such conformations during oxygenation.

Little evidence for structural rearrangement of arthropod Hc upon oxygenation has been reported (3). In the case of tarantula Hc, difference-spectra of the oxy- and the deoxy form (Figure 5) confirm differences in the near-ultraviolet region as observed with CD spectroscopy (17). Two experiments support the existence of a concerted transition for the 4x6-meric tarantula Hc (Figure 6). First, when all four copies of subunit *d* are labeled with a fluorescent dye at the same position, oxygenation alters the emission wavelength maximum (18).   The change in the emission maximum, in other words, the conformational transition, occurs in a stepwise rather than continuous manner, indicating a concerted transition. Second, when particular subunit types are replaced by modified types, the O_2 binding of the 4x6-meric tarantula Hc is drastically altered (19).

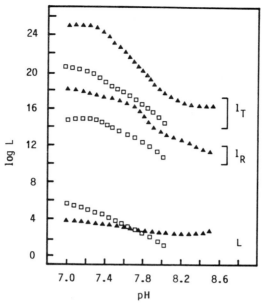

Figure 4*A*. Analysis of the O_2-binding of two cheliceratan Hcs according to the nesting model: pH-dependence of the different parameters. Spider *Eurypelma californicum* (▲), scorpion *Pandinus imperator* (□). The data are taken from references 13 and 14.

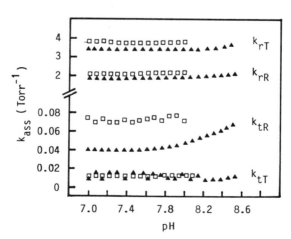

Figure 4*B*. Analysis of the O_2-binding of two cheliceratan Hcs according to the nesting model: pH-dependence of the different parameters. Spider *Eurypelma californicum* (▲), scorpion *Pandinus imperator* (□). The data are taken from references 13 and 14.

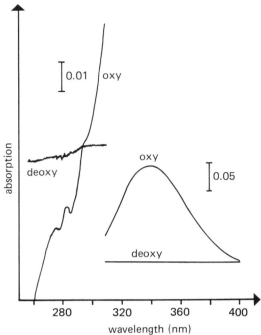

Figure 5. Difference absorption spectra of oxygenated and deoxygenated 24-meric spider Hc.

Tarantula Hcs with incorporated met-subunits have lower affinities than does the native oligomer. Thus, the conformational change induced by the modification of one subunit type is transferred to the other subunits within the oligomer.

How can a conformational transition be understood on the basis of the structure? Only for the hexameric Hc from the spiny lobster *Panulirus interruptus* is the structure known from x-ray analysis (4, 5). The hexamer is composed of two trimers in a sandwich-like orientation. However, counting the connections between the subunits, the hexamers should be regarded as a trimer of "tight dimers" rather than as a dimer of trimers. One subunit of the "tight dimer" belongs to the top trimer, the other one to the bottom trimer. A possible movement of the subunits within the hexamer is suggested by the "turning wheel" hypothesis (Figure 7).

The assumed change in the dimensions of each hexamer should also lead to a change in the space-diagonal of the 4x6-mer during oxygenation. This was recently shown by small angle x-ray scattering experiments with the 24-meric tarantula Hc (20). The oxyHc is longer than the deoxyHc by about 20 Å. This corresponds to 7% of the space-diagonal of the 24-mer. "Breathing" of Hcs exists and may also be detected in the molluscan Hc from *Helix pomatia*, which has a completely different structure (21).

Do isolated subunits already have the potential to adopt different conformations which are then stabilized by the association to the oligomer? Or are the different conformations of the subunits only established during the aggregation to oligomers? Quenching experiments using acrylamide with single subunits from tarantula Hc recently revealed differences in the quenching constants for the oxy and deoxy forms by at least a factor of two (22). This means that different conformations exist and can already be detected for isolated subunits. These conformational differences probably result from structural rearrangements at the active center caused by the change in the valency from Cu^+ (in the deoxy form) to Cu^{2+} (in the oxy form) (23, 24).

Size of Allosteric Units. It is not immediately understandable why, in applying the nesting model to the 4x6-meric tarantula Hc, the half-molecule rather than the most obvious structural unit, the hexamer, is regarded as the allosteric unit. However, arguments discussed elsewhere (3), in addition to preliminary results, might provide some insight into the relationship between the structure and the function of Hcs. Applying a thermodynamic statistical model (25) and maintaining the geometry between the four hexamers, the free energy for the interaction between the particular hexamers may be calculated by analyzing O_2-

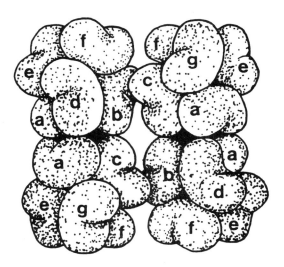

Figure 6. 24-meric spider Hc (*E. californicum*). The molecule is formed by two structurally identical 12-mers, each consisting of two closely connected hexamers. Each hexamer contains the subunit types *a, d, e, f, g* and *b*, or *c*.

binding curves. In the connection of tightly assembled hexamers, the amount of free energy involved is twice the amount of the free energy of interhexameric contacts. Thus, transitions within one hexamer should spontaneously induce conformational transitions in the tightly connected hexamer. This might explain why the half-molecules of the 4x6-meric scorpion and spider Hcs have to be considered as allosteric units when the nesting model is applied (12–14).

Conclusions

Arthropod Hcs with their complex structure appear to function in a simple way. Conformational transitions in one subunit are transferred to other subunits through the conservation of essential symmetries. Conformations are stabilized

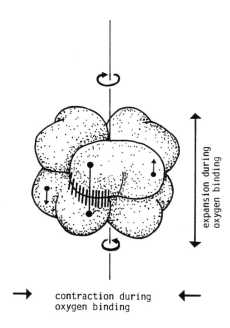

expansion during oxygen binding

contraction during oxygen binding

Figure 7. "Turning wheel" hypothesis (3): assumed movements of particular parts within a hexamer during the oxygenation process. The hexamer consists of two trimers. Each subunit of the top trimer is strongly associated with a particular subunit of the bottom trimer (26). A slight rotation of the two trimers against each other with respect to the three-fold axis would lead to a dislocation of some parts within a subunit as indicated by arrows. These parts seem mainly to belong to domain 3 (Figure 8). In order to avoid any gaps between the subunits of a trimer the subunits have to move closer to the three-fold axis. This results in a narrower but longer hexamer. Following this hypothesis, all six subunits are synchronized.

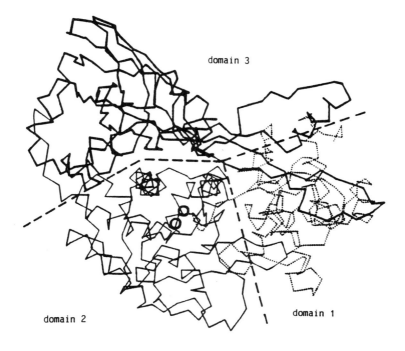

Figure 8. X-ray structure of a subunit of *P. interruptus* Hc. The three domains (--, --, ···) are closely packed, discrete structures. The three domains appear to be flexible with respect to each other, as is required in the "turning wheel" hypothesis (Figure 7). The two copper atoms in domain 2 are indicated by circles. The atomic coordinates were generously provided by W. G. J. Hol, Gronigen, FRG.

after self-assembly of subunits to the allosteric units, which seem to be the half-molecules of native Hcs. While the conformations are stabilized by the binding of the ligands, allosteric effectors shift the concentration ratios of the particular conformations, resulting in an appropriate O_2 affinity of the hemolymph.

Acknowledgements

The excellent technical assistance of Mrs. R. Drechsel is gratefully appreciated. I would like to thank Professor S. J. Gill and, especially, Reinhard Sterner for many fruitful discussions on the nesting theory, and Dagmar Hemmerich and Reinhard Sterner for their help with this manuscript. I am much indebted to Professor Bernt Linzen, who died two years ago, for his support and most valuable discussions on the structure-function relationship of Hcs. Financial support was received from the Deutschen Forschungsgemeinschaft (De 414/1-10).

References

1. Van Holde, K.E. and Miller, K.I. (1982) *Q. Rev. Biophys.* **15**: 1–129.
2. Ellerton, H.D., Ellerton, N.F. and Robinson, H.A. (1983) *Prog. Biophys. Molec. Biol.* **41**: 143–248.
3. Markl, J. and Decker, H. (1991) In *Advances in Comparative and Environmental Physiology*, ed. C. Mangum. In press.
4. Gaykema, W.P.J., Hol, W.G.J., Vereijken, J.M., Soeter, N.M., Bak, H.J. and Beintema, J.J. (1984) *Nature* **309**: 23–29.
5. Volbeda, A. and Hol, W.G.J. (1989) *J. Mol. Biol.* **209**: 249–279.
6. Markl, J. (1986) *Biol. Bull.* **171**: 90–115.
7. Salvato, B. and Beltramini, M. (1990) *Life Chem. Rep.* **8**: 1–47.
8. Decker, H., Savel, A., Linzen, B. and Van Holde, K.E. (1983) *Life Chem. Rep. Suppl. Ser.* **1**: 251–256.
9. Richey, B., Decker, H. and Gill, S.J. (1985) *Biochemistry* **24**: 109–117.
10. Decker, H., Robert, C.H. and Gill, S.J. (1986) In *Invertebrate Oxygen Carriers*, ed. B. Linzen, 383–388. Berlin: Springer-Verlag.
11. Robert, C.H., Decker, H., Richey, B., Gill, S.J. and Wyman, J. (1987) *Proc. Natl. Acad. Sci. U.S.A.* **84**: 1891–1895.
12. Decker, H., Robert, C.H. and Gill, S.J. (1988) *Biochemistry* **27**: 6901–6908.
13. Decker, H. and Sterner, R. (1990) *J. Mol. Biol.* **211**: 281–293.
14. Decker, H. (1990) *Biophys. Chem.* **37**: 257–261.
15. Decker, H. (1991) In *Invertebrate Dioxygen Carriers*, ed. G. Préaux. Leuven: Leuven University Press. In press.
16. Decker, H. and Sterner, R. (1990) *Naturwissenschaften.* **77**: 561–568.
17. Reisinger, P. (1986) In *Invertebrate Oxygen Carriers*, ed. B. Linzen, 203–206. Berlin: Springer-Verlag.
18. Leidescher, T. and Decker, H. (1990) *Eur. J. Biochem.* **187**: 617–625.
19. Decker, H., Savel-Niemann, A., Korschenhausen, D., Eckerskorn, E. and Markl, J. (1989) *Biol. Chem. Hoppe-Seyler* **370**: 511–523.
20. Decker, H., Sterner, R., Schwartz, E. and Pilz, I. (1991) In preparation.
21. Van Breemen, J.F.L., Ploegman, J.H. and Van Bruggen, E.F.J. (1979) *Eur. J. Biochem.* **100**: 61–65.
22. Boteva, R. and Decker, H. (1991) In preparation.
23. Brown, J.M., Powers, L., Kincaid, B., Larrabee, J.A. and Spiro, T. G. (1980) *J. Am. Chem. Soc.* **102**: 4210–4216.
24. Perutz, M.F. (1989) *Q. Rev. Biophys.* **22**: 139–236.
25. Di Cera, E. (1990) *Nuovo Cimento* **12**: 61–68.
26. Gaykema, W.P.J., Volbeda, A. and Hol, W.G.J. (1985) *J. Mol. Biol.* **187**: 255–275.

13
Subunits and Cooperativity of *Procambarus clarki* Hemocyanin

Nobuo Makino and Hiromi Ohnaka

Division of Biochemistry, Institute of Basic Medical Sciences, University of Tsukuba, Ibaraki 305, Japan

Introduction

Arthropod Hcs are comprised of six, or multiples of six subunits. For dodecameric and larger Hcs, it has recently been postulated that cooperative interactions between the hexameric structures must be taken into consideration in describing their cooperativity (1–3).

Among arthropods, many of the decapods (lobsters, crayfish and crabs) contain both hexameric and dodecameric Hcs (4). In studying the mechanism of Hc cooperativity, it is of interest to compare the function of the Hc molecules in each of the different association states.

Though the crayfish *Procambarus clarki* originates in North America, it is now most commonly found in Japan. The animal certainly seems to be a good source of Hc, but the protein is not yet well-characterized.

In this work, we have examined the subunit composition and O_2-binding isotherm on the basis of the allosteric model of Monod *et al.* (MWC model) (5). Results have been obtained that suggest the existence of cooperative interactions that extend beyond the hexameric units of the Hc.

Results and Discussion

Molecular Size of the Crayfish Hc. Hemolymph was obtained from 10 to 20 animals and was pooled. After the clot was removed, the Hc was precipitated with 55% ammonium sulfate and was stored at –5°C. Gel filtration (Figure 1) showed that the hemolymph contains two major components. The elution volume of the smaller component was close to those of *Panulirus japonicus* Hc

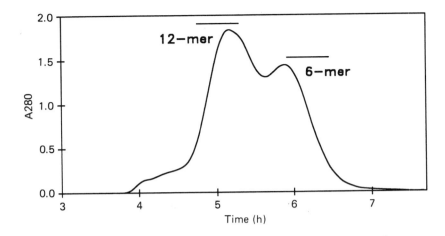

Figure 1. Gel filtration of *Procambarus* hemolymph. The hemolymph was dialyzed against 50 mM Tris/HCl pH 7.5 plus 10 mM $CaCl_2$ and was gel filtered on a Biogel A1.5m (Bio-Rad) column (1 x 110 cm) using the same buffer at the flow rate of 8 ml/h.

(6, 7) and horse spleen ferritin ($M_r = 440$ kDa), and was identified as a hexamer. The larger component was estimated to be a dodecamer. The gel filtration (Figure 1) indicated the presence of a still larger Hc component, though the content was small. In addition, another small peak was eluted at the void volume, but it showed no absorption band at 335 nm, which is characteristic of Hcs.

From the gel filtration profile of the pooled hemolymph, the hexamer and dodecamer contents were estimated at 40 and 55%, respectively. The larger Hc and non-Hc proteins account for the rest (about 5%). The larger Hc component was not studied in any more detail in this work.

The isolated dodecamer was stable and formed no smaller molecules during storage for seven days, as judged by gel filtration. The hexamer fraction, on the other hand, contained a small amount of a dodecamer (about 5% of the total protein), but the amount did not increase during storage.

The electronic spectra of the hexamer and dodecamer were indistinguishable, showing maxima at 278 and 335 nm. The $A_{335/278}$ ratio was 0.224 ± 0.007 (for six determinations).

Molecular Dissociation. Procambarus Hc was found to be very resistant to dissociation at alkaline pH. The conventional method for dissociation (overnight dialysis against EDTA at pH 8.5 to 9.5) did not produce a sufficient amount of the monomer. To obtain the monomer, the Hc was dialyzed for seven days against a buffer at pH 9.5 containing 10 mM EDTA and 1 M urea. About 30% of the protein was dissociated into the subunits (Figure 2A). Both hexamer and dodecamer could be dissociated into the monomers, though only partly (Figures 2B, C). The dodecamer was dissociated more easily. The $A_{335/278}$ ratio for the isolated monomers were 0.178 ± 0.021 (79% of the original value).

In the gel filtration of the dissociated whole Hc (Figure 2A), a molecular species was observed that is eluted between the hexamer and the monomer. It was also observed in the dissociated dodecamer (Figure 2C), though not in the hexamer (Figure 2B). From the alkaline PAGE pattern (Figure 3A), it was identified as a dimer. It is thought that the dimer-forming subunit is the "linker" which is necessary for association into the dodecamer (8).

Sulfhydryl groups do not seem to be involved in the dimerization, since 50 mM dithiothreitol (DTT) exerted no effect upon the gel filtration profile. This is in contrast to the properties of Hc from another crayfish *Cherax*, in which DTT caused dissociation of dimeric molecules (9).

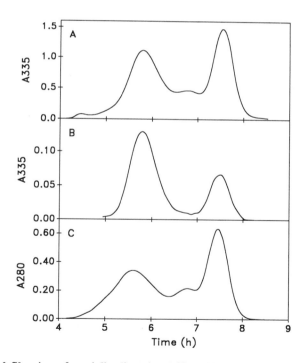

Figure 2. Gel filtration of partially dissociated Hcs. Hc was dialyzed against 50 mM ethanolamine/HCl buffer at pH 9.5 containing 10 mM EDTA and 1 M urea for 7 days, and was gel-filtrated with a Biogel A1.5m column (1 x 110 cm) at the flow rate of 8 ml/h. As an elution buffer, the same buffer at pH 9.5 containing 1 mM EDTA and 1 M urea was used. (A) The whole Hc, (B) the hexamer, and (C) the dodecamer.

As seen in Figure 3*A*, dissociation of the native whole Hc and the dodecamer produced another band, which moves slightly more slowly than the hexamer. It was not present in the dissociation product from the native hexamer (Figure 3*A*), and it is regarded as a heptamer, in which another molecule of the linker subunit is attached to the hexamer (10).

Fractionation of the Subunits. The monomer fraction obtained by gel filtration of the dissociated Hc was further fractionated by anion exchange chromatography (Figure 4) in the presence of 2M urea. Four major peaks were first eluted, followed by minor unresolved peaks. The major fractions were named one to four according to the order of elution. The amount of protein in the minor peaks varied with preparation, and probably consists of the reassociated subunits. Low $A_{335/278}$ ratios (0.02 to 0.10) of the subunit preparations indicate that they had lost most of the O_2-binding capacity. The subunits in fractions 1 and 2 showed a high tendency to reassociate, and readily formed hexamers upon the removal of urea at pH 9.5.

Figure 3*B* shows the SDS-PAGE patterns of the four DEAE-fractions thus obtained. The DEAE-fractions 1 and 2 were pure, but fractions 3 and 4 contained multiple components. In the electrophoretic patterns six different components

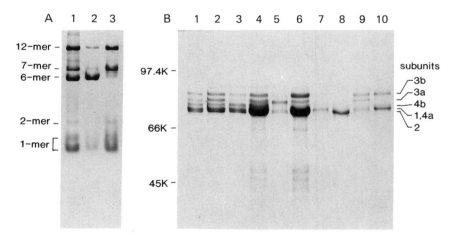

Figure 3. Electrophoresis of *Procambarus* Hc. (*A*) Alkaline PAGE of partially dissociated Hcs on 8% acrylamide gel (at pH 10.0) with a continuous buffer system (6). (1) The whole Hc, (2) the hexamer and (3) the dodecamer. (*B*) SDS/2-mercaptoethanol PAGE on 8% acrylamide gel. A low N,N'-methylene-bis (acrylamide) concentration (0.036%) was used to obtain a higher resolution. (1) Whole Hc, (2) the native dodecamer, (3) the native hexamer, (4) the hexamer (and also heptamer) fraction from dissociated native Hc, (5) the dimer fraction from dissociated Hc, (6) the monomer fraction from partially dissociated Hc, (7) DEAE-fraction 1, (8) DEAE-fraction 2, (9) DEAE-fraction 3 and (10) DEAE-fraction 4.

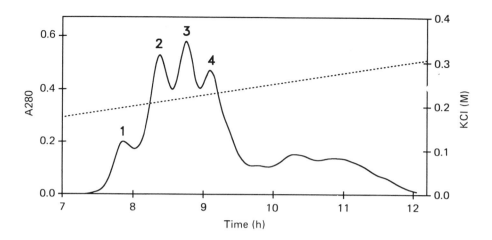

Figure 4. Anion-exchange chromatography of Hc monomers. The monomer fraction from the dissociation product of Hc was chromatographed on a DEAE-Toyopearlpak 650M column (2.2 x 20 cm, Toyo Soda, Tokyo), which was connected to a Pharmacia FPLC system. The protein was eluted with 0.05 M ethanolamine pH 9.5 containing 1 mM EDTA and 2 M urea with a KCl gradient from 0 to 0.5 M at the flow rate of 0.5 ml/min (———) A_{280}; (····) KCl concentration.

could be identified. As shown in Figure 3*B*, the subunit components were named 1, 2, 3*a*, 3*b*, 4*a* and 4*b*. Apparently, the four fractions cover all of the components in the native Hc. Subunits 1 and 4*a* showed such close mobilities that they were not separated in the SDS-PAGE of the native Hc. The native protein contains a considerable amount of subunit 4*b*, but the contents in fraction 4 were relatively small; the subunit is probably mainly present in the trailing part of DEAE-fraction 4. It is most likely that subunit 3*a* is the dimer-forming subunit, since it is absent in the native hexamer (Figure 3*B*). Except for the linker subunit, the subunit compositions of the native hexamer and dodecamer were similar.

Figure 5 shows the N-terminal sequence of subunit 1. It showed high homology to another crayfish Hc, *Astacus leptodactylus* Hc subunit *b* (11).

Oxygen-Binding Properties. Table 1 shows the values of P_{50} (partial pressure of O_2 at half saturation) and h_{max} (the maximum slope of the Hill plot) for the hexameric and dodecameric Hcs under various ionic conditions. As described above, the hexamer preparation contained a dodecamer, but its amount (about 5%) was small enough to distinguish the properties of the native hexamer and dodecamer.

Under physiological conditions (data sets 1 and 11 in Table 1), the P_{50} and h_{max} values were 4.2–4.3 mm Hg and 2.5–2.7, respectively. The values were in good agreement with those previously reported by Mangum (12) for the whole *Procambarus* Hc.

With respect to the cooperativity, it is interesting that the h_{max} values for the dodecamer were definitely higher than those of the hexamer at pH 7.5 or higher. At pH 6.5, their h_{max} values were similar.

Both $CaCl_2$ and NaCl enhanced the cooperativity of the Hc. In the presence of 10 mM $CaCl_2$ and 500 mM NaCl the h_{max} values of the hexamer and the dodecamer (data sets 6 and 16) were even higher than those observed in the physiological saline (data sets 1 and 11).

The dodecamer showed a larger Bohr effect in the presence of Ca^{2+}. At pH 6.5 the O_2 affinity of the dodecamer (data sets 12 and 15) was lower than that of the hexamer (data sets 2 and 5); but at pH 7.5 and 8.5, its affinity was higher because of the larger Bohr effect.

It is interesting that the Bohr effect was reversed in the absence of divalent cations in the pH range examined (data sets 8–10 and 18–20). No dissociation took place under the experimental conditions, as judged by gel filtration. It is suggested that the environment of the Bohr groups is greatly affected by the removal of the divalent cations. The effect is so large that it seems quite possible that the cations are directly bound to the ionizable groups.

Fitting of the O_2-Binding Curves to the MWC Model. In *Procambarus* Hc, as described above, the dodecamer showed a higher cooperativity than that of the hexamer; thus, it is of interest to see whether or not the interactions beyond the hexameric units are functioning in the dodecameric molecules. Here, as a first step of the quantitative analysis, we attempted a curve-fitting study of the O_2 binding using the original two-state model of Monod *et al.* (5).

According to the model, the fractional O_2 saturation of Hc (Y) is given as:

$$Y = \frac{LK_{TP}(1 + K_{TP})^{n-1} + K_{RP}(1 + K_{RP})^{n-1}}{L(1 + K_{TP})^n + (1 + K_{RP})^n} \tag{1}$$

where p and L are the pO_2 and allosteric constant, respectively. K_T and K_R are the O_2-binding constants for the T and R states, respectively and n is the number of O_2-binding sites in the cooperativity unit. In this work, n was taken as one of the variables to find the best-fit values.

Table 1 shows the results of the curve fitting. The fitting of the model was satisfactory; the values of RMSR (root mean square residuals) were of a similar magnitude to that of the experimental error (estimated as 0.3–0.4% in fractional O_2 saturation). It should be noted that n is not directly correlated with h_{max}. The approximate confidence limits (99%) for the fitted n values were determined by the F-test (13) and are also shown in Table 1. They are regarded as a statistical allowance for the estimated parameter values.

Table 1. O_2 equilibrium and fitted MWC parameters for *Procambarus* hexamer and dodecamer Hcs.[a] Numbers following 'E' show the power of 10.

Data set no.	pH	P_{50} (mmHg)	h_{max}	Fitted MWC Parameters				99% Confidence Limits for n[b]	1000 x RMSR[c]
				L	K_T	K_R	n		
6-mer, physiological saline[d]									
(1)	7.5	4.24	2.5	3.16E3	7.54E2	9.85E1	5.78	(5.00, 6.65)	4.12
6-mer, 10 mM $CaCl_2$									
(2)	6.5	4.35	1.5	4.18E1	1.01E1	3.29E1	8.54	(6.73, 10.55)	2.27
(3)	7.5	3.49	2.0	1.59E2	8.56E2	6.53E1	5.91	(4.59, 7.46)	4.86
(4)	8.5	2.80	2.2	1.99E2	9.92E2	7.44E1	6.58	(5.48, 7.82)	3.74
6-mer, 10 mM $CaCl_2$ + 500 mM NaCl									
(5)	6.5	6.50	1.7	1.38E2	7.30E2	3.85E1	5.52	(4.33, 6.86)	3.01
(6)	7.5	4.33	2.7	7.62E3	7.21E2	1.13E0	5.79	(5.14, 6.52)	3.77
(7)	8.5	2.51	2.3	4.55E2	1.29E1	1.39E0	4.97	(4.06, 6.02)	4.32
6-mer, 1 mM EDTA									
(8)	6.5	4.86	1.5	1.14E2	1.32E1	6.39E1	4.26	(2.52, 6.72)	2.53
(9)	7.5	8.51	1.8	4.41E1	3.86E2	3.44E1	3.49	(2.63, 4.54)	2.89
(10)	8.5	9.56	1.9	1.39E3	4.62E2	6.63E1	4.06	(3.41, 4.76)	2.60
12-mer, physiological saline[d]									
(11)	7.5	4.33	2.7	6.53E3	7.02E2	1.32E0	5.22	(4.46, 6.08)	4.55
12-mer, 10 mM $CaCl_2$									
(12)	6.5	6.53	1.5	1.34E2	9.68E2	2.80E1	8.48	(6.57, 10.65)	2.23
(13)	7.5	3.17	2.3	5.40E2	9.87E2	7.55E1	6.89	(6.21, 7.63)	2.48
(14)	8.5	2.44	2.5	2.02E3	1.42E1	1.50E0	6.13	(5.09, 7.31)	4.84
12-mer, 10 mM $CaCl_2$ + 500 mM NaCl									
(15)	6.5	8.04	1.7	2.08E2	6.37E2	3.05E1	6.15	(4.72, 7.80)	3.19
(16)	7.5	3.40	3.3	1.29E4	7.23E2	1.06E0	7.28	(6.60, 8.01)	3.06
(17)	8.5	2.37	3.0	1.20E0	1.15E1	2.15E0	5.90	(5.29, 6.58)	3.76
12-mer, 1 mM EDTA									
(18)	6.5	8.30	1.5	2.10E2	7.42E2	2.92E1	6.35	(4.58, 8.37)	2.73
(19)	7.5	12.76	1.9	8.90E2	4.61E2	3.11E1	5.31	(3.82, 7.00)	3.50
(20)	8.5	14.60	2.4	3.93E4	3.26E2	5.55E1	5.39	(4.51, 6.36)	3.91

[a]The O_2 equilibrium was measured with an Imai apparatus (16). Oxygenation of Hc was monitored at 335 nm.
[b]Approximate confidence limits from the F-test (14).
[c]RMSR of the fitting.
[d]Reproduces the cationic composition of the *Procambarus* hemolymph (13) (140 mM NaCl, 4 mM KCl, 7 mM $CaCl_2$ and 21 mM $MgCl_2$ in 50 mM Tris/HCl buffer pH 7.5).

```
            1        5        10        15        20
Pc 1    D-S-S-G-T-A-L-A-K-K-Q-Q-X-V-N-X-L-L-E-H-

Al b    D-A-S-G-A-T-L-A-K-R-Q-Q-V-V-N-H-L-L-E-H-
```

Figure 5. N-terminal sequence of *Procambarus* Hc subunit 1 (Pc 1). The subunit isolated by SDS-PAGE was transferred to Immobilon membrane (Millipore) and was sequenced with an Applied Biosystems 477A protein sequencer. The sequence of *A. leptodactylus* Hc subunit *b* (*Al b*) (11) is also aligned.

As seen from the confidence limits shown in Table 1, the fitting was not very sensitive to the *n* value, and in most cases the value of 6 fell within the statistically possible ranges. However, in some cases the *n* values clearly deviated from 6. In particular, it is notable that the *n* value of the dodecamer significantly exceeded 6 at pH 7.5 in the presence of Ca^{2+} (data sets 13 and 16 in Table 1). In the presence of 500 mM NaCl, both h_{max} and *n* values were high (data set 13). On the basis of the two-state concerted model, the subunit interactions extend beyond the hexameric unit, at least under certain conditions. Since the cooperativity of the dodecamer was generally higher than that of the hexamer, it is possible that the higher order of subunit interactions are generally effective, though not always appreciable. With the exception of the dimer-forming subunit, the subunit compositions of the dodecamer and the hexamer were not very different. Thus, it is unlikely that the functional difference is solely due to the subunit difference.

The *n* value for the dodecamer was also higher than 6 at pH 6.5 in the presence of Ca^{2+} (data set 12). In this case, however, the interpretation cannot be straightforward, since the Hill coefficient (1.5) was not high enough and also since the hexamer gave an *n* value higher than 6 under the same conditions (data set 2). Evidently, the model is not applicable in these particular cases.

When the divalent cations were removed with EDTA, the hexamer showed considerably lower *n* values at pH 7.5 and higher (data sets 9 and 10). This result may seem to indicate that the allosteric unit is smaller when the divalent cations are absent. However, it is also possible to explain the results on the assumption of *n* = 6 by modification of the MWC model, such as extensions to three affinity states (7) or to a hybrid state (14, 15).

In conclusion, the *Procambarus* Hc is regarded as one of the typical crustacean Hcs, but is unique in its resistance to molecular dissociation. Its associated molecules are especially stable, and at present it seems difficult to obtain fully active monomers. The results of the O_2 equilibrium studies suggest that interactions beyond the hexamer units contribute to the cooperativity in the dodecamer. Thus, the Hc may be rather suited to studies of the higher level of subunit interactions.

References

1. Savel-Niemann, A., Markl, J. and Linzen, B. (1988) *J. Mol. Biol.* **204**: 385–395.
2. Decker, H., Connelly, P.R., Robert, C.H. and Gill, S.J. (1988) *Biochemistry* **27**: 6901–6908.
3. Brouwer, M. and Serigstad, B. (1989) *Biochemistry* **28**: 8819–8827.
4. Van Holde, K.E. and Miller, K.I. (1982) *Q. Rev. Biophys.* **15**: 1–129.
5. Monod, J., Wyman, J. and Changeux, J.-P. (1965) *J. Mol. Biol.* **12**: 88–118.
6. Makino, N. and Kimura, S. (1988) *Eur. J. Biochem.* **173**: 423–430.
7. Makino, N. (1986) *Eur. J. Biochem.* **154**: 49–55.
8. Lamy, J., Lamy, J., Weill, J., Bonaventura, J., Bonaventura, J. and Brenowitz, M. (1979) *Arch. Biochem. Biophys.* **196**: 324–329.
9. Murray, A.C. and Jeffrey, P.D. (1974) *Biochemistry* **13**: 3667–3671.
10. Markl, J., Decker, H., Stöcker, W., Savel, A. and Linzen, B. (1981) *Hoppe-Seyler's Z. Physiol. Chem.* **362**: 185–188.
11. Schneider, H.-J., Voll, W., Lehmann, L., Grisshammer, R., Goettgens, A. and Linzen, B. (1986) In *Invertebrate Oxygen Carriers*, ed. B. Linzen, 173–176. Berlin: Springer-Verlag.
12. Mangum, C.P. (1983) *Mar. Biol. Lett.* **4**: 139–149.
13. Makino, N. (1989) *J. Biochem.* **106**: 418–422.
14. Brouwer, M., Bonaventura, C. and Bonaventura, J. (1978) *Biochemistry* **17**: 2148–2154.
15. Arisaka, F. and Van Holde, K.E. (1979) *J. Mol. Biol.* **134**: 41–73.
16. Imai, K. (1981) *Methods Enzymol.* **76**: 438–449.

14
Analysis of Oxygen Equilibrium Using the Adair Model for the Hemocyanin of *Limulus polyphemus* and the Hemoglobin of *Eisenia foetida*

Yoshihiko Igarashi,[a] Kazumoto Kimura,[b] Zhi-Xin Wang[c] and Akihiko Kajita[a]

[a]Department of Biochemistry and [b]Laboratory of Medical Sciences Section of Medical Engineering, Dokkyo University School of Medicine, Tochigi 321-02, Japan
[c]Institute of Biophysics, Academia Sinica, Beijing 100080, China

Introduction

Oxygen equilibrium curves of arthropod Hc and annelid Hb consisting of multiple subunits have been analyzed using the two-state model of Monod-Wyman-Changeux (MWC) and its extensions (1–4). However, it remains unclear whether these molecules undergo $R–T$ transition in their conformation. In the absence of direct structural evidence, it would seem somewhat controversial to apply the MWC model. The Adair scheme seems to be a better choice for representing the cooperativity of the highly assembled O_2-carrying proteins because it is a more general description of the ligand equilibria of O_2 carriers (5). This equation requires only that the binding sites in the consecutive steps of oxygenation be assumed to be functionally identical. However, only a few attempts have been made to apply the Adair model to giant respiratory proteins (6) because of the difficulty caused by the large number of O_2-binding sites. Relative to the MWC model, the Adair model has the advantage of showing cooperativity by Adair parameters at equilibrium without any consideration of R and T conformation of the O_2-binding protein.

In the present study, we attempted to describe the O_2 equilibria of *Limulus polyphemus* Hc and *Eisenia foetida* Hb by regarding the submultiple as a functional unit composed of 6 and 12 subunits, respectively. We succeeded in fitting the Adair model to the data, and obtained Adair constants which indicate that the increase in O_2 affinity occurs under physiological conditions during the later steps of the O_2 binding.

Materials and Methods

The hemolymph of the horseshoe crab *L. polyphemus* was provided by Dr. H. Sugita (Tsukuba University). The Hc was purified at $4°C$ according to Brenowitz *et al.* (7). The Hb was prepared from the earthworm, *E. foetida*. Oxygen equilibrium curves were determined at $25°C$ with an automatic oxygenation apparatus as described previously (8). The O_2 equilibrium data were analyzed according to the Adair model (5). N of the Adair model is defined as the number of O_2-binding sites of the functional units, and i denotes O_2-binding steps. K_i is defined as the ligand association constant for the ith step of the oxygenation. The parameters of the Adair model were obtained by the perturbation method of the least-squares minimization procedure with the aid of a DEC micro VAX II computer.

Results and Discussion

The Hill plot of the O_2 equilibrium of *L. polyphemus* Hc is given in Figure 1. The solid line and the broken line were drawn to represent the Adair model and the two-state MWC model, respectively; and the number of binding sites (N) was set at 6, as described above. Both curves appear to fit the measured points well, though a slight deviation is observed in the low saturation range of the MWC curve. Based upon the Adair model and the two-state MWC model, the sum of the square of residuals (SSR), which represents the accuracy of the data fitting the Hill plot, was 0.00544 and 0.104, respectively. The Scatchard plot (9) of the O_2 equilibrium of the Hc is given in Figure 2. The Adair model is represented by the solid line, and the MWC model by the broken line. The curve of the MWC model failed to coincide exactly with the data points, especially when below the 60% point of the O_2 saturation. These facts clearly indicate that the MWC model is an inadequate description of the Hc O_2 equilibrium state. As shown in the same figure, the solid line representing the Adair model fitted the data well. When N was fixed at 4, 8, and 12, the Adair equation adapted to the data as well as for when $N = 6$. As N was increased from

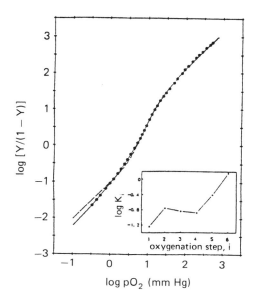

Figure 1. O_2 equilibrium curve of *L. polyphemus* Hc shown as a Hill plot. The solid and broken lines were calculated from the Adair model and MWC two-state model, respectively, for $N = 6$. The inset shows $\log K_i$ values as plotted against oxygenation step, *i*. K_i is given in mm Hg^{-1}. Conditions: Hc, 1.5 mg/ml; Bis-trispropane, 50 mM, pH 7.3; $CaCl_2$, 10 mM.

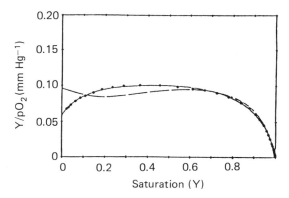

Figure 2. Scatchard plot of the O_2 equilibrium of *L. polyphemus* Hc. Conditions were the same as in Figure 1.

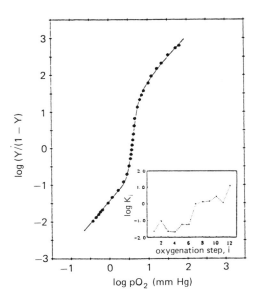

Figure 3. Hill plot of O_2 equilibrium of *E. foetida* Hb. The inset shows log K_i values as plotted against oxygenation step, *i*. K_i is given in mm Hg^{-1}. Conditions: Hb, 60 μM on a heme basis; Bis-trispropane, 50 mM, pH 7.5; $CaCl_2$ 20 mM.

4 to 6, the SSR of the Hill plot decreased to one tenth. However, when N was shifted from 6 to 8 or 12, the computed curves did not show much difference. At $N = 12$, there are 12 parameters from K_1 to K_{12} in the Adair equation, but the SSR of the Hill plots is not much smaller (0.00411) than it is when $N = 6$ (0.00544). Therefore, the model at $N = 6$ appears to be plausible. These calculated results indicate that the hexameric structures may be regarded as an allosteric unit. A similar conclusion about the allosteric unit has been suggested for *Tachypleus gigas* Hc under physiological conditions (1), while involvement of the 12-mer for *L. polyphemus* Hc as the allosteric unit, in addition to the 6-mer in their interacting cooperative unit (ICU) model (which allows for an equilibrium between the 12-mer and 6-mer), has also been postulated (4).

In the inset of Figure 1, the K_i values are obtained as described above and are plotted against *i*, the oxygenation step. As seen in the figure, K_i remained fairly constant for the binding of the first four molecules of O_2, and then increased gradually for the binding of the subsequent ligands. A similar change in the K_i value was observed in human Hb: cooperative interaction appears to be more dominant in later oxygenation steps than in the earlier steps (10).

Figure 3 illustrates the O_2 equilibrium curve of *E. foetida* Hb in a Hill plot where the simulated curve of the Adair model at $N = 12$ coincided well with the measured points. We have shown previously (2) that the O_2 equilibrium curves of *E. foetida* Hb could not be represented on the basis of the simple MWC model with parameters K_R, K_T and L. The curves could be described by the two-state model only when an interacting O_2-binding site (N) was varied within a range of $N = 6–10$. The native Hb may be dissociated into 1/12 molecules which are called submultiples, the cooperativity of which was found to be almost the same as that of the native Hb (11). Thus, the number of consecutive steps of oxygenation is considered to be 12 per allosteric unit, reflecting the involvement of the submultiple as a functional unit. The inset in Figure 3 shows the relation between the K_i values and the oxygenation steps indicated by i. The K_i values of *E. foetida* Hb remained constant until the half-oxygenation state ($i = 6$). Then, the K_i values increased gradually until the final step, $i = 12$.

All of these facts may be interpreted as suggesting that these O_2 carriers are in a deoxy state conformation during the early oxygenation steps and change to the oxy state conformation during the later steps.

References

1. Makino, N. (1989) *J. Biochem.* **106**: 418–422.
2. Igarashi, Y., Kimura, K. and Kajita, A. (1986) In *Invertebrate Oxygen Carriers,* ed. B. Linzen, 89–92. Berlin: Springer-Verlag.
3. Decker, H. and Sterner, R. (1990) *J. Mol. Biol.* **211**: 281–293.
4. Brouwer, M., and Serigstad, B. (1989) *Biochemistry* **28**: 8819–8827.
5. Adair, G.S. (1925) *J. Biol. Chem.* **63**: 529–545.
6. Makino, N. (1986) *Eur. J. Biochem.* **154**: 49–55.
7. Brenowitz, M., Bonaventura, C. and Bonaventura, J. (1984) *Biochemistry* **23**: 879–888.
8. Igarashi, Y., Kimura, K. and Kajita, A. (1985) *Biochem. Int.* **10**: 611–618.
9. Scatchard, G. (1949) *Ann. N. Y. Acad Sci.* **51**: 660–672.
10. Imai, K. (1982) *Allosteric Effects in Haemoglobin.* Cambridge: Cambridge University Press.
11. Igarashi, Y., and Kajita, A. (1986) *Dokkyo Igakukai Zasshi* **2**: 31–40 (in Japanese).

15
Structural and Functional Characterization of the Hemoglobin from *Lumbricus terrestris*

Kenzo Fushitani* and Austen F. Riggs

Department of Zoology, University of Texas, Austin, TX 78712, USA

Introduction

The extracellular Hb of the earthworm *Lumbricus terrestris* has an M_m near 3,800 kDa (1) and is composed of four major kinds of O_2-binding chains: *a*, *b* and *c* which form a disulfide-linked trimer, and chain *d* which is not covalently linked to any other chain (2, 3). Amino acid sequencing has shown both the position of the intra-molecular disulfide bond in each of the four chains and the positions of interchain disulfide bonds in the trimer (4, 5). Analyses of the heme content of the intact Hb, the *abc* trimer, and of isolated subunits *a*, *b*, *c* and *d* indicate that the intact molecule has 2200 g of non-heme protein per mole of heme (6). This value corresponds to 35,200 g of non-heme protein per one twelfth subunit, a value which appears to be consistent with those for bands V and VI observed with SDS-PAGE (7). An emerging picture of the structure of the intact molecule is that it consists of 192 heme-containing chains and 12 non-heme chains, at least some of which are required for the assembly of the intact molecule (8).

---
* Present address: Department of Biochemistry, Kawasaki Medical School, Kurashiki, Okayama, 701-01, Japan

The high cooperativity in O_2-binding by the intact Hb is illustrated by Hill coefficients as high as 7.9 and is modulated by pH and Ca^{2+} ions (9). The O_2-linked binding of a Ca^{2+} ion is associated with the release of two protons. The Bohr effect vanishes in the absence of cations. We seek the minimal structural unit that has an oxygenation-linked Ca^{2+} binding site. Full understanding of O_2-binding by a molecule consisting of about 200 chains requires knowledge of the organization of subunits and the determination of O_2-binding properties of both the subunits and various intermediate states of their aggregation. We report spectral and O_2-binding characteristics of isolated chains, the trimer, the partially reassembled subunit, and the intact Hb and a preliminary structural characterization of a non-heme chain with an M_m of 33–35 kDa.

Materials and Methods

Hb was extracted from the earthworm *Lumbricus* and purified in the CO form as described previously (9). The intact molecule thus purified was subjected to chromatography on Ultrogel ACA44 with a 0.1 M borate buffer, pH 9.3 and 1 mM EDTA to isolate both trimer and monomer. The three constituent chains *a, b* and *c* of the trimer were isolated by DEAE cellulose chromatography in the presence of dithiothreitol (DTT) (2). The monomer fraction was chromatographed in the same way. These procedures were done at 4°C and with CO-saturated buffers. All Hb samples to be used for spectral analysis were first converted to the oxy form by strong illumination under an atmosphere of O_2 at a temperature of 0°C. Aliquots were used to record spectra and for analyses of heme content by the pyridine hemochromogen method using a millimolar extinction coefficient of 32.0 at 557 nm (6). The absorption spectra of the CO forms were determined after reduction by sodium dithionite. Oxygen equilibria were measured with the Gill cell at 15°C between pH 6.4 and 9.0 either in the presence or absence of Ca^{2+} ions (10). Isolation of a non-heme containing chain of 35 kDa was performed as follows. The intact Hb was reduced with DTT, carboxymethylated, and chromatographed on Sephacryl S-200 with 0.05 M sodium acetate pH 6.5 with 6 M guanidine hydrochloride. Pooled fractions under the leading shoulder (Figure 1A) were dialyzed and subjected to reverse phase HPLC. The NH_2-terminal amino acid sequence was determined by a gas-liquid phase sequencer model 477A (Applied Biosystems) of the Protein Sequencing Center of the University of Texas.

Results and Discussion

The millimolar extinction coefficients at the absorption maxima in the ultraviolet and visible regions are given in Table 1 for all Hb samples. Ratios of absorbance between the Soret and UV regions (272-274 nm) for the CO-derivatives of the whole molecule, trimer chains *a, b, c* and *d* are 3.9, 4.6, 4.2,

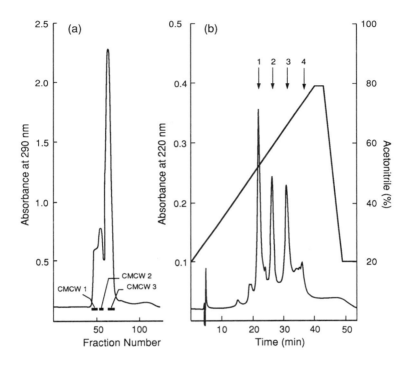

Figure 1. (*A*) Chromatogram from gel filtration of carboxymethylated whole Hb on Sephacryl S-200 in the presence of 6 M guanidine hydrochloride, 2.1 ml/tube, monitored at 290 nm. (*B*) Chromatogram from HPLC with a reverse phase column using 0.1% TFA/water-0.1% TFA/CH3Cn. Sample: CMCW2 from Figure 1*A*: 1, CMCW2-1; 2, chain *b*; 3, chain *c*; 4, chain *a*.

4.5, 5.1 and 5.4, respectively. The value for the whole molecule is lower than those of the others, suggesting that non-heme containing chains might be present in the whole molecule. Although these spectral analyses predict the possible presence of non-heme chains, components corresponding to non-heme chains were not isolated from either ACA44 gel or DE cellulose chromatography. Therefore, gel chromatography in the presence of 6 M guanidine hydrochloride with reduced and carboxymethylated Hb was carried out (Figure 1). Three fractions were subjected to reverse phase HPLC and to SDS-PAGE. Peak CMCW3 gave four chains, *a, b, c* and *d*, as expected. However, peak CMCW2 revealed a new peak (CMCW2-1) as shown in Figure 1*B*, which

Table 1. Determination of wavelength and millimolar absorption coefficients for *Lumbricus* Hb, the trimer and the constituent chains. Wavelengths are in nm. Values in parentheses give millimolar extinction coefficient (based on pyridine hemochromogen method, 32.0 at 557 nm).

		Whole Molecule	Subunit Trimer	Chain a	Chain b	Chain c	Chain d
HbCO	UV	273	274	272	273	273	273
		(47.0)	(38.7)	(47.2)	(40.4)	(40.6)	(37.9)
HbCO	Soret	419.7	419.8	421.1	420.6	419.2	419.6
		(183.3)	(178.0)	(198.2)	(181.8)	(207.1)	(204.7)
HbCO	β	537.8	537.5	538.0	538.5	537.0	537.2
		(13.5)	(12.9)	(15.3)	(13.6)	(14.2)	(13.2)
HbCO	α	569.0	569.0	570.0	569.1	568.5	567.3
		(13.7)	(12.6)	(14.6)	(13.4)	(13.6)	(12.6)
HbO_2	UV	278	279	278	280	280	279
		(47.6)	(39.0)	(47.6)	(41.7)	(40.5)	(38.0)
HbO_2	β	540.0	540.0	541.5	540.0	539.5	540.0
		(13.9)	(12.5)	(13.9)	(13.4)	(13.1)	(13.5)
HbO_2	α	575.8	576.5	576.5	576.2	576.5	574.2
		(14.6)	(12.5)	(12.9)	(13.3)	(12.8)	(13.4)

differed from chains a, b, c and d. SDS-PAGE of the peak CMCW2-1 showed a major band at the position corresponding to band V (described in reference 7) and two minor bands, one of which corresponds to chain a and the other which is positioned near chain d.   The two minor chains were removed by rechromatography on HPLC. The amino acid sequence of the first 28 residues of the major component of CMCW2-1 was found to be Ala-Ser-Asp-Pro-Tyr-Gln-Glu-Arg-Arg-Phe-Gln-Tyr-Leu-Val-Lys-Asn-Gln-Asn- Leu-Leu-Ile-Asp-Tyr-Leu-Ala-Lys-Lys-Leu (6). This sequence corresponds to that of chain D1A for which the first 13 residues had been reported (11). This chain must be considered as a constituent of the Hb in addition to the four heme-containing chains, a, b, c and d. SDS-PAGE (7) indicates that a somewhat larger non-heme chain ($\approx$36-38 kDa) also exists, and is probably present in CMCW1 (Figure 1A). This is consistent with the finding of two "linker" chains in the extracellular Hb of the related annelid, *Tylorrhynchus heterochaetus* (12). Isolation of these chains in native form would make possible experiments to clarify its structural and functional roles.

Although chains a and b were too unstable to permit measurement of the O_2 equilibria, chains c and d were relatively stable. Chain c was cooperative in O_2-binding (Hill coefficient, $\approx$ 1.6) and had a low O_2 affinity (log $P_{50} \approx$ 0.7 to

0.8), whereas chain d showed no cooperativity and a high affinity (log $P_{50} = -2.0$ to -0.05). The O_2 equilibria of both chains were independent of pH and Ca^{2+}. Cooperativity of the trimer was low: Hill coefficients of 1.4 were found but modulation of the O_2 affinity by pH and Ca^{2+} ions was similar to that of the intact molecule. These results indicate that the groups responsible for the Bohr effect and the site for oxygenation-linked Ca^{2+} binding must be present in the trimer. The calcium binding site may be formed between the negatively charged, NH_2-terminal extensions of chains a and c which are joined by disulfide bonds in the trimer (5). The complex formed by the addition of chain d to the trimer in equimolar proportions had a greatly enhanced cooperativity, though it was generally not as high as that of the intact molecule, except at low pH. Oxygenation of the complex was modulated by both pH and Ca^{2+} much as observed in the whole molecule. Cooperativity of oxygenation in the reassembled molecule was far greater than in either the trimer or chain d. This observation indicates that subunit d plays an essential functional role in the complex. This conclusion presumably also holds for the intact molecule. Measurements of the apparent M_m of chain c by gel filtration and of the reassembled complex by measurement of sedimentation velocity suggested the equilibria, $2c \rightleftharpoons c_2$ and $2(abcd) \rightleftharpoons (abcd)_2$. The conformation of the reassembled complex may be such that the dimerization of the $(abcd)$ unit depends on chain c–chain c interactions. These observations lead to a proposal of a possible assembly sequence: (1) $a + b + c \rightarrow abc$ (the disulfide-linked trimer), (2) $(2abc \rightarrow (abc)_2) + 2d \rightarrow (abcd)_2$, and (3) 24 $(abcd)_2$ + assembly chains $\rightarrow$ intact molecule.

Acknowledgements

This work was supported by National Science Foundation grants DMB8502857 and DMB88-10828, Welch Foundation grant F-0213 and National Institutes of Health grant GM 35847.

References

1. Vinogradov, S.N. (1985) *Comp. Biochem. Physiol. B* **82**: 1–15.
2. Fushitani, K., Imai, K. and Riggs, A.F. (1980) In *Invertebrate Oxygen Carriers*, ed. B. Linzen, 77–79. Berlin: Springer-Verlag.
3. Shishikura, F., Mainwaring, M.G., Yurewicz, E.C., Lightbody, J.J., Walz, D.A. and Vinogradov, S.N. (1986) *Biochim. Biophys. Acta* **869**: 314–321.
4. Shishikura, F., Snow, J.W., Gotoh, T., Vinogradov, S.N. and Walz, D.A. (1987) *J. Biol. Chem.* **262**: 3123–3131.
5. Fushitani, K., Matsuura, M.S.A. and Riggs, A.F. (1988) *J. Biol. Chem.* **263**: 6502–6517.

6. Fushitani, K. and Riggs, A.F. (1988) *Proc. Natl. Acad. Sci. U.S.A.* **85**: 9461–9463.

7. Shlom, J.M. and Vinogradov, S.N. (1973) *J. Biol. Chem.* **248**: 7904–7912.

8. Kapp, O.H., Mainwaring, M.G., Vinogradov, S.N. and Crewe, A.V. (1987) *Proc. Natl. Acad. Sci. U.S.A.* **84**: 7532–7536.

9. Fushitani, K., Imai, K. and Riggs, A.F. (1986) *J. Biol. Chem.* **261**: 8414–8423.

10. Dolman, D. and Gill, S.J. (1978) *Anal. Biochem.* **87**: 127–134.

11. Walz, D.A., Snow, J., Mainwaring, M.G. and Vinogradov, S.N. (1987) *Fed. Pro.* **46**: 2266.

12. Suzuki, T., Takagi, T. and Gotoh, T. (1990) *J. Biol. Chem.* **265**: 12168–12177.

16
Interaction of Divalent Metal Ions with the Hemoglobin of *Glossoscolex paulistus*: an EPR Study

Marcel Tabak, Maria H. Tinto, Hidetake Imasato and Janice R. Perussi

Institute of Chemistry and Physics, University of São Carlos, CP369, 13560 São Carlos, SP, Brazil

Introduction

Glossoscolex paulistus is an annelid found in the calcareous region of the state of S. Paulo, Brazil. Its Hb has a M_m of 3,100 kDa (1), and has a subunit structure similar to that of other annelid Hbs. Divalent ions like Ca^{2+} and Mg^{2+} act as effective modulators of O_2 affinity and cooperativity in these giant Hbs (2, 3), replacing the phosphates that have this role in tetrameric Hbs. In addition to this functional role, a stabilization of the oligomeric structure has been described as another important effect of divalent ions. It is known that divalent ions prevent the alkaline dissociation of annelid Hbs (4). Despite work in recent years on the role of divalent metal ions in the oxygenation of giant Hbs, very little is known about the binding of these ions to these molecules.

In our work, spectroscopy of Mn^{2+} is used as a tool to obtain information on the binding of Mn^{2+} to the Hb of *G. paulistus*, as well as on the effects of Ca^{2+} and Mg^{2+} upon the binding of Mn^{2+}. The number of binding sites is obtained under different experimental conditions and a classification of these sites is performed on the basis of the association constants.

Materials and Methods

Standard solutions of manganese sulfate of different concentrations in the range of 0.1–20 mM were used. Two types of experiments were performed in which either the Mn^{2+} concentration was kept constant and the Hb concentration varied, or *vice versa*. The Hb concentration was measured through the absorbance at 415 nm so that it is expressed as the concentration of heme, and the calculation of the number of sites are normalized to heme. Hb concentration varied in the range of 30–200 μM in heme. Equal volumes of Hb solution and aqueous solutions of ions were mixed and incubated overnight at 4°C to allow the equilibration of divalent ions, and the EPR spectra were measured at 25°C the next day. A Varian E-109 x-band spectrometer and a quartz flat cell were used.

Results and Discussion

In all experiments, the typical well-resolved six-line EPR spectrum was observed for Mn^{2+}. The hyperfine splitting was 92 Gauss and did not vary significantly. The intensity of the second component in the spectrum was used to obtain the Mn^{2+} concentration by using an appropriate calibration curve for

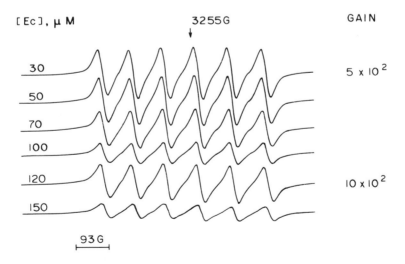

Figure 1. EPR spectra for solutions containing variable concentrations of Hb (shown on the spectra on the left, μM) and fixed concentrations of Mn^{2+} (2.3 x 10^{-3} M) at 25°C. $P = 20$ mW; Mod. Amp. = 4.0 g; $\Delta H = 1000$ G. Spectrometer gain is shown in the spectra on the right.

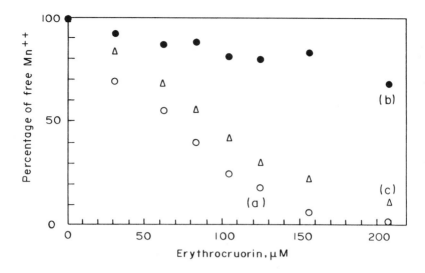

Figure 2.   Percentage of free Mn^{2+} relative to control in the absence of protein as a function of Hb concentration (heme basis). (*a*) Mn^{2+} alone. (*b*) 10 mM Ca^{2+}. (*c*) 10 mM Mg^{2+}.

Mn^{2+} in aqueous solution measured under the same conditions. In Figure 1, the EPR spectra are shown for solutions with a fixed concentration of Mn^{2+} (2.3 x 10^{-3} M) and variable Hb concentrations in the range 30–150 μM. A monotonic decrease in the intensity is observed upon increase in Hb concentration. The loss of intensity is related to the asymmetric coordination of the manganous ion and its time-dependent dynamics (5): the complexation to protein groups reduces the cubic symmetry of the hydrated ion, and large, time-dependent zero-field splittings are induced so that its modulation makes the EPR spectra very broad and undetectable (5). In this way, the only remaining EPR spectrum observed in the figure is due to the free Mn^{2+} in solution.

Figure 2 shows the plot of the percentage of free Mn^{2+} relative to the control in the absence of the Hb as a function of Hb concentration, obtained from the intensity of the EPR spectra. Results obtained in the presence of Ca^{2+} and Mg^{2+} are presented together, permitting the comparison of the binding of Mn^{2+} alone with the competitive binding in the presence of these diamagnetic ions. The presence of divalent ions leads to an increase in the concentration of free Mn^{2+} in solutions through direct competition for the available binding sites in the protein. The number of Mn^{2+} ions bound to the Hb was estimated from this figure by dividing the total variation of free Mn^{2+} in solution by the final Hb concentration. This number is 15 for Mn^{2+} alone, 4 in the presence of Ca^{2+}, and 12 in the presence of Mg^{2+}. In order to obtain additional information on the binding of Mn^{2+} our data were treated using Scatchard plots (6). Knowing the initial Mn^{2+} added in all solutions, the free Mn^{2+} was estimated for each Hb concentration; the bound Mn^{2+} was obtained in a straightforward manner. The Scatchard plot for Mn^{2+} binding is presented in Figure 3. This plot is due to the presence of two different classes of binding sites, one of high affinity and a second one of lower affinity. The abscissa provides the number of sites at the intersection ($v/L = 0$). We obtain 16 high affinity sites and an association constant of $5 \times 10^5 \, M^{-1}$; the number of lower affinity sites is 10 and their association constant is $3.6 \times 10^3 \, M^{-1}$.

Experiments were also performed at constant Hb concentration and variable Mn^{2+} concentration. The dependence of the concentration of free Mn^{2+} in solution as a function of the concentration of total added Mn^{2+} is a straight line that intercepts the abscissa at a finite concentration of total added Mn^{2+}. Two Hb concentrations were used: $30 \times 10^{-6} \, M$ and $700 \times 10^{-6} \, M$ on a heme basis. The intercept gives the concentration of added Mn^{2+} up to which level all Mn^{2+} are tightly bound so that no free Mn^{2+} is observed in solution. Dividing this concentration by the Hb concentration gives the number of ions bound per heme or the number of binding sites for Mn^{2+}. In the absence of divalent diamagnetic ions these numbers depend upon the Hb concentration: at the higher Hb

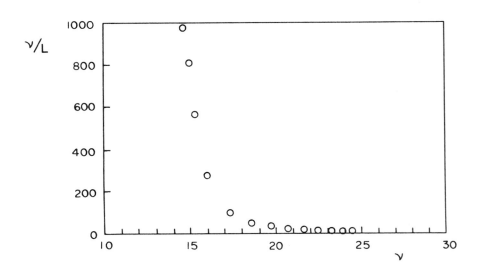

Figure 3. Scatchard plot for data in Figure 2A for the binding of Mn^{2+} to the Hb; v is the ratio of bound Mn^{2+} to protein and L is the free Mn^{2+} concentration for each Hb concentration.

concentration it is 12, and at the lower concentration it is 6. In addition, in the presence of Ca^{2+} and Mg^{2+} the number of sites are reduced to 1 and 5, respectively, at the higher Hb concentration, and to zero at the lower Hb concentrations.

Treatment of binding data through the Scatchard plot formalism (6) produced results consistent with those described above. In the case of the binding of Mn^{2+} alone, a total of 13 higher affinity sites are obtained with an association constant of 6.5×10^9 M^{-1}, and 11 lower affinity sites are obtained with an association constant of 1.2×10^3 M^{-1} at the higher Hb concentration of 70×10^{-6} M. These results are interesting in view of the fact that generally the effect of divalent ions upon oxygenation and cooperativity occurs at high concentration (50–100 mM), while the stabilization of the oligomeric structure is effective at a lower concentration (millimolar range). It is clear from our data that quite large association constants are observed for the higher affinity sites and that the number of sites obtained per heme is also quite high. This would suggest that the binding of divalent ions to the Hb is not as specific as it is in smaller proteins; but the binding is quite tight as deduced from the association constants. It might prove useful to further examine the possibility that the binding occurs as it does in polyelectrolytes where clusters of divalent ions interact with the macromolecule in a subtle manner.

Acknowledgements

The authors appreciate the support of Brazilian agencies CNPq, FAPESP and CAPES.

References

1. Costa, M.C.P., Bonafe, C.F.S., Meirelles, N.C. and Galembeck, F. (1988) *Braz. J. Med. Biol. Res.* 21: 115–118.
2. Fushitani, K., Imai, K. and Riggs, A.F.   (1986)   *J. Biol. Chem.* 261: 8414–8423.
3. Tsuneshige, A., Imai, K., Hori, H., Tyuma, I. and Gotoh, T. (1989) *J. Biochem.* 106: 406–417.
4. Kapp, O.H., Polidori, G., Mainwaring, M.G., Crewe, A.V. and Vinogradov, S.N. (1984) *J. Biol. Chem.* 259: 628–639.
5. Burlamacchi, L., Martin, G., Ottaviani, M.F. and Romaneli, M. (1978) *Adv. Mol. Relax. Processes* 12: 145–163.
6. Cantor, C.R. and Schimmel, P.R.   (1980)   *Biophysical Chemistry of Macromolecules*, Part III, 489–497. San Francisco: W.H. Freeman and Company.

17
Oxidation of the Extracellular Hemoglobin of *Glossoscolex paulistus*

Janice R. Perussi,[a] Adalto R. de Souza,[a] Maria H. Tinto,[a] Hidetake Imasato,[a] Nilce C. Meirelles[b] and Marcel Tabak[a]

[a]Institute of Physics and Chemistry, University of São Carlos, CP 369, 13560 São Carlos, SP, Brazil
[b]Institute of Biology, University of Campinas, 13081 Campinas, SP, Brazil

Introduction

The O_2 affinity of the extracellular Hb of *Glossoscolex paulistus* depends upon pH and the presence of ions; in particular, the presence of Cu ions decreases the cooperativity of this macroprotein (1, 2). The oxidation of the ferro Hb with potassium ferricyanide has been studied (3), and some unusual properties have been observed: the met derivative is formed at pH 7 with characteristic spectral properties of the aquomet form, but is easily converted to hemichrome away from pH 7. An ESR study (4) has also observed that hemichrome is formed at pH values different from 7 and has suggested that two different heme environments co-exist in the molecule. Another study used spectrophotometry and ESR spectroscopy (5) to confirm these results, and suggested the existence of some non-heme polypeptide chains. The aim of this work is to investigate the oxidation of iron in the extracellular Hb of *G. paulistus* under the influence of different chemical agents, of potassium ferricyanide and Cu ions, in particular. Since the number of heme groups in this macroprotein is over one hundred, the study of the kinetics of the oxidation of iron could be a useful tool in the characterization of different kinds of iron in terms of their exposure to the solvent.

Results and Discussion

The oxidation was monitored in a spectrophotometer, following the spectral changes as a function of time at 577 nm. The kinetics of oxidation are slower than are those for human Hb, in which there is a rapid phase in the millisecond range characterized by at least two phases. Kinetic data were analyzed as first order processes and the constants for the rapid and slow phases were obtained (Figure 1): $k_{slow} = 0.01$ min^{-1} and $k_{rap} = 0.3\text{-}1.0$ min^{-1}, depending upon the protein concentration. Using the graphic procedure presented in Figure 1, an extrapolation was made to initial absorbance ($t = 0$), with the assumption that the sum corresponds to the total number of oxidizable heme groups. In this way, the percentage of heme groups that react in each phase was calculated. The rapid phase corresponds to about 50% of the total iron with a half-life of 1–5 min and is essentially independent of protein concentration. The other 50% of the heme groups are oxidized in the slow phase with a half-life of about 160 min. This behavior is different from that of human Hb, in which the hemes are oxidized so rapidly that 100% of them are oxidized in about 1–2 min. Figure 2

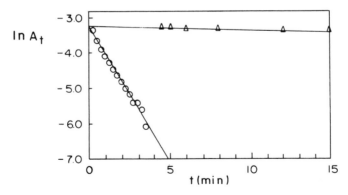

Figure 1. The determination of the pseudo-first order rate constants (min^{-1}) for the reaction of extracellular Hb (7 μM in heme) and ferricyanide (130 μM), pH 7, 25°C. (O) rapid phase, ($\triangle$) slow phase.

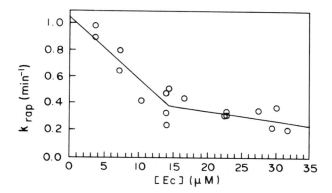

Figure 2. A plot of the pseudo-first order rate constant as a function of the Hb concentration (μM in heme); ferricyanide concentration = 130 μM, pH 7, 25°C.

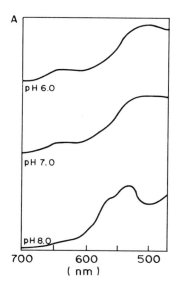

Figure 3. Optical spectra of Hb solution, 30 μM in heme, oxidized with 300 μM ferricyanide for one hour in 0.05M phosphate buffer at different pHs.

shows the dependence of the rapid phase constant as a function of protein concentration. It may be seen that the pseudo-first order constant of the rapid phase decreases with an increase in the protein concentration; above 15 μM it remains practically constant. This result may be interpreted as a dissociation of the oligomer which occurs at concentrations lower than 15 μM, making the heme groups more accessible and the oxidation reaction more rapid. The pseudo-first order constant of the rapid phase as a function of ferricyanide concentration at 14 μM heme shows that the dependence is linear and suggests a first order reaction relative to ferricyanide. The specific rate constant is estimated to be 2.6×10^3 min^{-1}M^{-1}.

A basic difference between this extracellular Hb and human Hb is the fact that in the reaction of the former with ferricyanide, in addition to a rapid initial phase that occurs during the first minute of reaction and which is similar to a reaction with human Hb, a second additional slower phase is apparent. Optical spectra obtained for oxidized forms (Figure 3) at pH $\leq$ 7.0 present maxima at about 500 and 630 nm, suggesting the formation of the aquomet derivative. At pH values higher than 7.5, a maximum at about 530 nm indicates the formation of the hemichrome.

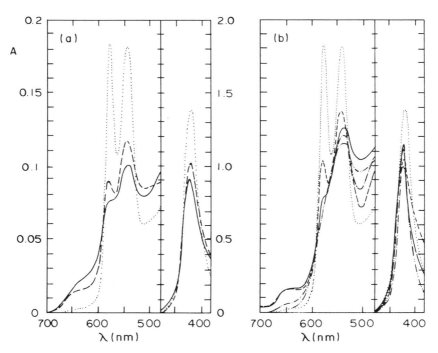

Figure 4. The reaction of extracellular Hb, 12 μM with respect to (*a*) Cu ions and (*b*) ferricyanide, pH 7, 25°C (····) control; (---) 1 : 1, (—) 1 : 3, (–·–··) 1 : 5, (–··–··) 1 : 10. Reaction time = 23 h.

Figure 4*a* presents the spectra obtained in the reactions of extracellular Hb and Cu ions after 23 h that suggest that the reaction is distinct as compared to the ferricyanide reaction (Figure 4*b*). The band at 630 nm is more pronounced and presents a characteristic profile. In the mixture of protein and metal in the proportion of 1 : 2, the band of a Cu(II)-protein complex appears in the region of 600 nm. In the presence of Cu in proportions higher than 1 : 2, it seems that some denaturation reaction and/or protein precipitation also occurs. The reaction with ferricyanide in the molar ratio of 1 : 1 is slow. The increase in absorbance at 630 nm is relatively more rapid than the decrease of the α and β bands at 577 and 540 nm, respectively. After 23 h of reaction it can be verified that the oxidation at ratios of 1 : 3, 1 : 5 and 1 : 10 was substantial and that, as a consequence of the reaction, there is a decrease in the absorbance of the Soret band (418 nm). This is different from that observed for human Hb in which the intensity increases with the oxidation with a simultaneous shift of about 10 nm to lower wavelengths. The greater increase in the intensity of the band at 630 nm compared to the decrease of the α and β bands could be related to the formation of the hemichrome as the oxidation proceeds. The effect of Cu which reduces the cooperativity of the Hb is certainly related to its oxidation effect. Further studies are in progress to characterize the interaction of Cu with Hb in detail.

Acknowledgements

The authors appreciate the support of the Brazilian agencies CNPq, FAPESP and FINEP.

References

1. Meirelles, N.C., Oliveira, B., Paula, E., Marangoni, S. and Marques, M.R.F. (1985) *Comp. Biochem. Physiol.* **82**B: 203–205.
2. Meirelles, N.C., Oliveira, B., Oliveira, A.R., Paula, E., Marangoni, S. and Rennebeck, G.M. (1987) *Comp Biochem. Physiol.* **88**A: 377–379.
3. Ascoli, F., Rossi Fanelli, M.R., Chiancone, E., Vecchini, P. and Antonini, E. (1978) *J. Mol. Biol.* **119**: 191–202.
4. Desideri, A., Verzili, D., Ascoli, F., Chiancone, E. and Antonini, E. (1982) *Biochim. Biophys. Acta* **708**: 1–5.
5. Tsuneshige, A., Imai, K., Hori, H., Tyuma, I. and Gotoh, T. (1989) *J. Biochem.* **106**: 406–417.

18

Kinetic Evidences for Slow Structural Changes in the Chlorocruorin from *Spirographis spallanzanii*

Rodolfo Ippoliti, Andrea Bellelli,
Eugenio Lendaro and Maurizio Brunori

Department of Biochemical Sciences and CNR Center for Molecular
Biology, University of Rome "La Sapienza," 00185 Rome, Italy

Introduction

Chlorocruorin (Chl), the respiratory protein from the blood of the polychaete worm *Spirographis spallanzanii*, is a giant hemeprotein ($M_w \approx 2,800$ kDa) characterized by interesting structural and functional properties (1, 2). Analysis of the O_2-binding isotherm (3, 4), which displays very high heme-heme interactions ($n > 5$), has led to the suggestion that the minimum cooperative unit contains twelve or six hemes within the larger protein matrix made up of 72 (5) or, possibly 96 (6) subunits.

In order to investigate more thoroughly the interactions between the minimum cooperative units and the completely assembled particle, we have carried out an extensive characterization of the kinetics of O_2- and CO-binding to Chl from *S. spallanzanii* for which only a very preliminary characterization is available (7).

In agreement with previous results on O_2-binding, equilibrium isotherms may be satisfactorily described by the MWC allosteric model (8), assuming a cooperative unit of six (4) or twelve (3) O_2-binding sites. Starting from these premises, we analyzed stopped-flow- and flash-photolysis kinetic data and found indications for the presence of very slow (quaternary?) relaxation(s), which lead to the accumulation of metastable intermediates not seen in similar experiments carried out with human Hb.

Materials and Methods

Chl from *S. spallanzanii* was prepared from live specimens from the Naples Bay as described by Antonini *et al.* (1, 2). Kinetic experiments, carried out by stopped-flow- and flash-photolysis, were performed following standard procedures (9).

Results and Discussion

We have determined the combination of Chl with CO by stopped-flow- and flash-photolysis at different protein and ligand concentrations, as well as at different degrees of photodissociation. The time course of the reaction, measured by mixing deoxy Chl with CO in the stopped-flow apparatus, conforms to a single exponential under pseudo-first order conditions. Figure 1, which reports the second order rate constant obtained at various CO concentrations, yields a value of $1' = 7 \times 10^4$ M^{-1}s^{-1}; we assume that this value represents the

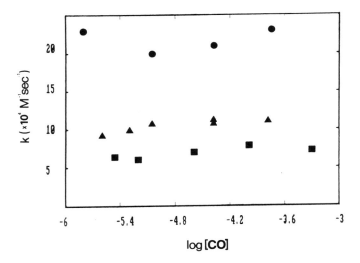

Figure 1. Second order rate constant for the binding of CO to Chl, determined at different ligand concentrations. Experimental conditions: pH 7.4, 0.05 M Tris/0.05 M Bis-Tris, 0.2 M NaCl, 5 mM CaCl$_2$, 5 mM MgCl$_2$; temperature = 20°C. Observation at 449 nm in all cases. (■) experiments by stopped-flow, Chl concentration 1.95 μM after mixing; (●) experiments by flash photolysis with partial (25%) photodissociation, Chl concentration from 0.33 to 0.66 μM; (▲) experiments by flash photolysis with complete (85-95%) photodissociation, Chl concentration from 0.33 to 0.66 μM.

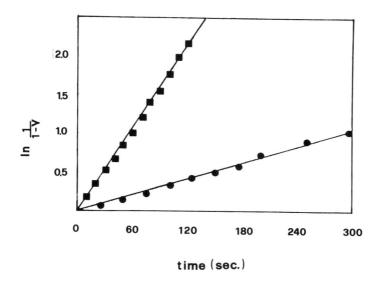

Figure 2. Time course of CO dissociation from ChlCO (●), as observed at 440 nm after mixing with excess microperoxidase. Chl concentration 7.5 μM; CO concentration 17.5 μM; microperoxidase concentration 50 μM. An experiment with HbA CO (■) is reported for comparison. Experimental conditions as in Figure 1.

combination rate constant for the T state, deoxy Chl. The partial photolysis experiments, also carried out at different CO concentrations, yield a value for the second order rate constant of $1' = 2.2 \times 10^5$ $M^{-1}s^{-1}$ (Figure 1); this is a minimum estimate for the reactivity of R state Chl, given that the fraction of ligand photolyzed is 25%.

Upon complete photolysis (85 to 95%), the time course for CO recombination yields a second order rate constant which, at CO concentrations above 20 μM, is twice as large as that measured by stopped-flow (see Figure 1). This unexpected finding cannot depend on the heterogeneity of the sample and implies that, in the short time in which the light pulses (150 μs), the ligand dissociates though the molecule does not fully relax to a stable deoxy conformation (unliganded T state).

To substantiate this hypothesis, we carried out more photolysis experiments at ligand (and protein) concentrations lower than 20 μM in order to slow down the recombination and to eventually allow enough time to achieve complete protein relaxation. Indeed, it was observed that 1) the time course of recombination becomes non-exponential, and 2) at the lowest concentrations (0.33 μM Chl and 2.5 μM CO), the average rate constant tends to approach the

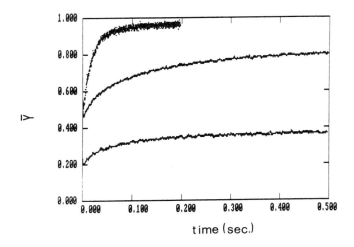

Figure 3. Time course of O_2 combination to deoxy Chl as followed in the stopped-flow apparatus at 455 nm. Experimental conditions as in Figure 1; O_2 concentrations (from top to bottom) = 670, 135 and 70 μM after mixing.

value observed by stopped-flow. This is also indicated by the data given in Figure 1. Therefore, we conclude that the state of the molecule populated immediately after complete photolysis does not correspond to either the T or the R state as shown by its intermediate reactivity, and that a slow protein relaxation takes place with a rate constant of approximately 1 s^{-1}.

The time course of CO dissociation from fully saturated ChlCO was determined by mixing the ChlCO with an excess of microperoxidase, following the method of Sharma *et al.* (10). The results, shown in Figure 2, indicate that the overall dissociation corresponds to a single exponential process with a rate constant of 1 = 0.0035 s^{-1}. Although the experiment yields fully deoxygenated Chl as the final product, we take this rate constant to be equal or close to that characteristic of R state Chl.

Taking the above values of 1' and 1 for R state Chl, the minimum estimate for the CO-binding constant to the high affinity state is $L_R = 6.5 \times 10^7$ M^{-1} (corresponding to $P_{50} = 0.011$ mm Hg).

Additional evidence for the presence of slow structural relaxations comes from the study of the O_2 combination time course. O_2 binding to Chl conforms to a single exponential for the R state, as followed by partial flash photolysis; the corresponding rate constant is $k' = 1.5 \times 10^6$ M^{-1}s^{-1}. On the other hand, the time course for the combination to the unliganded protein is unusually heterogeneous (see Figure 3); in fact, it is biphasic, does not follow second order behavior, and the observed rate constants, based on the equilibrium results and the kinetics of ligand dissociation (as measured independently by mixing Chl with O_2 in the presence of dithionite, following (11)), are much smaller than expected.

These experimental observations demonstrate that it is not possible to describe the kinetic behavior of Chl within a simple two-state model with rapid conformational changes. A plausible scheme to account for these results is as follows:

$$T_{Chl} \quad \rightleftharpoons \quad X_{Chl} \quad \rightleftharpoons \quad R_{Chl}$$

$$\updownarrow \qquad\qquad \updownarrow \qquad\qquad \updownarrow$$

$$T_{Chl}\,L \quad \rightleftharpoons \quad X_{Chl}\,L \quad \rightleftharpoons \quad R_{Chl}\,L \ .$$

The combination of CO with the T or the R state molecule is bimolecular, and the respective rate constants yield at least a four-fold increase in reactivity in going from T to R. A supposed intermediate species, X, is produced during the transition from the T to the R state, and vice versa, and its detection depends on experimental conditions. While the transition from X to T in unliganded Chl has been deduced from the flash photolysis experiments on Chl CO (see Figure 1), the reverse reaction ($T \rightarrow X$) could not be studied directly. By analogy with human Hb (12), we expect this transition to be first order and its rate constant to depend upon ligand saturation. In fact, we believe that this transition is responsible for the heterogeneous time course of O_2 combination (Figure 3).

This model for Chl implies that the pathways for ligand binding and release are different and depend upon the rates and concentrations of any given ligand; during O_2 binding, T state Chl is partially saturated and further O_2 binding is rate-limited by the structural transition to liganded X and R. In the case of CO recombination, the rate-limiting step usually is not the quaternary transition unless rapid photodissociation of the ligand is used to initiate the reaction; in this case, the X state, populated *via* the R state, may either rebind the ligand or decay to the T state (depending upon ligand concentration).

At present, it is not possible to envisage the nature of the intermediate state since it has been observed only in kinetic experiments and is not necessarily demanded by the equilibrium behavior of the protein. However, it may be of interest to suggest that the slow protein relaxation observed in the course of this work may correspond to quaternary transitions involving the reorganization of the entire molecule. This is not implausible considering that the present model describes the equilibrium data with the assumption of allosteric interactions exclusively involving the sites within each cooperative unit (with either six or twelve hemes), but does not take into account that the quaternary state of the assembled macromolecule may exert a significant constraint on the binding properties of the cooperative units. We propose, as a working hypothesis, that the slow relaxation process(es) described above are the first indications of quaternary conformational changes constraining the reactivity of the hemes in the whole molecule.

Acknowledgements

We express our thanks to Mr. P. Sansone (from the Zoological Station in Naples, Italy) for kindly providing the *S. spallanzanii*; and to Dr. A. Brancaccio (Rome) for skillful assistance. Work was partially supported by a grant from the Ministero dell'Universita' e della Ricerca Scientifica e Tecnologica of Italy.

References

1. Antonini, E., Rossi-Fanelli, A. and Caputo, A. (1962) *Arch. Biochem. Biophys.* **97**: 343–350.
2. Antonini, E., Rossi-Fanelli, A. and Caputo, A. (1962) *Arch. Biochem. Biophys.* **97**: 336–342.
3. Colosimo, A., Brunori, M. and Wyman, J. (1974) *Biophys. Chem.* **2**: 338–345.
4. Imai, K., Yoshikawa, S., Fushitani, K., Takizawa, H., Handa, T. and Kihara, H. (1986) In *Invertebrate Oxygen Carriers*, ed. B. Linzen, 367–374. Berlin: Springer-Verlag.
5. Guerritore, D., Bonacci, M.L., Brunori, M., Antonini, E., Wyman, J. and Rossi-Fanelli, A. (1965) *J. Mol. Biol.* **13**: 234–237.
6. Mezzasalma, V., Zagra, M., Di Stefano, L. and Salvato, B. (1986) In *Invertebrate Oxygen Carriers*, ed B. Linzen, 65–68. Berlin: Springer-Verlag.
7. Brunori, M., Guerritore, D., Antonini, E., Wyman, J. and Rossi-Fanelli, A. (1965) *Estr. Pubblic. Staz. Zool. (Napoli)* **34**: 521–526.
8. Monod, J., Wyman, J. and Changeux, J.-P. (1965) *J. Mol. Biol.* **15**: 88–118.
9. Brunori, M. and Giacometti, G.M. (1981) *Methods Enzymol.* **76**: 582–595.
10. Sharma, V.S., Schmidt, M.R. and Ranney, H.M. (1976) *J. Biol. Chem.* **251**: 4267–4272.
11. Gibson, Q. H. (1973) *Proc. Natl. Acad. Sci. U.S.A.* **70**: 1–4.
12. Sawicki, C. and Gibson, Q. H. (1976) *J. Biol. Chem.* **251**: 1533–1544.

19
Scapharca inaequivalvis Hemoglobins: Novel Cooperative Assemblies of Globin Chains

Emilia Chiancone,[a] Daniela Verzili,[a]
Alberto Boffi,[a] William E. Royer, Jr[b] and
Wayne A. Hendrickson[c]

[a]CNR Center of Molecular Biology and Department of Biochemical Sciences, University of Rome"La Sapienza," 00185 Rome, Italy
[b]Program in Molecular Medicine, University of Massachusetts Medical School, Worcester, MA 01605, USA
[c]Department of Biochemistry and Molecular Biophysics, Columbia University, New York, NY 10032, USA

Introduction

The crystal structures of the Hbs from the clam *Scapharca inaequivalvis* (HbI, a homodimer, and HbII, a heterotetramer) (1–3), and from the "fat innkeeper" worm *Urechis caupo* (a homotetramer) (4) have revealed novel assemblages of Mb-folded chains that differ markedly from that characteristic of the $\alpha_2\beta_2$ vertebrate Hb tetramer (5).

The *U. caupo* homotetramer is a non-cooperative molecule (6). On the other hand, *S. inaequivalvis* HbI and HbII both exhibit cooperative O_2 binding (7), and thus are of special interest in understanding the evolution of cooperativity in

Hbs. In particular, the highly cooperative homodimeric HbI (Hill coefficient, n = 1.5) represents the first exception to the rule, based on the behavior of vertebrate Hbs, that cooperativity is restricted to tetrameric molecules constructed from two different types of chains (5).

Relative to vertebrate Hbs, the most striking feature in the subunit arrangement of *S. inaequivalvis* HbI is the direct communication between the two heme groups, which are brought into contact by the involvement of both the E and F helices in the interface (HbII is a dimer of heterodimers assembled like HbI) (1-3). This feature provides a short pathway for transferring information about the ligation state of one heme group to the other. In the *U. caupo* tetramer, where only the E helix (but not the F helix) forms intersubunit contacts, this possibility is precluded, and the result is a non-cooperative molecule (4).

Cooperativity in the binding of O_2 requires that the molecule be able to alternate between distinct quaternary structures that are characteristic of the deoxygenated and liganded protein and are endowed with different ligand affinities. In vertebrate Hbs, a major role in the transmission of information from one heme group to the rest of the molecule, and, hence, in triggering the quaternary structural change, is played by the F helix. Thus, O_2-binding displaces the iron atom relative to the heme plane; in turn, this movement pulls the proximal His, $F8$, promotes a shift of the whole F helix and thereby results in a disruption of the bonds that stabilize the deoxygenated structure leading to major subunit rearrangements (8, 9).

The proximal His in the F helix is one of the few structural elements common to all globin chains (10). Hence, one may expect rearrangements of the F helix to be part of the triggering mechanism that permits heme-heme communication in all cooperative Hbs. The structural transitions that accompany ligand binding in *S. inaequivalvis* HbI will be discussed within this framework.

Results and Discussion

The HbI Homodimer. In the structure of *S. inaequivalvis* HbI at 2.4 Å resolution (2), the involvement of the hemes at the interface is the most striking, novel feature. One charged propionate group from each heme is intimately involved in the contact by forming an ionic interaction with Lys-96' (the prime indicates the symmetry-related subunit), and a hydrogen bond with Asn-100', two residues very close to the proximal His-101'. A water molecule is in a position to simultaneously form a hydrogen bond to propionate O_2s from both hemes. In addition to this heme–F region of the interface, interactions in several other regions contribute to the stability of the contact. A largely hydrophobic E-E area is formed where both E helices cross each other near their centers. Tyr-75 from each subunit is most extensively involved in this region and interacts across the interface with the aliphatic portion of Asn-79' and, *via a*

L   R

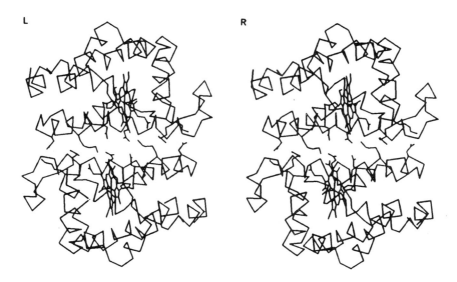

Figure 1. Stereo α-carbon plot of *S. inaequivalvis* HbI plotted down the molecular diad. Near the right and left ends of the diagram are the salt bridges formed between Lys-30 (at the beginning of the *B* helix) and Asp-89' (at the beginning of the *F* helix). Near the center, the interactions of the propionate groups from the hemes are shown including intrasubunit salt bridges with Arg-53 and -104 and interactions across the interface with Lys-96' and Asn-100'. Also shown are the side chains of the proximal His's. Reprinted with permission from *J. Biol. Chem.* **264**: 21057.

hydrogen bond, with Asp-82'. Two symmetric hydrophobic *E-F* patches are formed where the second and third turns of the *E* helix from one subunit interact with the second and third turns of the *F* helix of the other subunit. Involved in contacts are Gly-68, Ile-71, and Thr-72, and, respectively, Cys-92', Val-93' and Phe-97'. Finally, two symmetrically related salt-bridges are formed between Lys-30 at the beginning of the *B* helix (B2), and Asp-89' at the beginning of the *F* helix (F2).

The recent determination of the deoxyHbI structure at the same resolution (3) has revealed the transitions between liganded and unliganded states that occur in this molecule. In contrast to mammalian Hbs which undergo a major quaternary change upon ligand binding (the two $\alpha\beta$ dimers rotate relative to each other by 15° and translate by 0.8 Å), in *S. inaequivalvis* HbI, the overall quaternary change is rather subtle as one subunit rotates relative to the other subunit by only 3.4° and translates by 0.1 Å. On the other hand, tertiary changes that affect the interface are quite remarkable. The most dramatic difference in conformation between the two structures is observed for Phe-97, which is one helical turn from the proximal His. In the deoxy structure, the phenyl group packs tightly against the heme on the proximal side, but is extruded into the interface in the CO derivative, where it is in close contact with the methyl group of Thr-72'. This difference is related to the characteristic change in the position of the iron atom: whereas this lies 0.5 Å out of the mean plane of the heme in the deoxy porphyrin, it lies within 0.1 Å of the heme plane in the CO liganded one, and there is no longer sufficient space for the phenyl ring in the heme pocket. Another dramatic difference between the liganded and unliganded structures is the position of the hemes: in the CO derivative the heme is located about 0.6 Å deeper into the subunit than in the deoxy derivative. As a result, the iron atoms are 18.4 Å apart in CO HbI, but only 16.7 Å apart in deoxy HbI, and the propionate groups adopt different conformations at the interface. Moreover, Lys-96, which forms a trans-interface salt bridge with one propionate in the CO structure, switches to form a salt bridge with the other propionate and a hydrogen bond with the distal His-69' in the deoxy structure.

These structural data indicate that crucial roles in the cooperativity mechanism of HbI are played by Phe-97, which appears to keep the deoxy hemes in a low affinity conformation (by preventing the heme groups from being planar and the iron atom from being in that plane), and by the heme propionates, which are intimately involved in bonding interactions across the interface. The role of such interactions in the transmission of information between the two subunits is being investigated in proteins reconstituted with modified hemes.

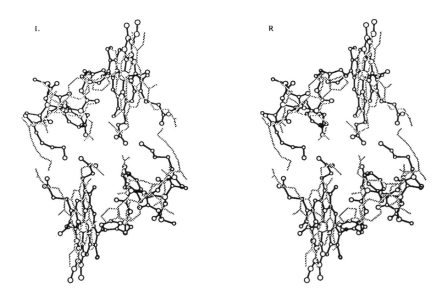

Figure 2. Stereo diagram showing deoxy (stippled bonds) and CO (open circles, dark bonds) *S. inaequivalivs* HbI models for two turns of the *F* helix and the hemes for both subunits. Note the movement of the hemes away from each other on ligand binding and rearrangements of side chains for Lys-96 and Phe-97.

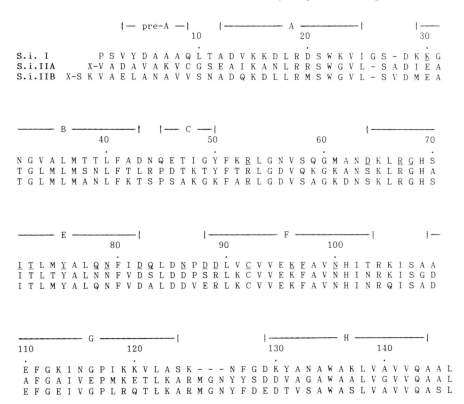

Figure 3. Comparison of the amino acid sequences of the constitutive *S. inaequivalvis* Hb chains. The numbering and the helix notation refer to the sequence of HbI. The sequences of HbII are taken from reference 18. Residues involved in contacts across the interface of HbI are underlined (2).

Two general considerations are in order regarding the structural basis of cooperativity in HbI. The insensitivity to allosteric effectors may be explained in terms of the tight coupling between the two heme groups, which are practically in direct communication. Moreover, the vicinity of the two heme groups and the major importance of ligand-linked tertiary conformational changes relative to quaternary ones explain the rapid rate of heme–heme communication. This is evidenced by data on the nanosecond geminate recombination of HbI with O_2 that indicate that the transfer of information between the two heme groups occurs within this time range, and, hence, is faster than the quaternary conformational change which takes place in microseconds (11), a behavior in marked contrast with that characteristic of vertebrate Hbs (12, 13).

Lastly, it is of interest that, despite the different structural bases, the manifestations of cooperativity in ligand binding in *S. inaequivalvis* HbI are the same as those in vertebrate Hbs. Thus, at equilibrium, cooperativity is primarily entropic in origin with O_2 as ligand (14); kinetically, it is apparent in the accelerating time course of CO-binding and in the rates of O_2 dissociation (15). However, a distinctive characteristic of *S. inaequivalvis* HbI is the cooperativity displayed in reactions that are non-cooperative in vertebrate Hbs. Cys-92 (*F*4), which is located at the subunit interface, reacts with bulky mercurials in a cooperative fashion, thus indicating that alterations in a limited part of the *E-F* interface are transmitted readily to the symmetrically related region. Another distinctive feature of HbI is the cooperative rupture of the bond between the N_ε of the proximal His and the heme iron which occurs at low pH in both deoxy- and NO-HbI (17). In human Hb, this bond represents the only structural link between the heme group and the protein; hence, it plays a critical role in the transmission of quaternary conformational changes (5). Cooperativity in this reaction thus provides an experimental proof of the transmission of conformational changes *via* a new pathway, the most likely candidate being the network of interactions at the heme–*F* interface.

The HbII Tetramer. The low resolution x-ray structure shows that HbII is a dimer of heterodimers constructed like HbI (1); thus, as expected, the constituent polypeptide chains of HbII display a high sequence similarity with the HbI chain, especially in the region of the *E* and *F* helices (18). Of all the residues involved in contacts across the dimer interface, about one-third are changed in the two tetramer chains. However, the amino acids crucial to the direct pathway of heme-heme communication, like Lys-96 and Asn-100, which interact with the propionates of the neighboring subunit, and Phe-97, which upon ligation swings out of the heme pocket into the interface, are all conserved. The substitutions involving Lys-30 (*B*2) and Asp-89 (*F*2) are intriguing as they involve two residues that form a salt bridge across the interface. In both tetramer chains, an aspartate and an arginine replace Lys-30 and Asp-89, respectively; hence, a salt bridge may still be formed, albeit with a sign reversal, a feature that is likely to direct the assembly of the two Hb components.

From a functional point of view, the presence of two dimeric building modules in HbII endows it with sensitivity to allosteric effectors, a condition which is lacking in HbI. This property is due to the complex interplay between the binding of O_2, protons and anions, which, in turn, also determines the O_2-linked polymerization of the protein (19).

Conclusions

The structural and functional characterization of the Hbs from *S. inaequivalvis* has revealed that cooperativity in O_2-binding may be achieved by means of subunit arrangements radically different from that of vertebrate Hbs. In the *S.*

inaequivalvis homodimer, cooperativity is based on hemes communicating directly across a novel subunit interface formed by the *E* and *F* helices. Upon ligand binding, the iron displacement relative to the heme plane triggers the transition toward the structure of the liganded derivative, just as in vertebrate Hbs. It, however, is the only feature common to the two mechanisms of cooperativity; thus, in *S. inaequivalvis* the changes in the interactions that link the heme groups result in more striking alterations in tertiary structure but in smaller variations of quaternary structure than in mammalian Hbs.

References

1. Royer, W.E., Jr., Love, W.E. and Fenderson, F.F.   (1985)   *Nature* **316**: 277–280.
2. Royer, W.E., Jr., Hendrickson, W.A. and Chiancone, E. (1989) *J. Biol. Chem.* **264**: 21502–21061.
3. Royer, W.E., Jr., Hendrickson, W.A. and Chiancone, E. (1990) *Science* **249**: 518–522.
4. Kolatkar, P.R., Meador, M.E., Stanfield, R.L. and Hackert, M.L.   (1988) *J. Biol. Chem.* **263**: 3462–3465.
5. Perutz, M.F. (1979) *Annu. Rev. Biochem.* **48**: 327–386.
6. Garey, J.R. and Riggs, A.F. (1986) *J. Biol. Chem.* **261**: 16446–16450.
7. Chiancone, E., Vecchini, P., Verzili, D., Ascoli, F. and Antonini, E. (1981) *J. Mol. Biol.* **152**: 577–592.
8. Perutz, M.F. (1980) *Proc. Roy. Soc. Lond., B* **208**: 135–162.
9. Baldwin, J. and Chothia, C. (1979) *J. Mol. Biol.* **129**: 175–220.
10. Dickerson, R.E. and Geis, I. (1983) *Hemoglobin*, 68–73. Menlo Park, CA: Benjamin/Cumming.
11. Chiancone, E. and Gibson, Q.H. (1989) *J. Biol. Chem.* **264**: 21062–21065.
12. Olson, J.S., Rohlfs, R.I. and Gibson, Q.H.   (1987)   *J. Biol. Chem.* **262**: 12930–12938.
13. Hofrichter, J., Henry, E.R., Sommer, H., Deutsch, Ikeda-Saito, A.M., Yonetani, T. and Eaton, W.A. (1985) *Biochemistry* **24**: 2667–2679.
14. Ikeda-Saito, M., Yonetani, T., Chiancone, E., Ascoli, F., Verzili, D. and Antonini, E. (1983) *J. Mol. Biol.* **170**: 1009–1018.
15. Antonini, E., Ascoli, F., Brunori, M., Chiancone, E., Verzili, D., Morris, R.J. and Gibson, Q.H. (1984) *J. Biol. Chem.* **259**: 6730–6738.
16. Boffi, A., Gattoni, M., Santucci, R., Vecchini, P., Ascoli, F. and Chiancone, E. (1987) *Biochem. J.* **241**: 499–504.
17. Coletta, M., Boffi, A., Ascenzi, P., Brunori, M. and Chiancone, E. (1989) *J. Biol. Chem.* **265**: 4828–4830.
18. Petruzzelli, R., Boffi, A., Barra, D., Bossa, F., Ascoli, F. and Chiancone, E. (1989) *FEBS Lett.* **259**: 133–136.
19. Boffi, A., Vecchini, P. and Chiancone, E.   (1989)   *J. Biol. Chem.* **265**: 6203–6209.

20
High and Low Spin Forms of Oxidized Dimeric *Scapharca inaequivalvis* Hemoglobin

Carla Spagnuolo,[a] Alessandro Desideri,[b] Francesca Polizio[b] and Emilia Chiancone[a]

[a]CNR Center of Molecular Biology and Department of Biochemical Sciences, University "La Sapienza," 00185 Rome, Italy
[b]Department of Biology, University "Tor Vergata," 00173 Rome, Italy

Introduction

The intracellular dimeric Hb (HbI) from the Arcid clam *Scapharca inaequivalvis* is made up of two identical subunits that exhibit cooperative interaction in O_2 binding (1). In a previous study on the oxidized protein, sedimentation velocity experiments showed that this derivative undergoes a reversible pH-dependent dissociation into monomers; monomer formation is more marked at acid pH values. Parallel optical spectroscopy measurements indicated that dissociation is accompanied by a high to low spin transition. The low spin species is characterized by the presence of absorption bands, typical of hemichromes in which the sixth heme ligand is contributed by the protein. The high spin species displays an unusually high absorption at 600 nm (2).

An EPR study of oxidized HbI has been performed to gain deeper insight into the correlation between changes in the heme environment and changes in quaternary structure as a function of pH.

Results and Discussion

The X-band EPR spectrum of HbI obtained by oxidation by potassium hexacyanoferrate at pH 9.4 is shown in Figure 1. The spectrum is characterized by a rhombic high spin signal ($g_1 = 6.18$, $g_2 = 5.89$, $g_3 = 2.0$) and by a small percentage of the hydroxyl low spin species ($g_1 = 2.60$, $g_2 = 2.17$, $g_3 = 1.80$); it is independent of protein concentration consistent with the observed stability of the dimeric form at this pH value over the protein concentration range covered (2.6×10^{-4} - 5×10^{-5} M heme) (2). The rhombicity of the high spin signal cannot be ascribed to a specific structural feature of the heme, since such signals have been observed in many hemoprotein derivatives (3–5). It may be envisaged that the protein moiety puts a constraint upon the heme which destroys the electronic equivalence of the porphyrin ring in all directions, thereby causing the splitting of the EPR absorption derivatives in the $g = 6$ region. The percentage of rhombicity, R, varies with the degree of constraint and is defined as

$$R = 100 \left(\Delta g / 16 \right)$$

where Δg is the absolute difference in g values between the two components near $g = 6$ (4). For HbI, a value of R around 2% may be calculated; similar values have been reported for the isolated α chains of HbA (6) and for the low pH form of cyt c (5). The presence of a constrained structure at the heme site had never been observed in a liganded derivative of HbI, although several spectroscopic techniques indicated the presence of constraints in the deoxy derivative (1, 7, 8).

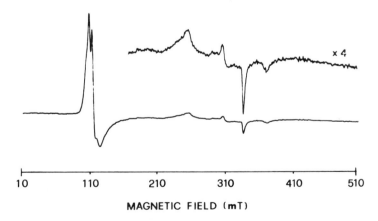

Figure 1. X-band EPR spectrum of oxidized *S. inaequivalvis* HbI at pH 9.4. Protein concentration: 2.6×10^{-4} M (heme); oxidation obtained with the addition of 1.5 equivalents of potassium ferricyanide per heme at pH 9.4. Setting conditions: microwave frequency, 9.42 GHz; microwave power, 15 mW; modulation amplitude, 1.0 mT; temperature, 14.5°K.

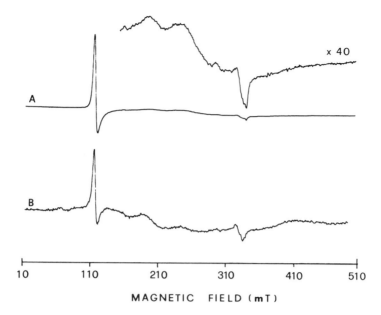

Figure 2. X-band EPR spectra of oxidized *S. inaequivalvis* HbI at pH 7.1. Protein concentration: (*A*) 2.6 x 10^{-4} M (heme), (*B*) 0.54 x 10^{-4} M (heme); oxidation obtained by addition of 1.5 equivalents of potassium ferricyanide per heme at pH 7.1. Temperature: 5°K. Other setting conditions as in Figure 1.

At pH 7.0, the EPR signals change as a function of protein concentration, as expected from the significant shift of the monomer-dimer equilibrium of oxidized HbI towards dissociation (2). At the higher concentration studied (Figure 2*A*), the EPR spectrum is characterized by a high spin signal with an axial symmetry and a linewidth of 5.1 mT in the $g = 6$ line, which are ascribable to an aquomet heme, and by several broad lines in the low magnetic field region, which are ascribable to low spin species. In particular, three of the low spin lines are better defined and their corresponding g values are 3.4, 2.9, and 2.3. The g values at 2.9 and 2.3 could be attributed to the first two absorption derivatives of a low spin form that has been identified in other heme compounds as a bisimidazole : heme complex with both imidazoles protonated at N_δ in other heme compounds. This form is characterized by three lines, usually $g_1 = 2.9$, $g_2 = 2.3$, and $g_3 = 1.5$ (9–10). The broad line with $g_1 = 3.4$ has been observed in cyt c at alkaline pH values (11), and in *Glycera* Hb (12) where it corresponds to an ε-amino : heme iron : imidazole structure. When the protein concentration decreases to 5 x 10^{-5} M, the intensity of the high spin signal decreases by 25% with a concomitant increase in the intensity of the low spin species (Figure 2*B*). These EPR results in combination with the published sedimentation velocity

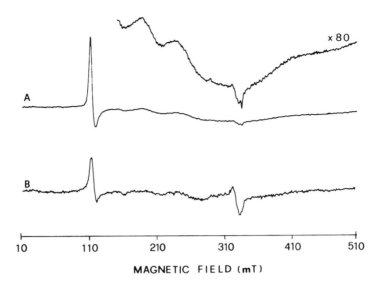

Figure 3. X-band EPR spectra of oxidized *S. inaequivalvis* HbI at pH 5.9. Protein concentration: (*A*) 2.6 x 10^{-4} M (heme), (*B*) 0.54 x 10^{-4} M (heme); oxidation obtained by addition of 1.5 equivalents of potassium ferricyanide per heme at pH 5.9. Temperature, 5°K. Other setting conditions as in Figure 1.

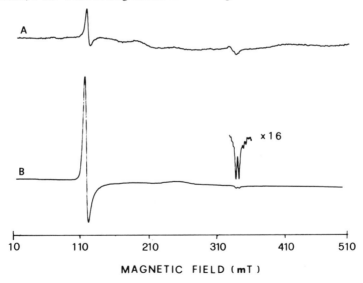

Figure 4. X-band EPR spectra of oxidized *S. inaequivalvis* HbI at pH 7.0, in the absence (*A*) or presence (*B*) of a molar excess of sodium fluoride.   Protein concentration: (*A*) 0.54 x 10^{-4} M (heme), (*B*) 0.70 x 10^{-4} M (heme); oxidation obtained with the addition of 1.5 equivalents of potassium ferricyanide per heme at pH 7.0. Temperature, 5°K. Other setting conditions as in Figure 1.

data (2) suggest that the oxidized heme is in the aquomet form when the protein is dimeric and that upon dissociation into monomers it transforms itself into hemichrome.

The EPR experiments performed at pH 5.8 confirm this scenario. At this pH value, the monomer-dimer equilibrium is shifted towards dissociation more than it is at pH 7.0 (2). Accordingly, a change in protein concentration from 2.6 x 10^{-4} to 5 x 10^{-5} M results in a bigger decrease of the high spin EPR signal (50%), than it does at pH 7.0 (Figures 3A, B). It is noteworthy that the linewidth of the $g = 6$ line is broader at pH 5.8 than at pH 7.0 (7.0 mT), indicating that the symmetry of the heme is lower.

Lastly, the EPR spectrum of the fluoride adduct was measured to study the effect of an external ferric heme ligand on hemichrome formation. Figure 4B shows the spectrum of HbI oxidized at pH 7.0 after the addition of a molar excess of NaF. The presence of a doublet with a splitting of about 4.6 mT in the $g = 2$ region, due to the interaction of the ^{19}F nucleus with the unpaired electron of the metal (4), evidences a direct binding of the anion to the iron atom. Binding of the anion prevents the formation of the low spin hemichrome species, as indicated in the comparison of the two spectra given in Figure 4.

In conclusion, the present EPR study on oxidized HbI indicates that in the dimer the heme is an aquomet form characterized by an axial high spin signal at neutral and slightly acid pH values. In contrast, the heme in the monomer is characterized by at least two environments which may be identified as a bisimidazole and an ε-amino : imidazole complex. At alkaline pH values, the axial high spin species undergoes a transition toward an unusual rhombic form, which is indicative of the presence of a constraint at the heme site.

Acknowledgements

We wish to thank Francesca de Martino for her skillful assistance in the preparation of the Hb samples.

References

1. Chiancone, E., Vecchini, P., Verzili, D., Ascoli, F. and Antonini, E.   (1981) *J. Mol. Biol.* **152**: 577–592.
2. Spagnuolo, C., Ascoli, F., Chiancone, E., Vecchini, P. and Antonini, E.   (1983) *J. Mol. Biol.* **164**: 627–644.
3. Gurd, F.R.N., Falk, K.E., Malmström, B.G. and Vänngård, T.   (1967) *J. Biol. Chem.* **242**: 5724–5730.
4. Peisach, J., Blumberg, W.E., Ogawa, S., Rachmilewitz, E.A. and Oltzick, R.   (1971) *J. Biol. Chem.* **246**: 3342–3355.
5. Peisach, J. and Gersonde, K. (1977) *Biochemistry* **16**: 2539–2545.
6. Peisach, J., Blumberg, W.E., Wittenberg, B.A., Wittenberg, J.B. and Kampa, L.   (1969) *Proc. Natl. Acad. Sci. U.S.A.* **63**: 934–939.

7. Verzili, D., Santucci, R., Ikeda-Saito, M., Chiancone, E., Ascoli, F., Yonetani, T. and Antonini, E. (1982) *Biochim. Biophys. Acta* **704**: 215–220.
8. Inubushi, T., Yonetani, T. and Chiancone, E. (1988) *FEBS Lett.* **235**: 87–92.
9. Peisach, J. and Mims, W.B. (1977) *Biochemistry* **16**: 2795–2799.
10. Ikeda, M., Iizuka, T., Takao, H. and Hagihara, B. (1974) *Biochim. Biophys. Acta* **336**: 15–24.
11. Brautigan, D.L., Feinberg, B.A., Hoffman, B.M., Margoliash, E., Peisach, J. and Blumberg, W.E. (1977) *J. Biol. Chem.* **252**: 574–582.
12. Seamonds, B., Blumberg, W.E. and Peisach, J. (1972) *Biochim. Biophys. Acta* **263**: 507–514.

21
Aplysia Myoglobin: Involvement of Two Kinds of Carboxyl Groups in the Autoxidation Reaction

Ariki Matsuoka and Keiji Shikama

Institute of Biology, Tohoku University, Sendai 980, Japan

Introduction

In the usual Mbs, the distal (*E*7) His is known to play a key role in the stability properties of the bound O_2; most of the autoxidation reaction of MbO_2 to metMb is explained by proton-catalyzed processes involving the distal His as a catalytic residue *via* its imidazole ring and a proton-relay mechanism (1), as well as the distal His stabilizing the bound O_2 by hydrogen bond formation (2).

On the other hand, there are a few interesting Mbs in which the usual distal His is lacking. This new type of Mb was first isolated by Rossi-Fanelli and Antonini (3) from *Aplysia limacina*, a gastropod mollusc from the Mediterranean. The crystallographic structure of this molecule has now become available at 1.6 Å resolution, its distal (*E*7) residue being recognized as Val at position 63 (4).

Recently, we have succeeded in isolating native MbO_2 directly from the radular muscle of *Aplysia kurodai*, a common species around the Japanese coast, and have examined its stability properties. When compared with sperm whale MbO_2 as reference, *Aplysia* MbO_2 is found to be extremely susceptible to autoxidation, and its pH dependence is also unusual.

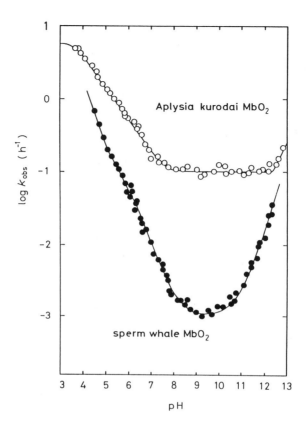

Figure 1. The pH profile for the stability property of *Aplysia* MbO$_2$ in 0.1 M buffer at 25°C. The logarithmic values of the observed rate constant, k_{obs}, for the autoxidation of MbO$_2$ to metMb are plotted against the pH of the solution. The computed curve (—) was obtained by a least-squares fitting to the experimental points (O) over the whole range of pH studied, based on equation (4) for a "three-state model." For comparison, the pH profile of sperm whale MbO$_2$ (●) is shown from the previous paper (6). Concentration: 25 μM for *A. kurodai* MbO$_2$; 50 μM for sperm whale MbO$_2$.

Results and Discussion

Stability Properties of MbO$_2$. Under air-saturated conditions, MbO$_2$ is oxidized easily to metMb with generation of the superoxide anion as

$$\overset{k_{obs}}{MbO_2 \rightarrow metMb + O_2^-} \qquad , \qquad (1)$$

where k_{obs} represents the first order rate constant observed at a given pH (5). Therefore, the rate of the autoxidation is written as

$$- d\,[\,MbO_2\,]/dt \; = k_{obs}\,[\,MbO_2\,] \qquad . \qquad (2)$$

This process of autoxidation was followed by a plot of $- \ln(\,[MbO_2]_t\,/[MbO_2]_o)$ *versus* time t, where the ratio of MbO_2 concentration after time t to that at time $t = 0$ may be monitored by the absorbance changes at 578 nm (α-peak of *Aplysia* MbO_2). The observed first order rate constant, k_{obs} in h^{-1}, was thus determined from the slope of each straight line.

If the values of k_{obs} are plotted against the pH of the solution, a profile of the stability of MbO_2 may be obtained. Figure 1 shows such a typical profile for *Aplysia* MbO_2 over the wide range of pH 4–13 in 0.1 M buffer at 25°C. It is clear that *Aplysia* MbO_2 is extremely susceptible to autoxidation over the whole range of pH studied. Its rate is, for instance, 100 times higher at pH 9.0, and its pH dependence is unusual and much less steep when compared with sperm whale MbO_2 as a reference (6). All these properties of *Aplysia* MbO_2 are in sharp contrast to the stability of sperm whale MbO_2, which depends strongly upon the pH of the solution.

In a previous paper (6) we concluded that the unusual pH profile for the autoxidation rate of *Aplysia* MbO_2 is best explained in terms of a "three-state model." In this scheme, it is assumed that two different kinds of dissociable groups, AH with pK_1 and BH with pK_2, are involved in the reaction. Consequently, there are three forms of the MbO_2, represented by A, B and C, at molar fractions of α, β, and $1 - \alpha - \beta$, respectively, which are in equilibrium with each other, but which differ in dissociation states for the groups AH and BH (see equation (3)). These forms may be oxidized to metMb by the displacement of O_2^- from MbO_2 by an entering H_2O molecule, and at extremely high pH by an entering OH^- ion. Therefore, the autoxidation reaction of *Aplysia* MbO_2 may be written as

$$MbO_2(\,AH, BH\,) \;\underset{}{\overset{K_1}{\rightleftharpoons}}\; MbO_2(\,A^-, BH\,) \;\underset{}{\overset{K_2}{\rightleftharpoons}}\; MbO_2(\,A^-, B^-\,),$$

$$\qquad \Big\downarrow k_o^A \qquad\qquad\qquad \Big\downarrow k_o^B \qquad\qquad\qquad \Big\downarrow k_o^C \quad \Big\downarrow k_{OH}^C \qquad (3)$$

$$\quad\;\; \text{metMb} \qquad\qquad\qquad \text{metMb} \qquad\qquad\qquad\;\; \text{metMb}$$

where, for each form of MbO_2, k_o is the rate constant for the displacement by H_2O, and k_{OH} is the rate constant for the displacement by OH^-.

For this reaction, the observed rate constant, k_{obs} in equation (2), was thus given by

$$k_{obs} = \{k_o^A[H_2O]\}(\alpha) + \{k_o^B[H_2O]\}(\beta) + \{k_o^C[H_2O] + k_{OH}^C[OH^-]\}(1 - \alpha - \beta) \quad (4)$$

where

$$\alpha = \frac{[H^+]^2}{[H^+]^2 + K_1[H^+] + K_1K_2} \quad ,$$

$$\beta = \frac{K_1[H^+]}{[H^+]^2 + K_1[H^+] + K_1K_2} \quad , \qquad (5)$$

$$1 - \alpha - \beta = \frac{K_1K_2}{[H^+]^2 + K_1[H^+] + K_1K_2} \quad .$$

With the use of a computer, we have carried out iterative least-squares calculations for various values for K_1 and K_2, the adjustable parameters in equation (5). The best fit to more than 70 experimental values of k_{obs} was obtained as a function of pH as shown in Figure 1. In this way, the rate constants and the acid dissociation constants involved in the autoxidation reaction of *A. kurodai* MbO_2 were established in 0.1 M buffer at 25°C as summarized in Table 1.

In these kinetic formulations, one of the most remarkable features is that *Aplysia* MbO_2, lacking the usual distal His, does not show the proton-catalytic term $k_H[H_2O][H^+]$ that may play a dominant role in the autoxidation reaction of sperm whale MbO_2 (6).

Thermodynamic Characterization of the Groups AH *and* BH. By analyzing the pH dependence for the autoxidation rate of *Aplysia* MbO_2 at 15, 25 and 35°C, we have studied the effect of temperature on the dissociation processes involved. Figure 2 shows a van't Hoff plot for the resulting values of K_1 and K_2, the acid dissociation constants of the groups *AH* and *BH*, respectively, in 0.1 M buffer.

Table 1. Rate constants and acid dissociation constants obtained from the pH-dependence for the autoxidation reaction of *A. kurodai* MbO_2 in 0.1 M buffer at 25°C.

State of MbO$_2$	A(AH, BH)	K_1 ⇌	B(A⁻, BH)	K_2 ⇌	C(A⁻, B⁻)
k_o (h⁻¹M⁻¹)	0.11		0.013		0.0018
k_{OH} (h⁻¹M⁻¹)	—		—		0.83
pK		4.3		6.1	

From the slope of each straight line, the enthalpy change ($\Delta H°$) of ionization for each group was determined to be practically zero kcal·mol⁻¹, both associated with a large negative value of $\Delta S°$ of the order of –20 cal·mol⁻¹·K⁻¹ (7). These thermodynamic parameters obtained for each group are all those to be expected for the ionization of a carboxyl group. On the basis of these results, and also taking into account the fact that *Aplysia* Mb contains only a single His residue corresponding to the heme-binding proximal residue, we may unequivocally conclude that the two kinds of dissociable groups, *AH* with pK_1 = 4.3 and *BH* with pK_2 = 6.1, must both be carboxyl groups, and that the protonation of these groups is responsible for an increase in the autoxidation rate in the acidic pH range.

Titration Behavior of metMb. Our sequence data show that *A. kurodai* Mb has nine aspartic and five glutamic acids distributed almost equally along the amino acid sequence (8, 9). Out of these 14 carboxyl groups, it is extremely difficult to determine which residue is responsible for *AH* or *BH*. However, any carboxyl group that is in a location where it may affect the molecular O_2 bound to the heme iron ought to be of primary importance. In this respect, it should be

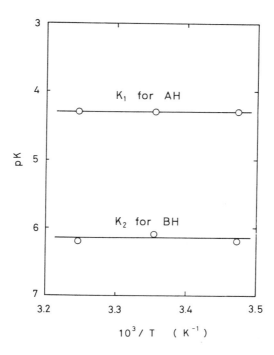

Figure 2. The van't Hoff plot for the acid dissociation constants K_1 and K_2 of the groups *AH* and *BH* involved in the autoxidation reaction of *Aplysia* MbO₂ in 0.1 M buffer.

noted that the two protoheme propionic acid side-chain carboxyl groups may not be ruled out as possible candidates for *AH* and/or *BH*. The protonation of the propionic acid side-chain groups may also provide some effects on the electronic configuration of the heme iron.

Here, it seems of great interest to note that an unusual increase in the Soret peak of *Aplysia* metMb may be observed in the acidic pH range. Figure 3 shows the change in absorbance monitored at 408 nm by adding 0.1 M HCl or KOH in increments of 50 μl to a 5 ml solution of the protein (10.5 μM) in 0.1 M KCl at 25°C. On the more acidic side, below pH 3.5, there was an abrupt decrease in the extinction coefficient, probably due to acid denaturation of the protein.

We have then carried out a curve-fitting analysis with the use of a computer and have found that a single dissociation process is insufficient to explain all the parts of the titration curve. Instead, the best fit to the experimental data was obtained by the equation:

$$\varepsilon_{408} = \varepsilon_A(\alpha) + \varepsilon_B(\beta) + \varepsilon_C(1 - \alpha - \beta) \tag{6}$$

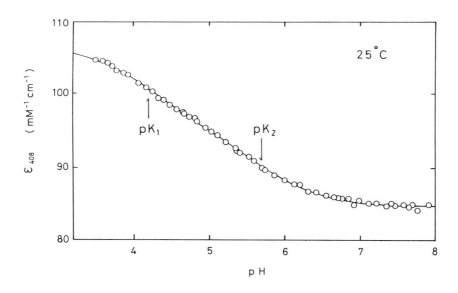

Figure 3. Absorbance changes in the Soret-peak of *Aplysia* metMb as a function of pH in 0.1 M KCl at 25°C. The computed curve (—) was obtained by a least-squares fitting to the experimental points (O), based on equation (6). MetMb concentration: 10.5 μM.

This is completely analogous to equation (4), with the involvement of two kinds of dissociable groups, AH with pK_1 and BH with pK_2, from equation (5). The resulting values in 0.1 M KCl at 25°C were as follows: $pK_1 = 4.2$, $pK_2 = 5.7$, $\varepsilon_A = 106.7$ mM^{-1}cm^{-1}, $\varepsilon_B = 95.3$ mM^{-1}cm^{-1} and $\varepsilon_C = 84.7$ mM^{-1}cm^{-1}. Furthermore, the thermodynamic parameters obtained for these dissociation processes are all characteristic of a carboxyl group, and are practically identical with those for the functional groups AH and BH assumed in the autoxidation reaction of *Aplysia* MbO$_2$ (10).

In conclusion, at the present time we do not know which carboxyl residues are responsible for AH and BH. However, our present data clearly indicate that both groups must be located at a position very close to the heme iron so as to affect its electronic configuration.

References

1. Shikama, K. (1988) *Coord. Chem. Rev.* **83**: 73–91.
2. Phillips, S.E.V. and Schoenborn, B.P. (1981) *Nature* **292**: 81–82.
3. Rossi-Fanelli, A. and Antonini, E. (1957) *Biokhimia* **22**: 335–344.
4. Bolognesi, M., Onesti, S., Gatti, G., Coda, A., Ascenzi, P. and Brunori, M. (1989) *J. Mol. Biol.* **205**: 529–544.
5. Gotoh, T. and Shikama, K. (1976) *J. Biochem.* **80**: 397–399.
6. Shikama, K. and Matsuoka, A. (1986) *Biochemistry* **25**: 3898–3903.
7. Matsuoka, A. and Shikama, K. (1988) *Biochim. Biophys. Acta* **956**: 127–132.
8. Suzuki, T., Takagi, T. and Shikama, K. (1981) *Biochim. Biophys. Acta* **669**: 79–83.
9. Takagi, T., Iida, S., Matsuoka, A. and Shikama, K. (1984) *J. Mol. Biol.* **180**: 1179–1184.
10. Matsuoka, A. (1989) Ph. D. Thesis, Tohoku University, Sendai, Japan.

22
Aplysia limacina Myoglobin: Molecular Bases for Ligand Binding

Martino Bolognesi,[a] Francesco Frigerio,[a] Claudia Lionetti,[a] Menico Rizzi,[a] Paolo Ascenzi[b] and Maurizio Brunori[b]

[a]Department of Genetics and Microbiology, University of Pavia, Via Taramelli 16, 27100 Pavia, Italy
[b]Department of Biochemical Sciences and CNR Center of Molecular Biology, University of Rome "La Sapienza," 00185 Rome, Italy

Introduction

The application of protein engineering techniques to hemoproteins has allowed the exploration of the role of distal residues and their ligand-binding functions in O_2-carrying proteins (1). In the field of globins, a wealth of crystallographic, structural and functional information is available (2) such that, until suitable site-directed mutagenesis experiments can be performed, we may establish a correlation between the nature of the distal residues observed in different O_2 carriers and their detailed function in ligand(s) (de)stabilization processes (3). In the present communication we describe some of the structural details of the mollusc *Aplysia limacina* Mb that pertain to the process of ligand binding, considering the crystallographic analyses conducted at atomic resolution on the native ferric protein, and on its azide, imidazole and fluoride complexes. Consideration of the thermodynamic and kinetic parameters for ligands binding to *A. limacina* Mb, to monomeric *Chironomus thummi thummi* Hb (type III),

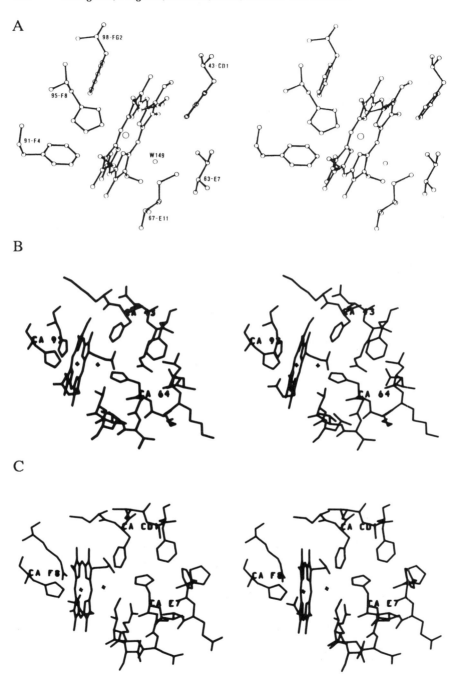

Figure 1.   Stereo views of the heme surroundings in the "aquomet" forms of: (A) *A. limacina* Mb; (B) sperm whale Mb; (C) *C. th. thummi* Hb.

to monomeric *Glycera dibranchiata* Hb (type II) and, by comparison, to sperm whale Mb allows us to relate the functional properties of these globins to the structural environment of the distal site heme pocket.

Materials and Methods

A. limacina Mb was prepared from the buccal muscle of the gastropod as detailed elsewhere (4). *C. th. thummi* Hb was a generous gift of Prof. Robert Huber (Max Planck Institut für Biochemie, Martinsried bei München, FRG). Ferric sperm whale Mb was purified from commercial hemoprotein preparations (type II, from Sigma Chemical Co., St. Louis, MO, USA), as detailed (5). All chemicals (from Merck AG, Darmstadt, FRG) were of analytical grade and used without further purification.

The crystal structure of ferric *A. limacina* Mb was refined at 1.5 Å resolution to a crystallographic *R* factor of 15%, following the procedures described previously (4). The three-dimensional structures of the *A. limacina* Mb : azide, : imidazole and : fluoride complexes were all refined at 2.0 Å resolution, employing crystals of the derivatives obtained by soaking in solutions containing saturating levels of the respective ligand. Data collection was performed by single crystal diffractometry; processing of the collected intensities as well as crystallographic refinements followed the procedures described elsewhere (6). The 2.0 Å resolution structures of *A. limacina* Mb : azide, :imidazole and :fluoride complexes were refined to final crystallographic *R* factors of 15.4, 15.6 and 13.6%, respectively. Using quite similar crystallographic procedures the structure of sperm whale Mb : imidazole adduct has been refined to a crystallographic *R* factor of 14.8%, at 2.0 Å resolution (7). All refined structures show ideal stereochemistry, with approximate rms deviations from ideal bond lengths of 0.020 Å, and from ideal bond angles of 3.0°. The three-dimensional models of *C. th. thummi* Hb and of sperm whale Mb have been obtained from the Brookhaven Protein Data Bank (8) and inspected by means of an Evans and Sutherland PS330 graphics workstation.

Kinetic parameters for imidazole binding to *A. limacina* Mb were estimated by rapid-mixing stopped-flow experiments, mixing the imidazole derivative of the hemoprotein with azide (0.2 M) in an attempt to monitor the substitution of imidazole by azide (as expected on the basis of the equilibrium constants (9)). The half-time of the imidazole dissociation was estimated to be less than 7 ms, because the substitution was lost in the dead-time of the stopped-flow apparatus; thus, the dissociation rate constant, k_{off}, should be faster than 100 s^{-1}. From values of the equilibrium constant ($K = 1.0 \times 10^1$ M^{-1}) and of k_{off}, the value of the second order association rate constant, k_{on}, was estimated to be greater than 1.0×10^3 M^{-1} s^{-1}. Kinetic parameters for azide binding to *C. th. thummi* Hb were determined by temperature-jump experiments as detailed previously (9). The value of the association equilibrium constant for azide binding to *C. th. thummi* Hb was determined spectrophotometrically as reported elsewhere (9).

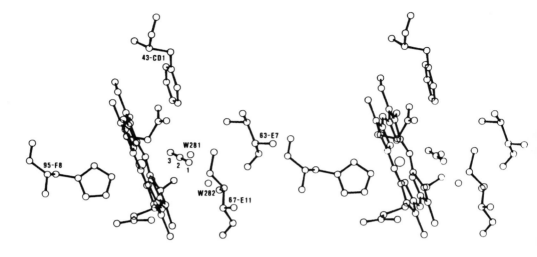

Figure 2*A*. Stereo view of the heme region of azide derivatives in *A. limacina* Mb.

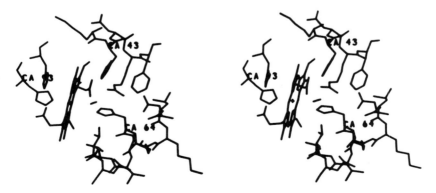

Figure 2*B*. Stereo views of the heme region of azide derivatives in sperm whale Mb.

Results and Discussion

Molecular Structures. The tertiary structure of *A. limacina* Mb closely conforms to the conventional globin fold, with local modifications, such as kinks of about 30° in helices *A* and *G* at residues Pro-16(*A*14) and Pro-113(*G*13). In the native ferric Mb the heme iron is penta-coordinated, such that the unique water molecule present in the heme pocket (*W*149) is not bound to the metal, but is hydrogen bonded to the carbonyl O_2 of residue Val-63 (*E*7), 4.6 Å away from the iron (Figure 1). The heme pocket appears fairly open to the solvent, due to the replacement of the conventional His *E*7 residue with the smaller Val-63. The residue at position *E*11 is Ile-67. In the *CD* corner *A. limacina* Mb shows an Asp residue at position 45(*CD*3), the negative charge of which is compensated by the side chain of residue Lys-59(*E*3).

Figure 2 shows a view of the distal site of *A. limacina* Mb in the presence of the azide ligand. It may be appreciated that the azide (at pH 7.0) binds to the iron, forming an angle of 115° normal to the heme. The free end of the linear anion (atom 1 in Figure 2) is pointing towards the outside of the heme pocket. The Fe-azide coordination bond is 2.1 Å, and the Fe–*N*3 direction makes an angle of 9° with respect to the heme normal. The iron atom moves toward the porphyrin plane, and the proximal coordination bond between His-95(*F*8) *NE*2 and Fe is 2.1 Å, slightly longer than in the native structure (2.05 Å). The negative charge of azide, and its physical dimensions may play a role in causing the displacement of propionate group IV (lower in Figures 1, 2), which drastically alters its conformation with respect to the native protein, moving to the opposite side of the heme. The *N*1 atom of azide falls in a position which is almost coincident with that occupied by *W*149 in the "aquomet" structure at pH 7.0, indicating that this site is readily available to accommodate polar molecules.

Figure 3 represents the structure of *A. limacina* Mb : fluoride complex. In this adduct, the major observed conformational readjustment is centered at residue Arg-66(*E*10), which appears to be disordered in the solvent, and not visible in the Fourier maps of all the other derivatives of *A. limacina* Mb. The conformation adopted by Arg-66(*E*10), together with the contribution of three ordered, bridging H_2O molecules (*W*192, *W*266 and *W*328, in Figure 3) and of the heme propionate III (upper in Figure 3), allows the bound ligand to stabilize in an extended network of hydrogen bonds that stretches from Arg-66(*E*10) to the *CD* region residues Phe-42(*C*6), Asp-45(*CD*3), and to Lys-59(*E*3). The fluoride ion is 2.2 Å from the heme iron, which is essentially in the heme plane with the proximal bond being 2.3 Å (6). Inspection of the refined model of *A. limacina* Mb : imidazole adduct (Figure 4*A*) shows that the protein undergoes conformational readjustments in order to accommodate the unusually bulky ligand. Collisions of the imidazole ring mainly with residues Phe-43(*CD*1), Val-63(*E*7), and Ile-67(*E*11) are at the basis of side chain readjustments and of a displacement of about 2.6 Å of the *CD* region of the molecule. In Figure 4*A* the structure of the "aquomet" form of *A. limacina* Mb (*C*α trace) has been overlaid for comparison with that of the residues surrounding the imidazole site.

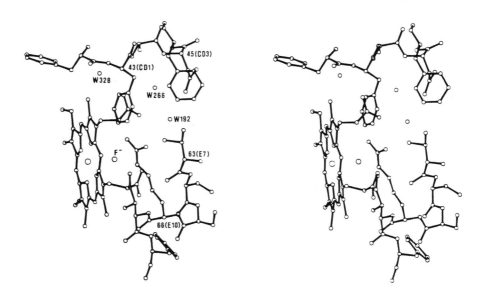

Figure 3. Stereo view of a region of the heme pocket in the *A. limacina* Mb : fluoride complex.

The ligand binds to the heme iron with a coordination bond of 2.3 Å, is in a plane tilted by 16° with respect to the porphyrin normal, and forms an angle of −21° with respect to the heme $N2$–$N4$ direction. The His-95($F8$) $NE2$-Fe proximal bond is 2.2 Å long.

In sperm whale Mb : imidazole complex the structural perturbations brought about by the imidazole ligand are quite different from those observed in *A. limacina* Mb : imidazole. As shown in Figure 4*B*, binding of imidazole does not significantly alter the geometry of the polypeptide backbone. On the other hand, it requires a substantial alteration of the conformations of residues His-64(*E7*), which swings out of the heme pocket, of Arg-45(*CD3*) and of Asp-60(*E3*). The distal Fe-imidazole coordination bond is 2.2 Å, and a hydrogen bond connects the ligand $NE2$ atom to $ND1$ of His-64(*E7*). The imidazole ring is tilted by 16° with respect to the heme normal, and forms an angle of −41° with respect to the heme $N2$–$N4$ direction.

Structure-Function Relationships in Monomeric Hemoproteins.   Values of the kinetic and thermodynamic parameters for azide-, fluoride- and imidazole-binding to ferric *A. limacina* Mb, *C. th. thummi* Hb, *G. dibranchiata* Hb and sperm whale Mb are summarized in Table 1. Analyses of the functional data and of the structural details allow the following conclusions to be drawn.

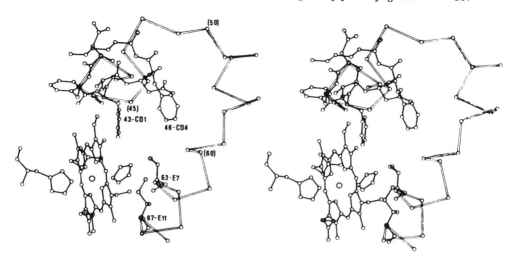

Figure 4A. Stereo view of the heme surroundings in the imidazole derivative of *A. limacina* Mb.

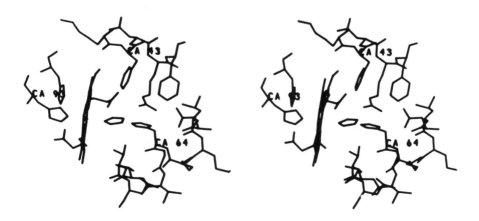

Figure 4B. Stereo view of the heme surroundings in the imidazole derivative of sperm whale Mb.

Second order combination rate constants for azide, fluoride and imidazole-binding to *A. limacina* Mb, *C. th. thummi* Hb and *G. dibranchiata* Hb are faster, by at least 20-fold, than those reported for the formation of the sperm whale Mb : ligand complexes (see Table 1). Such a finding may be related to the structure of the distal site of these globins; indeed, only ferric sperm whale Mb shows an Fe-coordinated water molecule. Removal of this ligand from the Fe sixth coordination position impairs ligand-binding processes in sperm whale Mb. On the other hand, a more direct access to the ligand coordination site is granted in globins the distal site of which does not contain an "endogenous" H_2O molecule (10), possibly because there is a lack of sufficient stabilization by the E7 residue, apolar in *A. limacina* Mb and in *G. dibranchiata* Hb, and of a non-productive orientation in *C. th. thummi* Hb.

Dissociation rate constants of *A. limacina* Mb : azide, : imidazole, and *G. dibranchiata* Hb : fluoride complexes are also faster by at least 20-fold than those reported for sperm whale Mb : azide, : imidazole and : fluoride adducts as well as for *A. limacina* Mb : fluoride and *C. th. thummi* Hb : azide complexes. This finding may reflect the stabilization of the exogenous ligand by hydrogen bonding to distal residues. In the sperm whale Mb : azide adduct (11) (Figure 2), the ligand is located in the inner part of the heme crevice, and the nitrogen atom

Table 1.   Values of kinetic and thermodynamic parameters for the binding of azide, fluoride and imidazole to sperm whale Mb, *A. limacina* Mb, *C. th. thummi* Hb and *G. dibranchiata* Hb.

Hemoprotein	Ligand	k_{on} (M^{-1}s^{-1})	k_{off} (s^{-1})	K (M^{-1})
Sperm whale Mb[a]	Azide	2.5×10^4	1.0	2.5×10^4
	Fluoride	3.6	6.0×10^{-2}	6.0×10^1
	Imidazole	5.0×10^1	4.8	1.0×10^1
A. limacina Mb	Azide[b]	1.8×10^6	7.5×10^2	3.1×10^3
	Fluoride[c]	1.5×10^3	1.0×10^1	1.0×10^2
	Imidazole	$> 1 \times 10^3$ [d]	$> 1 \times 10^2$ [d]	1.0×10^1 [e]
C. th. thummi Hb[d]	Azide	1.8×10^6	2.0	8.5×10^5
G. dibranchiata Hb	Fluoride	$> 1 \times 10^2$ [c]	$> 1 \times 10^2$ [c]	1.0 [f]

[a]pH = 6.0, 0.1 M phosphate buffer; T= 25°C (14).
[b]pH = 6.1, 0.1 M phosphate buffer; T= 25°C (9).
[c]pH = 7.0, 0.1 M phosphate buffer; T= 20°C (6,15).
[d]pH = 6.1, 0.1 M phosphate buffer (present study).
[e]pH = 6.0, 0.1 M phosphate buffer; T= 20°C (5).
[f]pH = 6.85, 0.1 M phosphate buffer; T= 22°C (16).

coordinated to the heme iron is hydrogen bonded to the His-64 (*E7*) *NE2* atom, as, based on model building studies which used the sperm whale Mb aquomet structure as a reference (12), is believed to occur in the :fluoride complex. In the sperm whale Mb : imidazole complex His-64(*E7*) *ND*1 atom is hydrogen bonded (3.0 Å) to the *NE2* atom of the bound ligand molecule (7) (see Figure 4).

Inspection of the crystallographic structural data, summarized in Figures 2 and 4, indicates that the ligand stabilizing interactions are lacking in both *A. limacina* Mb : azide and : imidazole complexes. On the other hand, model building studies based on the structure of ferric *C. th. thummi* Hb (13) do not allow us to evaluate a preferential orientation of the linear azide anion bound to the heme iron. Ligand stabilization may, however, be achieved through conformational readjustment(s) of the His-57(*E7*) side chain. Concerning the unexpectedly slow dissociation rate constant of the *A. limacina* Mb : fluoride complex as compared to the *G. dibranchiata* Hb : fluoride adduct, Figure 3 shows that ligand stabilization may be brought about by a residue different from *E7* (*i.e.*, Arg-66(*E*10)), which is polar, long and flexible enough to reach the distal site and provide extensive hydrogen bonding capabilities. The values of the association equilibrium constants for ligand binding reported in Table 1, indeed reflect the different kinetic behavior of the monomeric hemoproteins considered.

As a whole, the present study indicates that, even though different monomeric hemoproteins display an evident overall structural homology, their detailed ligand-binding mechanisms are quite diverse, despite the "apparent" near coincidence of the thermodynamic parameters.

References

1. Perutz, M.F. (1979) *Trends Biochem. Sci.* 14: 42–44.
2. Brunori, M., Coletta, M., Ascenzi, P. and Bolognesi, M. (1989) *J. Mol. Liq.* 42: 175–194.
3. Rohlfs, R.J., Matthews, A.J., Carver, T.E., Olson, J.S., Springer, B.A., Egeberg, K.D. and Sligar, S.G. (1990) *J. Biol. Chem.* 265: 3168–3176.
4. Bolognesi, M., Onesti, S., Gatti, G., Coda, A., Ascenzi, P. and Brunori, M. (1989) *J. Mol. Biol.* 205: 529–544.
5. Bolognesi, M., Cannillo, E., Ascenzi, P. Giacometti, G.M., Merli, A. and Brunori, M. (1982) *J. Mol. Biol.* 158: 305–315.
6. Bolognesi, M., Coda, A., Frigerio, F., Gatti, G., Ascenzi, P. and Brunori, M. (1990) *J. Mol. Biol.* 213: 621–625.
7. Lionetti, C., Guanziroli, M.G., Frigerio, F., Ascenzi, P. and Bolognesi, M. (1990) *J. Mol. Biol.* 217: 409–412.
8. Bernstein, F.C., Koetzle, T.F., Williams, G.J.B., Meyer, Jr., E.F., Brice, M.D., Rodgers, J.R., Kennard, O., Shimanouchi, T. and Tasumi, M. (1977) *J. Mol. Biol.* 112: 535–542.
9. Giacometti, G.M., Da Ros, A., Antonini, E. and Brunori, M. (1975) *Biochemistry* 13: 1584–1588.

10. Giacometti, G.M., Ascenzi, P., Brunori, M., Rigatti, G., Giacometti, G. and Bolognesi, M. (1981) *J. Mol. Biol.* **151**: 315–319.
11. Stryer, L., Kendrew, J.C. and Watson, H.C. (1964) *J. Mol. Biol.* **8**: 96–104.
12. Takano, T. (1977) *J. Mol. Biol.* **110**: 537–568.
13. Steigemann, W. and Weber, E. (1979) *J. Mol. Biol.* **117**: 309–338.
14. Antonini, E. and Brunori, M. (1971) *Hemoglobin and Myoglobin in Their Reactions with Ligands*, 219–234. Amsterdam: Elsevier North-Holland Publishing Co.
15. Giacometti, G.M., Ascenzi, P., Bolognesi, M. and Brunori, M. (1981) *J. Mol. Biol.* **146**: 363–374.
16. Seamonds, B., Forster, R.E. and George, P. (1971) *J. Biol. Chem.* **246**: 5391–5397.

23
Biophysical Characterization of Constituents of the *Glycera dibranchiata* Oxygen Transport and Utilization System: Erythrocytes and Monomer Hemoglobins

James D. Satterlee

Department of Chemistry, Washington State University, Pullman, WA, 99164-4630, USA

Introduction

Glycera dibranchiata, the common blood worm, contains multiple Hbs in nucleated erythrocytes. Without a highly developed, structural vascular system, these cells apparently circulate in the central body cavity in response to the organism's movements. The total Hb content of the erythrocytes may be separated into high and low M_w fractions (1), and the constituents of these two fractions have been the subjects of recent inquiry.

In this paper we present results from our initial ^{31}P NMR studies of living, fully intact specimens of *G. dibranchiata* and of the intact erythrocytes. We also summarize the current state of affairs with respect to the characterization of the protein components of the low M_w Hb fraction.

The *G. dibranchiata* erythrocytes are of interest in the context of understanding the overall O_2 utilization of this evolutionarily less-developed organism. *G. dibranchiata* is found buried in intertidal areas on the northeastern United States coast, and this habitat implies that the organism's O_2 delivery system must be capable of functioning under a variety of pO_2s (2).

The Hb content of these erythrocytes is of interest for precisely the same reason. Multiple Hbs may be necessary to sustain the organism in an environment offering varying O_2 availability. In this respect, the low M_w Hb fraction has been shown to consist primarily of monomer Hbs (1, 3–10). Between 1971 and 1974, pioneering work by Padlan and Love (11) and by Riggs and co-workers (12, 13) resulted in a crystal structure and primary sequence data for components of the total monomer Hb fraction. That work revealed that at least some components of the monomer Hb fraction lacked the commonly found His that corresponds to the primary sequence position *E*7 in Mb. This "distal" His lies closest to the ligand binding site and influences ligation dynamics. This substitution is exceptional: whereas other wild-type Mbs and Hbs exhibit *E*7 substitutions (*e.g.*, elephant Mb, opossum Hb, human Hb Zurich), in *G. dibranchiata*, unlike in virtually all other naturally occurring *E*7 mutants in which the replacement is by a polar or charged amino acid, the substitution is by Leu, which results in a nonpolar side chain positioned next to the heme ligand binding site.

Although the original (12, 13) and more recent (14) sequencing studies revealed low primary sequence homology with Mb, x-ray crystallography indicates a three-dimensional structure similarity of one of the *G. dibranchiata* monomer Hbs to Mbs (11, 15).

The combination of an exceptional *E*7 substitution and structural similarity to Mb manifested in a fully functional monomer Hb led to several ligation studies, the most notable being that of Parkhurst *et al.* (7), which revealed comparatively enhanced CO affinity and reduced O_2 affinity. That work also gave the first indication of significant protein heterogeneity within the monomer Hb fraction. However, the differing isolation and purification techniques, and the lack of a purity standard meant that preparations from different laboratories were undoubtedly different—a fact that has made the rationalization of dynamics results virtually impossible. Consequently, applying high resolution analytical biochemistry to elucidate easily measured purity criteria and to define the extent of protein heterogeneity using high resolution methods (6, 9, 10) has constituted a significant step towards the standardization of future studies.

Materials and Methods

Specimens of blood worms were purchased from Maine Bait Co., Wiscasset, ME, USA. Each specimen was taxonomically classified, then incised and the fluid collected over ice. Erythrocytes were collected by centrifugation at 4°C,

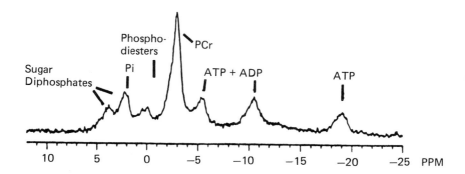

Figure 1. Whole-body ^{31}P NMR spectrum of a live specimen of *G. dibranchiata*. Observed shifts are reported relative to external phosphoric acid. Suggested assignments are shown by analogy to other work.

washed three times with a solution of 0.5M KCl, 0.01 M MgCl$_2$ and 0.02 M glucose. Although not buffered, the solution pH was 7.12. The buffy coat was completely removed by aspiration.

Cells were immediately used for NMR spectroscopy by suspending them in the wash solution that contained 15% D$_2$O for lock to a hematocrit level of approximately 50%. ^{31}P NMR spectra were obtained on a GE 360 spectrometer operating at 146 MHz, at a calibrated probe temperature of 22°C. Phosphorous shifts are reported relative to external H$_3$PO$_4$, assigned 0 ppm. Spectra of live specimens of the annelid were obtained by packing a suitably sized specimen into a 12 mm NMR tube. All spectra were obtained over periods of 15 min to 1 hr. The spectrum presented in Figure 1 was obtained from a specimen that was alive at the end of the experimentation.

Results and Discussion

Figure 1 presents the ^{31}P NMR spectrum of a single live specimen of *G. dibranchiata*. This whole body spectrum is dominated by a phosphocreatine peak (PCr) and includes ATP and ADP resonances. With the exception of the PCr peak, it is similar to the ^{31}P NMR spectrum of the intact *G. dibranchiata* erythrocytes shown in Figure 2, indicating that the PCR is due to the body of the specimen. This is not unreasonable since the worm resembles a hollow tube, closed at each end and filled with erythrocyte-containing coelomic fluid.

The spectrum of the erythrocytes (Figure 2) is very similar to that of other nucleated cells. It differs from the spectrum of non-nucleated erythrocytes, such as human red blood cells, by displaying relatively larger ATP and ADP resonances upfield. This is a typical situation for nucleated cells. The erythrocytes in Figure 2 also display a larger number of other phosphorous-containing compounds than do human erythrocytes, which are assigned in these figures by analogy to their peak positions in the spectra of other nucleated cells (16).

These spectra indicate the ease with which this complete organism and its erythrocytes may be studied by multinuclear NMR spectroscopy. To date we have also begun high resolution ^{31}P NMR assignments of the lysed erythrocyte contents and we have detected ^{23}Na and ^{35}Cl NMR signals from the intact erythrocytes, indicating the feasibility of measuring transmembrane ion exchange rates.

Concerning the question of heterogeneity of the monomer Hb fraction obtained as the result of gel filtration separation of the erythrocyte Hb content, two laboratories have shown that it contains three major monomer Hbs. Originally, we detected these three monomer Hbs using a combination of isoelectric focusing (IEF) and low pressure ion exchange chromatography (5). Three individual Hbs were also initially detected by proton NMR spectroscopy (5), and we subsequently isolated three major monomer Hbs and reported their individual UV-visible and NMR spectra (6). This work was independently corroborated about one year later using a slightly different ion exchange technique (8). That work, the multiple times we have reproduced our isolations, and our success in protein sequencing and in creating a cDNA library (14, 17) all combine to provide strong evidence that the three individual monomer Hbs are authentic.

Several properties of these individual Hbs are interesting. First, they are not distinguishable by size. All three of the monomer Hbs perform identically on size exclusion chromatography (on Sephacryl S200, Sephadex G75 and G-50) and SDS-PAGE. Some discrimination between the three components is achieved with IEF (9, 10). Components III and IV behave identically, showing doublet bands at approximately $pI = 6.8$ and $pI = 7.3$. The lower pI band corresponds to reduced protein whereas the higher pI band is the oxidized (met) Hb. The apo-protein of each of these monomers focuses as a single tight line at approximately $pI = 7.7$. IEF bands for component II are in approximately the same positions, but are shifted a detectable amount to a lower pI, thereby being distinguishable from the bands of components III and IV. This shift corresponds to ~1 mm on the gel and represents a real, but only slight, isoelectric point shift. What is unusual about component II is that solutions of the holoprotein also yield an IEF band corresponding to the apo-protein, apparently indicating lower affinity of the heme for component II globin.

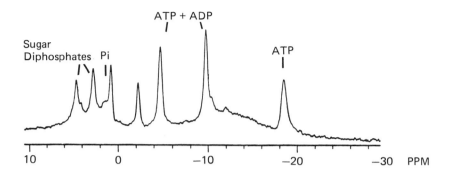

Figure 2. ^{31}P NMR spectrum of packed erythrocytes from *G. dibranchiata*. Spectra were obtained at 146 MHz at 20°C. Assignments are made by analogy to other work on nucleated cells.

Proton NMR spectroscopy has revealed two other unique characteristics of the monomer Hbs (18, 10). First, in the ferric native form of each component the heme exists in two possible orientations, 180° reversed from each other relative to the globin matrix. The major orientation consists of 80–90% of the total (19). Second, in each component the major orientation in both the ferrous, CO-ligated form (18) and the ferric native form (19), is reversed 180° from that found in Mb. This NMR work was the first to detect this phenomenon in the monomer Hbs and it has been confirmed by the recent refinement of one of the monomer Hb structures (15). Recent assignments of selected proton resonances for each cyanide ligated monomer on the Mb also establish this point (19).

The last significant, anomalous property of these monomer Hbs that we have detected thus far is their unusually slow cyanide-binding kinetics (21, 22). Compared to other ferriheme proteins whose rates (k_1 *(app)*) range from 1 x 10^2 to 1 x 10^5, the monomer metHb components exhibit the following rates (k_1 *(app)*): component II = 0.491 ($M^{-1}S^{-1}$); component III = 0.302 ($M^{-1}S^{-1}$); component IV = 1.82 ($M^{-1}S^{-1}$). Whereas such rates may be fast on the evolutionary time scale, they are anomalously slow in comparison to other heme proteins. Our recent sequencing work (14) offers ideas for why such slow cyanide binding occurs. First, the distal His (a polar residue) that occurs at position *E*7 is replaced by the non-polar side chain of Leu, thereby lowering the polarity at the heme ligand binding site. Furthermore, in the two full sequences completed thus far (14) both monomer Hbs display a heme pocket polarity opposite that of Mb. Both of these microevnironmental differences probably contribute to the anomalous kinetics. Details such as these will be elucidated using site-directed mutants constructed as described elsewhere in this volume.

Acknowledgements

This work was supported by grants from the National Institutes of Health (DK30912, AM30912 and HL01758 (Research Career Development Award)).

References

1. Vinogradov, S.N., Machlik, C.A. and Chao, L.L. (1970) *J. Biol. Chem.* **245**: 6533–6538.
2. Hoffman, R.J. and Mangum, C.P. (1970) *Comp. Biochem. Physiol.* **36B**: 211–228.
3. Seamonds, B., Forster, R.E. and Gottlieb, A.J. (1971) *J. Biol. Chem.* **246**: 1700–1705.
4. Weber, R.E., Sullivan, B., Bonaventura, J. and Bonaventura, C. (1977) *Comp. Biochem. Physiol.* **58B**: 183–187.
5. Kandler, R.L. and Satterlee, J.D. (1983) *Comp. Biochem. Physiol.* **75B**: 499–503.
6. Kandler, R.L., Constantinidis, I.C. and Satterlee, J.D. (1984) *Biochem. J.* **225**: 131–138.
7. Parkhurst, L.J., Sima, P. and Gross, D.J. (1980) *Biochemistry* **19**: 2688–2692.
8. Cooke, R.M. and Wright, P.E. (1985) *Biochim. Biophys. Acta* **832**: 357–364.
9. Constantinidis, I.C. and Satterlee, J.D. (1987) *Biochemistry* **26**: 7779–7786.
10. Constantinidis, I.C., Kandler, R.L. and Satterlee, J.D. (1989) *Comp. Biochem. Physiol.* **92B**: 619–622.
11. Padlan, E.A. and Love, W.E. (1974) *J. Biol. Chem.* **249**: 4067–4078.
12. Imamura, T., Baldwin, T.O. and Riggs, A.F. (1972) *J. Biol. Chem.* **247**: 2785–2797.
13. Li, S.L. and Riggs, A.F. (1971) *Biochim. Biophys. Acta* **236**: 208–210.
14. Simons, P.C. and Satterlee, J.D. (1989) *Biochemistry* **28**: 8525–8530.
15. Arents, G. and Love, W.E. (1989) *J. Mol. Biol.* **210**: 149–161.
16. Hollis, D.P. (1980) In *Biological Magnetic Resonance*, eds. L.J. Berliner and J. Reuben, vol. 2, 1–44. New York: Plenum Press.
17. Simons, P.C. and Satterlee, J.D. (1991) This volume.
18. Cooke, R.M. and Wright, P.E. (1985) *Biochim. Biophys. Acta* **832**: 365–372.
19. Constantinidis, I.C., Satterlee, J.D., Pandey, R.K., Leung, H.K. and Smith, K.M. (1988) *Biochemistry* **27**: 3069–3076.

20. Mintorovitch, J., Satterlee, J.D., Pandey, R.K., Leung, H.K. and Smith, K.M. (1990) *Inorg. Chim. Acta* **170:** 157–159.
21. Mintorovitch, J. and Satterlee, J.D. (1988) *Biochemistry* **27:** 8045–8050.
22. Mintorovitch, J., Van Pelt, D. and Satterlee, J.D. (1989) *Biochemistry* **28:** 6099–6104.

24
Oxygen Equilibrium Characteristics of Hemerythrins from the Brachiopod, *Lingula unguis*, and the Sipunculid, *Siphonosoma cumanense*

Kiyohiro Imai,[a] Hideo Takizawa,[b, c] Takashi Handa[b] and Hiroshi Kihara[c]

[a]Department of Physicochemical Physiology, Osaka University School of Medicine, Osaka 565, Japan
[b]Department of Chemistry, Faculty of Science, Science University of Tokyo, Tokyo 162, Japan
[c]Department of Physics, Jichi Medical School, Tochigi 329-04, Japan

Introduction

Hr occurs in various aggregation states, such as monomer (myoHr), trimer, tetramer and octamer. The subunits of these oligomers have a common three-dimensional structure which is identical or very similar to the compact bundle of four antiparallel α-helices of monomeric myoHr which is related to Hr just as Mb is to Hb. Each subunit has a pair of nonequivalent Fe atoms, to one of which O_2 binds reversibly. The O_2 affinity of Hr varies over a relatively small range depending upon the source used. Cooperativity in O_2 binding is totally absent in the Hrs examined to date, except for two octameric Hrs from the brachiopods, *Lingula unguis* (1) and *Lingula reevii* (2). These two Hrs also exhibited a Bohr effect (pH dependence of O_2 affinity), whereas none of the Hrs from the other species showed that. The structure and function of the Hrs and the myoHr were reviewed by several investigators (3–7).

In view of the current structural and functional data for Hr, little is known about the molecular mechanism of O_2 binding, such as the regulation of O_2 affinity and the stereochemistry of homotropic and heterotropic interactions. It is also entirely unknown why some octameric Hrs exhibit cooperativity while others do not, and what functional significance the oligomeric states have. Recently, Satake *et al.* (8) reported that *Lingula* Hr is a hetero-octamer which consists of an equal number of two distinct subunits (α and β) and the whole molecule may be of the form, $\alpha_4\beta_4$. Further, recent studies using *Lingula* Hr and a trimeric Hr from the sipunculid, *Siphonosoma cumanense*, revealed that the oligomeric state is necessary to prevent autoxidation of the Fe atoms (9, 10). These findings provide useful information regarding the significance of oligomerization in Hr.

In the present study, we investigated O_2 equilibrium properties of two Hrs; one, the octameric, cooperative Hr from *L. unguis*, and the other, the trimeric, non-cooperative Hr from *Siphonosoma*, to obtain a better understanding of the structure-function relationships in Hr.   Preliminary results were reported elsewhere (11). Kinetic studies using the same Hrs were also published (12, 27).

Materials and Methods

L. unguis shells and *Siphonosoma* worms were collected at Ariake Bay and in Seto Inland Sea, respectively, of Japan, and their Hrs were prepared according to Joshi and Sullivan (13) and Addison and Dougherty (14), respectively, with minor modifications.   In addition to the isolated Hrs, *Siphonosoma* hemerythrocytes, which were washed and suspended in isotonic buffer (0.25 M Na_2SO_4 + 0.05 M Tris-HCl (pH 7.3) + 10mM EDTA), and a hemolysate sample, which was dialyzed against distilled water, were also prepared. Oxygen equilibrium curves spanning a wide range of O_2 saturation were determined with an improved version (15, 16) of an automatic oxygenation apparatus (17).

Results and Discussion

Oxygen Equilibrium Under Various Solution Conditions. Figure 1 shows Hill plots of O_2 equilibrium data for *Lingula* Hr in 0.03 M bis-Tris-propane at different pH values (Side *A*), and for *Siphonosoma* hemolysate under various solution conditions (Side *B*). The shape of the Hill plot for *Lingula* Hr is pH dependent: the middle portion of the plot is steep at slightly alkaline pH values while it becomes less steep at more alkaline or acidic pH values. When pH decreases from 8.18 to 6.53, the lower asymptote of the plot moves in parallel toward the right and gradually approaches a certain position, followed by shifts to the right of the upper asymptote. Similar behavior of the Hill plot was also observed for human Hb (HbA) (see Figure 6.2 of reference 16). This result indicates that, exactly as is the case in Hb, the cooperativity of *Lingula* Hr is pH dependent and the oxygenation of *Lingula* Hr is associated with a transition between two alternate states (T and R) whose affinities are independent of pH. In contrast to *Lingula* Hr, the O_2 equilibrium curve of *Siphonosoma* hemolysate is insensitive to the kind of buffer used, the pH value and the presence of Cl^- or $CaCl_2$ and the Hill plot is always linear with a slope of unity.

The values of P_{50}, and of the maximal slope of the Hill plot, h_{max}, for *Lingula* Hr as obtained from the equilibrium curves in Figure 1, are listed in Table 1, which also includes the parameter values obtained under other solution conditions. The parameter values for *Siphonosoma* hemolysate, together with those for Hr solution and hemerythrocyte suspension are listed in Table 2. These parameter values are plotted against pH in Figure 2. From these data the O_2 equilibrium characteristics of *Lingula* Hr are summarized as follows: 1) None of $CaCl_2$, NaCl or $KClO_4$ exerts significant influence on both overall O_2 affinity ($1/P_{50}$) and cooperativity (h_{max}); 2) at acidic pH values, 0.07 M potassium phosphate buffer containing 0.1 M NaCl produces P_{50} and h_{max} values identical with or similar to those in 0.03 M bis-Tris-propane buffer containing 0.064 M Cl^-; whereas at alkaline pH values, the former buffer

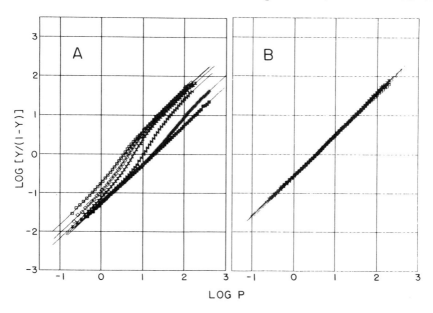

Figure 1. Hill plots of O_2-binding to *Lingula* Hr (*A*) and *Siphonosoma* hemolysate (*B*). *P*, partial pressure of O_2 in mm Hg; *Y*, fractional O_2 saturation. Experimental conditions for *Lingula* Hr: in 0.03 M bis-Tris-propane + 0.064 M Cl^-; Hr concentration, 50 *m*M binding site; 25°C; (☐) pH 8.18; (‡) pH 7.87; (⊠) pH 7.64; (⊠) pH 7.32; (✳), pH 6.93; (⊠) pH 6.53. Conditions for *Siphonosoma* hemolysate: Hr concentration, 100 µM site; 25°C; (☐) 0.05 M Tris-HCl (pH 8.1); (◊) 0.05 M bis-Tris (pH 7.4) + 0.1 M Cl^-; (⊠)0.1 M phosphate (pH 7.4); (⊠) 0.05 M bis-Tris (pH 7.4) + 25 mM $CaCl_2$.

system gives larger P_{50} values than does the latter buffer system. The buffer dependence of P_{50} is not attributable to the difference in Cl$^-$ concentration; 3) P_{50} shows strong pH dependences above pH 7. Both Bohr plots (log P_{50} vs. pH) in Figure 2 become steepest at pH 7.4, giving the Bohr coefficient ($\delta H^+ = \Delta \log P_{50}/\Delta pH$) values of -0.78 in 0.03 M bis-Tris-propane and -0.51 in 0.07 M phosphate; and 4) the h_{max} value is greater than unity and is pH dependent above pH 7, whereas it is almost unity below pH 7. It becomes maximal ($h_{max} = 1.8$) at pH 7.5 in 0.03 M bis-Tris-propane, and at pH 7.7 in 0.05 M phosphate. The present results agree with an earlier observation by Manwell (1) that $h_{max} = 1.8$ at pH 7.6 and $h_{max} = 1.0$ at pH 6.7. A similar pH dependence of cooperativity was also observed for Hr from *L. reevii* (2, 18).

Table 1. Oxygenation parameters of *L. unguis* Hr.

Conditions[a]	P_{50}[b]	h_{max}[c]	K_T[d]	K_R[e]	L_0[f]
In 0.03 M bis-Tris-propane + 0.064 M Cl$^-$					
pH 8.18	3.5	1.43	0.15	0.46	4.4×10^1
pH 7.87	4.5	1.66	0.089	0.46	3.6×10^2
pH 7.64	5.9	1.78	0.066	0.37	4.3×10^2
pH 7.42	8.2	1.79	0.07	0.35	3.2×10^3
pH 7.32	10.2	1.73	0.057	0.28	3.1×10^3
pH 6.93	17.1	1.26	0.053	0.15	1.4×10^3
pH 6.53	18.5	1.02	0.05	0.06	2.3
In 0.07 M potassium phosphate + 0.1 M NaCl					
pH 8.18	5.3	1.57	0.09	0.25	1.3×10
pH 7.64	8.9	1.82	0.066	0.41	2.1×10^4
pH 6.87	18.0	1.11	0.05	0.083	2.5×10
pH 6.28	18.6	1.06	0.044	0.064	4.0
In 0.03 M bis-Tris-propane (pH 7.4)					
no additive	8.9	1.76	0.072	0.38	9.0×10^3
+ 0.08 M CaCl$_2$	9.5	1.50	0.09	0.27	1.0×10^3
+ 0.63 M NaCl	9.3	1.66	0.066	0.4	2.3×10^4
+ 0.63 M KClO$_4$	8.2	1.75	0.07	0.24	1.6×10^2

[a]Other conditions: 25°C; Hr concentration, 35–50 μM site.
[b]P_{50} in mm Hg.
[c]Maximal slope of the Hill plot.
[d]O_2 association equilibrium constant for the T state in mm Hg^{-1}.
[e]Same as K_T but for the R state.
[f]The allosteric constant ($= [T_0]/[R_0]$).

Table 2. Values of P_{50} and h_{max} for *Siphonosoma* Hr, hemolysate and hemerythrocyte.

Conditions[a]	P_{50}[b]	h_{max}
Hemerythrin		
In 0.1 M Tris-acetate (pH 8.1)	3.0	1.06
In 0.05 M bis-Tris (pH 7.4) + 0.1 M Cl^-	3.2	1.05
In 0.1 M potassium phosphate (pH 7.4)	3.0	1.10
Hemolysate		
In 0.05 M Tris-HCl (pH 8.1)	3.4	1.05
In 0.05 M bis-Tris (pH 7.4) + 0.1 M Cl^-	3.5	1.09
In 0.05 M bis-Tris (pH 7.4) + 25 mM $CaCl_2$	3.5	1.12
In 0.03 M bis-Tris-propane (pH 7.4) + 0.06 M $KClO_4$	3.3	1.07
In 0.1 M potassium phosphate (pH 7.4)	3.3	1.11
Hemerythrocyte		
In 0.05 M Tris (pH 7.3) + 0.25 M Na_2SO_4 and 10 mM EDTA	3.2	1.0

[a]Other conditions: 25°C; Hr concentration, 100–140 μM site.
[b]In mm Hg.

P_{50} and h_{max} values are essentially identical for Hr solution, hemolysate and hemerythrocyte suspension from *Siphonosoma* (Table 2). Thus, the O_2 equilibrium curve of *Siphonosoma* Hr is little influenced by the pH, the kind of buffer system, additives such as Cl^-, $CaCl_2$ and $KClO_4$ and intracellular components in the hemerythrocyte. The Bohr coefficient is –0.04 or less (in absolute value), and h_{max} is smaller than 1.12. Thus, this Hr lacks both homotropic and heterotropic effects.

Interestingly, the O_2 affinity of *Lingula* Hr becomes similar to that of *Siphonosoma* Hr at a high pH (Figure 2). These two Hrs also showed similar O_2 association rate constant values at pH 7.6 (12). Upon a pH decrease to 6.8, the rate constant for *Lingula* Hr was reduced by a factor of 9, which is indicative of the stabilization of a low affinity state, while the rate constant for *Siphonosoma* Hr remained unchanged (12).

Heat of Oxygenation. Figure 3 shows log P_{50} vs $1/T$ plots (T is the absolute temperature) which were obtained from O_2 equilibrium data at four different temperatures, 15, 20, 25 and 30°C. From the slope of these Arrhenius plots, the heat of oxygenation (ΔH) was calculated to be –8.6 kcal/mol for *Lingula* Hr and –17.8 kcal/mol for *Siphonosoma* Hr.

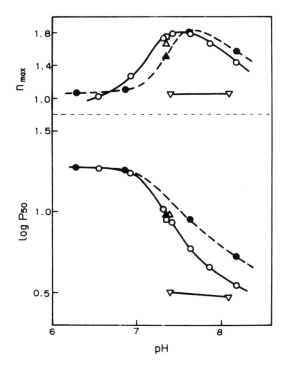

Figure 2. pH dependences of log P_{50} and h_{max} values for *Lingula* and *Siphonosoma* Hrs. Data in Tables 1 and 2 are plotted. ($\bigcirc$) *Lingula* Hr in 0.03 M bis-Tris-propane + 0.064 M Cl⁻; ($\bullet$) *Lingula* Hr in 0.07 M phosphate + 0.1 M NaCl; ($\square$), ($\triangle$) and ($\blacktriangle$) *Lingula* Hr with no additive, 0.63 M NaCl and 0.08 M CaCl₂, respectively, in 0.03 M bis-Tris-propane; ($\triangledown$) *Siphonosoma* Hr in 0.05 M bis-Tris or 0.1 M Tris-acetate.

Linear MWC Plots. To infer the size of the allosteric unit of *Lingula* Hr, the present oxygenation data were further analyzed within the framework of the Monod-Wyman-Changeux (MWC) allosteric model (19). Here, the allosteric unit is meant to be the minimum unit within the Hr molecule which substantially preserves the full cooperativity of the whole molecule. A linearized form of the MWC equation (20) is rewritten in the form (11)

$$\log Z = (N-1) \log X + \log L_0,$$

where $Z = (K_R - Q)/(Q - K_T)$, $X = (1 + K_T P)/(1 + K_R P)$, $Q = Y/(1 - Y)/P$; Y is the fractional O₂ saturation and P is the partial pressure of O₂. K_T and K_R are the intrinsic association equilibrium constants for the T and the R states, respectively, N is the number of O_2-binding sites per allosteric unit, and L_0 is the allosteric constant ($= [T_0]/[R_0]$). K_T and K_R values were obtained from the lower and upper asymptotes, respectively, of the experimental Hill plots (15, 16).

The present oxygenation data for *Lingula* Hr were plotted according to the above equation, producing linear plots (Figure 4). The K_T and K_R values used for these plots are listed in Table 1. It was found that these plots are well-expressed by straight lines with a slope of 7 (therefore, $N = 8$). The L_0 values in Table 1 were adjusted so that each line fitted to the corresponding experimental plot. From the present graphic analysis, it has been concluded that the allosteric unit of *Lingula* Hr is the one possessing eight O_2-binding sites, in other words, the whole molecule.

Maximal K_R/minimal K_T ($= 9$) for *Lingula* Hr (Table 1) is much smaller than the corresponding value ($= 950$) for HbA (16). L_0 is much smaller for *Lingula* Hr than for HbA ($= 3 \times 10^6$). The small K_R/K_T and L_0 values for *Lingula* Hr provide the reasons why this Hr shows less cooperativity than HbA, in spite of the larger size of its allosteric unit.

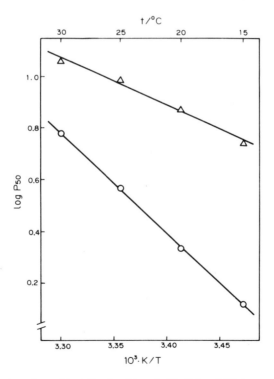

Figure 3. Arrhenius plots of log P_{50} for *Lingula* Hr (Δ) and *Siphonosoma* Hr ($\bigcirc$). In 0.03 M bis-Tris-propane (pH 7.4) + 0.06 M Cl⁻.

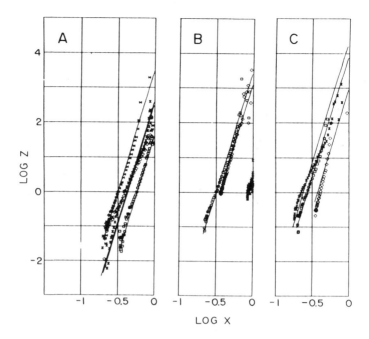

Figure 4.   Linear MWC plots of oxygenation data for *Lingula* Hr.   The MWC parameter values are listed in Table 1.   Straight lines were calculated with $N = 8$.   (*A*) in 0.03 M bis-Tris-propane + 0.064 M Cl⁻; (□) pH 8.18; (◊) pH 7.87; (⊠) pH 7.64; (|X|,) pH 7.42.   (*B*) in the same buffer as in *A*; (□) pH 7.32; (◊) pH 6.93; (⊠) pH 6.53.   (*C*) in 0.03 M bis-Tris-propane (pH 7.4); (□) no additive; (◊) + 0.08 M CaCl₂; (⊠) + 0.63 M NaCl.

Comparison of O₂-Binding Properties.   Table 3 compares the O₂-binding properties of the Hrs and the myoHr studied to date.   Only the two octameric *Lingula* Hrs exhibit allosteric properties (including cooperativity and the Bohr effect), while other Hrs, no matter whether octameric or trimeric, are non-allosteric.   Inversely, the effect of perchlorate which is observed in myoHr and some octameric Hrs is absent in *Lingula* Hr.   It is noteworthy that the O₂ association constant ($1/P_{50}$) for non-allosteric Hrs, with the exception of *Phascolosoma* Hr, ranges between 0.22 and 0.56 mm Hg⁻¹ and that these values are comparable to the affinity of the *R* state of *Lingula* Hr (see the K_R value at alkaline pH in Table 1).

The large ΔH value for *Siphonosoma* Hr is comparable to those observed in other non-allosteric Hrs.   The small ΔH value for the Hr from *L. unguis* may be characteristic of allosteric Hr, but we are still waiting for the determination of ΔH for the Hr from *L. reevii*.   Our interpretation for the small ΔH value of

Table 3. Comparison of O_2-binding properties of hemerythrins and myohemerythrin.

Species[a]	P_{50}[b]	h_{max}	Bohr effect[c]	Effect of ClO_4^-[d]	ΔH[e]	Conditions	Ref.
Lingula unguis (8)	3.5-18.5	1.02-1.79	-0.78	Absent	-8.6	0.03 M bis-Tris-propane + 0.06 M Cl-, pH 6.53-8.18, 25°C	Present study
Lingula reevii (8)	8-16	1.0-1.7	Present			Phosphate (Γ = 0.2), pH 6.7-7.6, 22-24°C	1
	5.1-15.0	1.1-2.0	-0.59			Phosphate (Γ = 0.2), pH 6.3-8.0, 22°C	18
Siphonosoma cumanense (3)	3.0-3.2	1.05-1.06	Absent	Absent	-17.8	0.05 M bis-Tris + 0.1 M Cl- or 0.1 M Tris-acetate, pH 7.4-8.1, 25°C	Present study
Siphonosoma ingens	1.8	1.00	Absent			Phosphate (Γ = 0.2), pH 6.03-8.78, 18.5°C	21
Sipunculus rudus	3.8	1.1	Absent		-13.5	0.15 M phosphate, pH 6-9.5, 20°C	22
Phascolosoma agassizii (3)	10.0	1.0	Absent		-17	Phosphate (Γ = 0.2), pH 5.8-7.9, 25°C	23
Phascolopsis gouldii (8)	3.5-4.0	1.0-1.1	Absent	Present (↓)	-18[f]	0.1 M Tris/SO_4^{2-}, pH 7.1-8.8, 23°C	18
Themiste zostericola coelomic Hr (8)	3.4-4.5	1.1	Absent	Present (↓)	-20	pH 5.8-8.0, 20-25°C	21, 24
myoHr (1)	1.08	1.0	Absent	Present (↑)		pH 7.6, 20°C	

[a]The number in parentheses indicates the number of subunits per molecule.
[b]In mmHg.
[c]Measured by the Bohr coefficient (= Δlog P_{50}/ΔpH).
[d]Downward and upward arrows in parentheses indicate decrease and increase, respectively, in oxygen affinity upon the addition of $KClO_4$.
[e]Heat of oxygenation in kcal/mol.
[f]Unpublished data of Peticolas *et al.* as quoted by Klotz and Klotz (25).

Lingula Hr is as follows. The observed ΔH value includes the heat of Bohr proton release associated with O_2 binding. If the ΔH value is corrected for the heat of proton release, which is an endothermic process, the ΔH value for *Lingula* Hr would become similar to ΔH values for other Hrs with no Bohr effect. The marked pH dependence of cooperativity appearing in the Bohr effect region (Figure 2) suggests that the Bohr groups of *Lingula* Hr participate in intersubunit ionic links which would contribute to the stabilization of the T quaternary structure, exactly in the same manner as in Hb (26).

Acknowledgements

The authors thank Professor Koji Hoshino, Hiroshima University, for a generous supply of *Siphonosoma* worms.

References

1. Manwell, C. (1960) *Science* **132**: 550–551.
2. Richardson, D.E., Reem, R.C. and Solomon, E.I. (1983) *J. Am. Chem. Soc.* **105**: 7780–7781.
3. Kurtz, D.M., Shriver, D.F. and Klotz, I.M. (1977) *Coord. Chem. Rev.* **24**: 145–178.
4. Klippenstein, G.L. (1980) *Am. Zool.* **20**: 39–51.
5. Wood, E.J. (1980) *Essays Biochem.* **16**: 1–47.
6. Wilkins, R.G. and Harrington, P.C. (1983) *Adv. Inorg. Biochem.* **5**: 51–85.
7. Toulmond, A. (1985) *Symp. Soc. Exp. Biol.* **39**: 163.
8. Satake, K., Yugi, M., Kamo, M., Kihara, H. and Tsugita, A. (1990) *Protein Sequences Data Anal.* **3**: 1–6.
9. Fuseya, M., Ichimura, K., Yamamura, T., Tachiiri, Y., Satake, K., Amemiya, Y. and Kihara, H. (1989) *J. Biochem.* **105**: 293–298.
10. Tambo, F., Ichimura, K., Fuseya, M., Takahashi, S., Yamamura, T., Kihara, H. and Satake, K. (1989) *Biochem. Int.* **18**: 311–317.
11. Imai, K., Yoshikawa, S., Fushitani, K., Takizawa, H., Handa, T. and Kihara, H. (1986) In *Invertebrate Oxygen Carriers*, ed. B. Linzen, 367–374. Berlin: Springer-Verlag.
12. Zimmer, J.R., Tachiiri, Y., Takizawa, H., Handa, T., Yamamura, T. and Kihara, H. (1986) *Biochim. Biophys. Acta* **874**: 174–180.
13. Joshi, J.G. and Sullivan, B. (1973) *Comp. Biochem. Physiol.* **44B**: 857.
14. Addison, A.W. and Dougherty, P.L. (1982) *Comp. Biochem. Physiol.* **72B**: 433–438.
15. Imai, K. (1981) *Methods Enzymol.* **76**: 438–449.
16. Imai, K. (1982) *Allosteric Effects in Haemoglobin*. London: Cambridge University Press.
17. Imai, K., Morimoto, H., Kotani, M., Watari, H., Hirata, W. and Kuroda, M. (1970) *Biochim. Biophys. Acta* **200**: 189–196.

18. Richardson, D.E., Emad, M., Reem, R.C. and Solomon, E.I. (1987) *Biochemistry* **26**: 1003–1013.
19. Monod, J., Wyman, J. and Changeux, J.-P. (1965) *J. Mol. Biol.* **12**: 88–118.
20. Decker, H., Savel, A., Linzen, B. and Van Holde, K.E. (1983) In *Life Chem. Rep. Suppl. Ser.* **1**: 251–256.
21. Manwell, C. (1960) *Comp. Biochem. Physiol.* **1**: 277–285.
22. Bates, G., Brunori, M., Amiconi, G., Antonini, E. and Wyman, J. (1968) *Biochemistry* **7**: 3016–3020.
23. Manwell, C. (1958) *Science* **127**: 592–593.
24. Chadwick, R.A. and Klippenstein, G.L. (1983) *Comp. Biochem. Physiol.* **74A**: 687–692.
25. Klotz, I.M. and Klotz, T.A. (1955) *Science* **121**: 477–480.
26. Perutz, M.F. (1970) *Nature* **228**: 726–734.
27. Tachiiri, Y., Ichimura, K., Yamamura, T., Satake, K., Kurita, K., Nagamura, T. and Kihara, H. (1990) *Eur. Biophys. J.* **18**: 9–16.

25
The Dynamics of Dioxygen Binding to Hemerythrin

Daniel Lavalette and Catherine Tetreau

INSERM Unit 219, Curie Institute, University of Paris South, 91405 Orsay, France

Introduction

Oxygen carrier heme-proteins have proven to be very convenient tools for investigating internal protein dynamics. When a short pulse of laser light is absorbed by an oxy-protein, the heme–O_2 bond is immediately disrupted. Since oxy- and deoxy-proteins display different absorption spectra, an optical signal is available for monitoring the sequence of events accompanying the return to the bound state. Because the energetics of rebinding are sensitive to the protein conformation, the reaction constitutes an internal time probe which is modulated by protein movements. Heme-proteins (*e.g.*, Mb or Hbs) have long been investigated in many laboratories. H. Frauenfelder and his co-workers pioneered the field of photolysis in glassy solvents at low temperatures; in other words, at practically infinite viscosity. They showed that proteins exist in a large number of conformational substates with slightly different reaction rates and that O_2 binding is governed by several energy barriers in sequence rather than just one. It was later recognized that some of the barriers and associated rate constants were also dependent upon viscosity in agreement with their assumed dynamic nature (1, 2).

Hrs are another class of respiratory proteins in which O_2 does not bind to an independent prosthetic group like a heme, but rather to a binuclear complex of two Fe atoms (for a recent review, see reference 3). They are found in some species of marine invertebrate and are not related in any way to heme-proteins. A preliminary study of the octameric Hr of the marine worm *Sipunculus nudus* revealed that, as with heme-proteins, HrO_2 can be photodissociated, thus opening a new way of studying this class of invertebrate O_2 carriers as well as illuminating new aspects of protein dynamics in general (4).

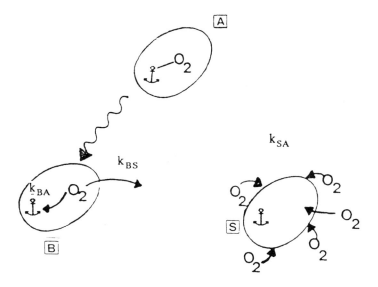

Figure 1. Simplified scheme of the fate of the protein-O_2 pair after photodissociation of the oxygenated protein (see text for further discussion of the protein states). In the present work, photodissociation was achieved by a 532 nm laser pulse of 20 ns duration, which was absorbed by the broad charge-transfer band of HrO_2 in the visible spectral region. This band is absent in Hr and, therefore, photodissociation is accompanied by a bleaching of the protein sample.

Results and Discussion

The basic laser-photolysis experiment is described schematically in Figure 1. The protein-O_2 complex is shown at A in its bound state. In state B, a photon has just been absorbed and the O_2-protein bond has been broken. The O_2 molecule may now rebind directly without leaving the protein, resulting in geminate recombinations at the first order rate k_{BA}. Alternatively, O_2 may escape into the solvent at the rate k_{BS}. At pH 7.4 and 5°C, about 50% of the photodissociated HrO_2 recombine within about 200 ns according to the geminate reaction path, which, of course, is independent of the concentration of O_2. The proteins which have not recombined after this rapid initial phase are shown in S; this corresponds to the situation in which the O_2 molecules have escaped into the solvent. The "empty" proteins are surrounded by O_2 molecules that compete for rebinding from the solvent at a bimolecular rate constant k_{SA}. This gives rise to a slow final rebinding phase with a rate proportional to the O_2 concentration.

To investigate the reaction in more detail, the three rate constants were measured in glycerol/water mixtures of different viscosities at a constant temperature of 5°C. Figure 2 shows that the escape rate k_{BS} decreased upon increase in the viscosity, whereas the geminate rate k_{BA} did not. An immediate consequence was that the proportion of O_2 escaping into the solvent severely decreased and practically vanished at the highest viscosity. Below 100 cP, rebinding from outside k_{SA} was also found to be independent of viscosity, showing that the rate of re-entry must be barrier-limited. A moderate decrease above 100 cP was found to correspond to a beginning of diffusion control of the bimolecular reaction, and not to an intrinsic viscosity dependence (5). From these apparently simple findings two fundamental questions emerge: how can a pure intramolecular first order reaction rate like k_{BS} depend upon viscosity and why does O_2 escape depend upon viscosity whereas its re-entry does not ?

That the internal rebinding rate k_{BA} does not depend upon viscosity is in agreement with the view that this rate actually corresponds to the ultimate bond formation step between O_2 and the binuclear Fe site of Hr. This is basically an electronic quantum process in which no viscosity effect is expected and which should be approximated by Eyring's theory of absolute reaction rates, commonly used in organic and inorganic chemistry. The rate constant follows an

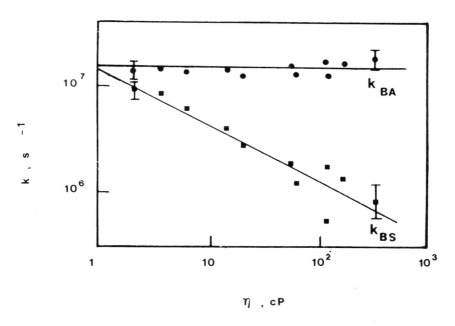

Figure 2. Viscosity dependence of the internal rate constants for O_2 geminate rebinding with Hr (k_{BA}) and escape into the solvent (k_{BS}). Experiments were performed at 5°C, pH 7.5, in glycerol/water mixtures.

$$o \neq \langle x\varrho \rangle \qquad\qquad o = \langle x\varrho \rangle$$

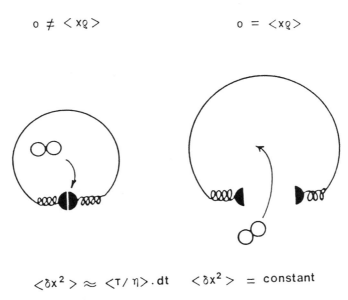

$$\langle \delta x^2 \rangle \approx \langle T/\eta \rangle \cdot dt \qquad \langle \delta x^2 \rangle = \text{constant}$$

Figure 3. Brownian-motion driven fluctuations of protein residues (symbolized by the black balls and the springs) have no effect upon the entry of O_2 if the gate may be considered as "open" on the average ($\langle dx \rangle \neq 0$, *right*). For a "closed" gate, ($\langle dx \rangle = 0$, *left*), the fluctuations may lead to an opening within a finite waiting time dependent upon temperature and local viscosity.

Arrhenius law with a pre-exponential factor and an activation energy depending only on the energy levels of the reactants. However, the transition state theory is unable to explain the viscosity dependence of the escape rate kBS. Since viscosity or friction imply that parts of a system are moving, a description of the escape process must be formulated in terms of internal motions of Hr.

The simplest model is to assume that O_2 cannot escape the protein unless some amino acid side-chain residue moves about in order to open a gate of width dx in the protein matrix. If the gate is closed most of the time (Figure 3, *left*), then the average displacement of the residue measured using the equilibrium position as origin will be zero. Due to the local brownian motion, however, the mean square displacement increases uniformly with time at a rate that is inversely proportional to viscosity (this follows from Einstein's formula for brownian motion). For an unbound O_2 molecule still inside the protein, there will be a finite waiting time and, hence, a finite rate before the gate opens enough for the molecule to escape into the solvent. The actual viscosity dependence of the rate may be complicated by the fact that local movements are constrained and cannot be described as easily as can those of a free brownian particle. This presumably accounts for the fractional power of viscosity observed in Figure 2. However, when the gate is assumed to be open on the average (Figure 3, *right*) nothing seriously prevents O_2 from entering or leaving the protein. The reaction rate may be decreased by a geometric probability factor but is not expected to depend upon viscosity. The situations shown in Figure 3 may indeed explain the results of the Hr experiments if we identify the process of Figure 3 (closed) with the escape rate k_{BS} and that of Figure 3 (open) with the rebinding from the outside k_{SA}. Thus, one single mechanism may, at least qualitatively, account for both processes. It implies, however, that Hr undergoes a noticeable conformational change upon deoxygenation, since it has to switch from a structure with a fluctuating but closed gate to another one with a permanently open gate. The question of the nature of the gate is not yet solved. Based upon x-ray diffraction studies, it has been observed that one possible path leading to the binuclear Fe center is more or less obstructed by His and Trp residues arranged in parallel planes (6). The present data are not yet sufficient to determine whether this observation is related to our "gate" mechanism.

The free-energy diagram which may be drawn to describe O_2 dissociating and rebinding with Hr should include both the internal, viscosity-independent energy barrier for bond formation (*i.e.*, the ordinary reaction coordinate) and the fluctuating, viscosity-dependent energy barrier for escape. Moreover, two different diagrams have to be drawn: one for the oxy-protein and one for the deoxy-protein. Clearly, one single reaction coordinate is not sufficient to even

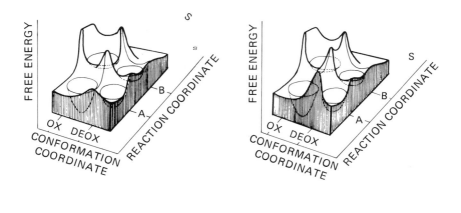

Figure 4. A possible free-energy surface for the HrO_2 system as a function of a "small-ligand reaction coordinate" and of a protein conformation coordinate. Viscosity-dependent barriers separate the oxy- and deoxy-conformations of the protein. The surface is shown at low viscosity (*e.g.*, 1 cP) on the left and at high viscosity (*e.g.*, 100 cP) on the right. Note the increased barrier height along the conformation coordinate.

approximately describe the real situation. We need a generalized protein conformation coordinate in addition to the reaction coordinate so that the system may move along these two mutually perpendicular coordinates on a free energy surface as shown in Figure 4.

The reaction coordinate is used to describe quantum effects according to transition state theory. The protein conformation coordinate parametrizes large conformational changes as well as small fluctuations in the positions of the protein residues. We shall assume that motions over the barriers along the conformation coordinate occur due to coupling with the solvent by friction forces and local brownian motion. Thus, transitions along the conformation coordinate will be viscosity-dependent; Kramer's theory provides the necessary theoretical background (7).

HEMERYTHRIN

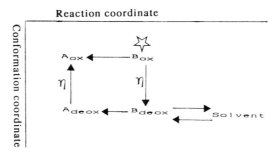

MYOGLOBIN

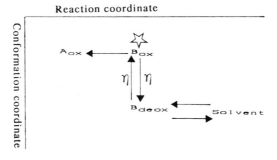

Figure 5. A possible common scheme for O_2-binding to heme-proteins and Hr. The states correspond to those of the free-energy surface of Figure 4 as viewed from above. The star denotes the initial state of the protein-O_2 pair immediately after photodissociation. If access to state A_{deox} is forbidden for some reason in heme-proteins, the reaction path will appear as sequential and reversible, in contrast to the quasi-cyclic reaction path displayed by Hr.

Our experiments suggest that Hr may be in the oxy- or deoxy-conformation. In each conformation, the O_2 molecule may bind to the Fe, but the ultimately stable state is, of course, A_{ox} (Figure 4). Similarly, there will be two B states, B_{ox} and B_{deox}. Since photodissociation is too fast for the protein conformation coordinate to change, the system at first becomes trapped in B_{ox}. If the protein were rigid (this would correspond to a very large barrier between B_{ox} and the solvent S), there would be no way out and O_2 would rebind directly to A_{ox}. However, during the time the system spends at B, brownian motion may, for a time, accumulate an excess energy into the protein coordinate and cause the system to move to B_{deox}. This is equivalent to opening the gate considered in the time-dependent picture. The probability, as discussed above, depends upon viscosity, and so does the barrier. Oxygen may then leave the protein in its B_{deox} conformation.

Rebinding from the solvent S occurs at first along the reaction coordinate of the deoxy-protein, presumably because the barrier between B_{deox} and A_{deox} is much smaller than the barrier with B_{ox}. No viscosity effects are expected along a reaction coordinate and the small barrier means that the "gate is open." Presumably, the optical spectrum of A_{deox} is not distinguishable from that of A_{ox}. Therefore, the final conformational relaxation towards A_{ox} cannot be detected, but should be viscosity-dependent! Other time-resolved spectroscopic techniques (*e. g.*, infrared or Raman) could possibly test the validity of the proposed reaction scheme by providing information on this final, yet undetectable step.

The separation into a reaction and a conformation coordinate provides us with a description of what is going on in the course of the reaction of O_2 with Hr. Is this representation also consistent with the findings on hemoproteins? In their study of Mb, Frauenfelder *et al.* found several barriers arranged in sequential order (1). The viscosity-dependence decreased with increased penetration into the protein, and the final heme–O_2 barrier showed no viscosity-dependence at all, presumably for similar reasons as in Hr. But in contrast to Hr, the sequential arrangement of the barriers in Mb implies that the rebinding of O_2 follows a path which is the inverse of the dissociation path. The number of barriers does not really matter at this stage of speculation. There is no reason to oppose the reversibility of one system to the apparent irreversibility of the other. Figure 5 shows how the system described in Figure 4 may be changed into one that would resemble Mb by simply assuming that A_{deox} is not accessible because the barrier viewed from B_{deox} might be too high. The only possibility would then be a return via the viscosity-dependent $B_{deox} \rightarrow B_{ox}$ transition. The system then appears as purely sequential, although both conformation and reaction coordinates are successively involved in the process.

The kinetics considered in this work were always exponential and the internal rebinding rate was found to follow the Arrhenius law $k_{BA} = A \exp(-H/RT)$ with $A = 3.8 \times 10^7 s^{-1}$, and $H = 2.5$ kJ.Mol^{-1}. However, proteins are subject to conformational fluctuations and these values are averages over an unknown distribution of conformational substates, each displaying a slightly different rate

(1, 5, 8). At viscosities considerably higher than those used here, conformational fluctuations may be slower than O_2 rebinding and the kinetics may become strongly non-exponential. In the accompanying paper, this effect is used to determine the underlying distribution of rate constants among the conformational substates and to deduce the real value of the pre-exponential factor and of the activation enthalpy for the internal O_2-binding reaction to Hr (8).

References

1. Austin, R.H., Beeson, K.W., Eisenstein, L., Frauenfelder, H. and Gunsalus, I.C. (1975) *Biochemistry* 14: 5355–5373.
2. Beece, D., Eisenstein, L., Frauenfelder, H., Good, D., Marden, M.C, Reinisch, L., Reynolds, A.H., Sorensen, L.B. and Yue, K.T. (1980) *Biochemistry* 19: 5147–5157.
3. Kurtz, D.M. (1986) In *Invertebrate Oxygen Carriers,* ed. B. Linzen, 25–36. Berlin: Springer-Verlag.
4. Alberding, N., Lavalette, D. and Austin, R.H. (1981) *Proc. Natl. Acad. Sci. U.S.A.* 78: 2307–2309.
5. Lavalette, D. and Tetreau, C. (1985) *Eur. J. Biochem.* 145: 555–565.
6. Stenkamp, R.E., Sieker, L.C., Jensen, L.H., McCallum, J.D. and Sander-Loehr, J. (1985) *Proc. Natl. Acad. Sci. U.S.A* 82: 713–716.
7. Kramers, H.A. (1940) *Physica* 7: 284–304.
8. Lavalette, D., Tetreau, C., Brochon, J.C. and Livesey, A.K. (1991) *Eur. J. Biochem.* Submitted.

26

Distribution of Hemerythrin's Conformational Substates from Kinetic Investigations at Low Temperature

Catherine Tetreau,[a] Daniel Lavalette[a] and Jean-Claude Brochon[b]

[a]INSERM Unit 219, Curie Institute and [b]Laboratory of Electromagnetic Radiation, University of Paris South, 91405 Orsay, France.

Introduction

In addition to their biologically essential macroscopic motions, proteins also present equilibrium fluctuations; in other words, they exist in a large number of slightly different structures that react at different rates. As a consequence, the protein ensemble is kinetically inhomogeneous and must be described by a continuous distribution function of rate constants, $f(k)$. At physiological temperature and low viscosity, the fluctuations are so rapid in comparison to most protein reactions that the differences in reaction rates are averaged out. Exponential kinetics are observed, but the corresponding rate constant is actually an average value, $<k>$. In contrast, at low temperature and high viscosity, each conformational substate (CS) remains frozen and reacts at its own rate; the kinetics turn into a superposition of an infinite number of exponentials. Such non-exponential kinetics are related to the probability distribution of the rate constants by $N(t) = \int f(k) \exp(-kt).dk$. These concepts have emerged from extensive flash photolysis investigations of liganded hemoproteins (1, 2) and, more recently, of Hr (3). Although there is no *a priori* reason why the average rate, $<k>$, measured under physiological conditions should obey a simple

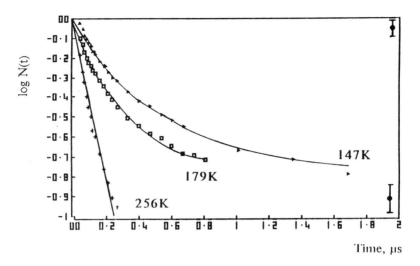

Figure 1. Semi-log plots of the geminate recombination of O_2 with Hr after laser photolysis in 60% w/w ethylene glycol/water (256 and 147°K) or in 79% w/w glycerol/water (179°K). $N(t)$ is the fraction of unrecombined Hr at time t after the maximum of the laser pulse. The solid lines are fits to the data calculated using the MEM.

Arrhenius's law as does any elementary reaction rate ($k = A \exp[-H/RT]$), it can be shown that this might be the case under sufficiently general conditions (3, 4), provided that A and H are replaced by effective parameters A_{eff} and H_{eff}. Our ultimate understanding of the significance of these effective values and of their correlation with the substates' parameters must be based on the knowledge of the distribution, $f(k)$. An ensemble of molecules in fast conformational relaxation cannot be distinguished from an ensemble of "static" proteins with a constant reaction rate since both display exponential kinetics. Information about the conformational substates (CS's) may be obtained only by slowing down the fluctuations. In the present work, we attempt to characterize the CS's of Hr by determining its rate constant spectrum $f(k)$ for the geminate recombination of O_2 (*i.e.*, the internal rebinding with its original site) following photodissociation. The usual approach to recovering $f(k)$ from the non-exponential kinetics observed at low temperature consists of assuming that CS's differ only in their activation enthalpy and that the pre-exponential factor A remains constant. Using a model to describe the enthalpy distribution function, $g(H)$, least-square procedures are applied to fit the experimental data. Here we use an entirely different and new approach, known as the Maximum Entropy Method (MEM), which permits the direct recovery of the rate constant spectrum $f(k)$ without using the above-mentioned restrictive assumptions.

The Maximum Entropy Method.   The MEM has been applied in a variety of fields, in particular in the analysis of lifetimes in pulse-fluorimetry (5). In the theory of communication (6), the entropy function measures the uncertainty of the occurrence of one particular event among a distribution of random events. Therefore, maximizing the entropy of a distribution of rate constants means that one will choose, among all conceivable $f(k)$ that fit $N(t)$ equally well, the least constrained one; in other words, the distribution of exponentials which reproduces $N(t)$ within the required accuracy without including an *a priori* knowledge about $f(k)$. In brief, the continuous $f(k)$ spectrum was approximated by a discrete set of rate constants k_i with individual weight $f(k_i)$. The superposition of the exponential decays was computed and the calculated kinetics, $N_{calc}(t)$, were compared to the data points, $N_{exp}(t)$. Starting from an equiprobable distribution of rates, the weights of the k_is were iteratively adjusted, with the dual constraints of fitting the data $N_{exp}(t)$ as closely as possible and simultaneously maximizing the entropy of the $f(k_i)$ probability set. Details of the procedure may be found in reference 7.

Results and Discussion

Figure 1 displays the geminate rebinding kinetics recorded in the two extreme regimes of fast and slow conformational relaxation. Strong deviation from an exponential may be noticed in 79% w/w glycerol/water (GLY) at 179°K, and in 60% w/w ethylene glycol/water (EGOH) at 147°K. Since both solvents are below their estimated freezing temperatures (4), experiments performed at low temperature correspond to the situation in which the CS's distribution is frozen, whereas at 256°K, the fluctuations are fast enough to lead to an average exponential rate. The rate constant spectra calculated by the MEM are given in Figure 2. The narrow spectrum at 256°K is in sharp contrast with the broad distributions associated with the kinetics in the rigid glasses. The temporal fits are given as solid lines in Figure 1. A noise-free exponential signal should yield a delta function spectrum. Noise factors, as well as the use of discrete instead of continuous k values, are likely to be responsible for the finite, though very narrow, width of the $f(k)$ curve at 256°K.

The shift of the spectra towards smaller k values as the temperature is lowered is in agreement with the fact that the kinetics slow down. Experimental results with Hr (3) and with hemoproteins (1), suggest that only H is distributed whereas A retains a constant value. If we make this additional hypothesis, it can be shown that a simple transformation permits us to obtain $g(H)$ provided that A is known. From the Arrhenius plot of the most probable rate of $f(k)$ below the freezing temperature of EGOH (4), a value of $A = 2 \times 10^{10}$ s^{-1} was obtained and $g(H)$ could be calculated (Figure 3). The distribution at 179°K has been determined in GLY and the change in $g(H)$ reflects the temperature dependence of the frozen CS distribution if there is no specific solvent dependence. Spectral

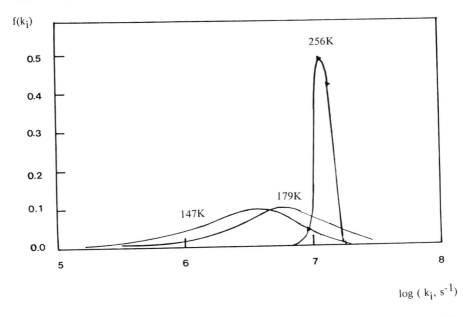

Figure 2. Rate constant spectra of the geminate recombination of O_2 with Hr recovered using the MEM.

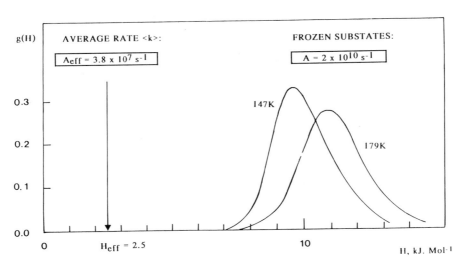

Figure 3. Arrhenius parameters for geminate O_2 rebinding with Hr: right part, frozen CS; left part, average protein rate under physiological conditions (fast fluctuations).

narrowing as the temperature is lowered appears reasonable since less CS are expected to be populated. Under physiological conditions, the average rate $<k>$ for Hr was found to follow an Arrhenius law (3). For comparison, the values of H_{eff} and A_{eff} are indicated in Figure 3. Strikingly, H_{eff} does not coincide either with the most probable nor with the average value of $g(H)$. Even the probability for any real protein to have an enthalpy of activation equal to the effective value H_{eff} goes to zero; moreover, A_{eff} differs by more than two orders of magnitude from the true value of A. This shows that a comparison of the thermodynamic parameters of similar reactions performed by different proteins requires a complete knowledge of the distribution of protein substates.

References

1. Austin, R.H., Beeson, K.W., Eisenstein, L., Frauenfelder, H. and Gunsalus, I.C. (1975) *Biochemistry* **14**: 5355–5373.
2. Beece, D., Eisenstein, L., Frauenfelder, H., Good, D., Marden, M.C., Reinisch, L., Reynolds, A.H., Sorensen, L.B. and Yue, K.T.   (1980) *Biochemistry* **19**: 5147–5157.
3. Lavalette, D. and Tetreau, C. (1988) *Eur. J. Biochem.* **145**: 555–565.
4. Lavalette, D., Tetreau, C., Brochon, J.C. and Livesey, A.K. (1991) *Eur. J. Biochem.* Submittted.
5. Livesey, A.K. and Brochon, J.C. (1987) *Biophys. J.* **52**: 693–706.
6. Shannon, C.E. and Weaver, W.   (1964)   *The Mathematical Theory of Commmunication*. Urbana: The University of Illinois Press.
7. Gull, S.F. and Skilling, J. (1984) *IEE Proceedings* **131**: 646–650.

Part III
Amino Acid and cDNA Sequences

27
Translation of the cDNA Sequence for the Polymeric Hemoglobin of *Artemia*

Clive N.A. Trotman,[a] Anthony M. Manning[a], Luc Moens,[b] Kerry J. Guise[a] and Warren P. Tate[a]

[a]Department of Biochemistry, University of Otago, Dunedin, New Zealand
[b]Department of Biochemistry, University of Antwerp, Antwerp 2610, Belgium

Introduction

The estimated M_w of at least 260 kDa for the complete Hb of *Artemia* (1) would be sufficient for 16 conventional globin domains. The molecule comprises two subunits (2), each thought to contain a series of concatenated domains resulting from the translation of a continuous messenger RNA (3). The subunits may be identical, as in HbI ($\alpha\alpha$) or HbIII ($\beta\beta$), or different ($\alpha\beta$), as in HbII (4). The cDNA sequence of nine continuous domains has recently been deduced (5). This preliminary analysis of the translation of the cDNA and the application of knowledge-based modeling techniques described in this paper indicate that all domains have a high probability of conforming to the globin fold. The *Artemia* domains have a very convincing homology with the vertebrate and invertebrate globins in the current database, but nevertheless display a few departures from convention including a novel and well-conserved linker sequence between domains.

Results and Discussion

Translation of the Artemia *Globin cDNA*. The translated sequence was subdivisible into nine successive globin-like domains (Table 1), which have 17–38 percent homology. Alignment of the nine sequences (translates $T1$ to $T9$) over their conserved sites strongly suggested that the *Artemia* domains possessed the classical globin helices A to H with their corresponding turns. Differences in length were mostly accommodated in the turns, as would be expected in variants on a conserved tertiary structure. Final refinement of the translation and adjustment of the alignment was performed with the aid of Template II of Bashford *et al*. (6), with an associated database of 226 sequences (the database to which we will refer unless otherwise stated). The sum of penalties per domain ranged from 3.4 to 8.3 with no violations of "Required" (class 2) sidechains.

Key conserved residues and characteristic types of sequences were recognized in the *Artemia* sequence. The expected invariant $CD1$ Phe and $F8$ His, and the almost invariant $E7$ His were primary assignments. Other striking features of the globin pattern generally included Gly-Leu-Ser at the equivalent of the N-terminus, Trp at $A12$, His-Pro-Glu at the start of the C-helix, and Val at $E11$. Further landmarks became apparent within the *Artemia* domain family itself, including a $B10$ Phe, $C4$ Tyr, $F5$ Gly and $G5$ Phe. A consensus sequence Val-Asp-Pro-Val-Thr-Gly-Leu emerged as diagnostic of the inter-domain linker.

One of the domains, $T3$, was recognizable as equivalent to the polypeptide $E1$ sequenced by Moens *et al*. (7). Differences between that and the present sequence were considered to be indicative of polymorphism within the *Artemia* genus, particularly since doubt often surrounds the true origin of commercial cysts. The $E7$ polypeptide sequence (8) was not found as such, although regions of similarity to it were present in $T6$, possibly also reflecting species variation.

Homology and Modeling at Specific Residues. Deviations characteristic of *Artemia* have been identified by measuring their penalties against Bashford's Template II and by detecting consistent departures from the consensus globin. In most cases the anomalies are represented in several of the domains, thus reducing the probability of their resulting from cloning or sequencing artifacts. Structural implications of these anomalies have been investigated through comparison of those structures in the Brookhaven data bank whose sequences were the most similar to *Artemia*, namely *Chironomus thummi thummi* deoxy HbIII (9), file $1ECD$, sperm whale deoxy Mb (1), file $1MBD$, and *Petromyzon marinus* cyano met HbV (11), file $2LHB$.

Region NA. The $NA2$ Leu is almost completely conserved throughout all globins; however, the replacements with Ile in $T1$ or $T8$, or Phe in $T9$, are conservative. This region is discussed in more detail with the inter-domain linker.

Table 1. *Artemia* globin domains *T1–T9* aligned with sperm whale Mb, *Chironomus* HbIII and part of *Petromyzon* globin V. The *Artemia* sequence from the N-terminus to position *A3* was obtained by amino acid sequencing, and from *A4* to the C-terminus is the translation of the cDNA sequence.

```
                 Linker        A     8  12 15        B    6  10          C2 4
Myoglobin             VL   SEGEWQLVLHVWAKVE  A   DVAGHGQDILIRLFKS   HPETLEK
Chironomus             L   SADQISTVQASFDKVK  G   DPVGILYAVFKA       DPSIMAK
Artemia T1         AEVSGI  LVSDKATIKRTWATVT      DLPSFGRNVFLSVFAA   KPEYKNL
Artemia T2   LRRQIDLEVTGL   SCVDVANIQESWSKVS  G  DLKTTGSVVFQRMING   HPEYQQL
Artemia T3   SSLKRVDPITGL   SGLEKNAILSTWGKVR  G  NLQEVGKATFGKLFTA   HPEYQQM
Artemia T4   RQADIVDPVTHL   TGRQKEMIKASWSKAR  T  DLRSLGQELFMRMFKA   HPEYQTL
Artemia T5   ATSEEADPVTGL   YGKEIVALRQAFAAVT  P  RNVEIGKRVFAKLFAA   HPEYKNL
Artemia T6   FQLGQVDSNT L   TALEKQSIQDIWSNLR  S  TGLQDLAVKIFTRLFSA  HPEYKLL
Artemia T7   LQLERINPITGL   SAREVAVVKQTWNLVK  P  DLMGVGMRIFKSLFEA   FPAYQAV
Artemia T8   QQSYKQDPVTGI   TDAEKALVQESWDLLK  P  DLLGLGRKIFTKVFTK   HPDYQIL
Artemia T9   IGLKEVNPQNAF   SAYDIQAVQRTWALAK  P  DLMGKGAMVFKQLFTD   HG YQPL


             CD   4   7   D          E    5  78 11           EF
Myoglobin    FDR  FKHLK  TEAEMKA   SEDLKKHGVTVLTALGAILK   KGHHEAE
Chironomus   FTQ  FAGKD  LESIKG    TAPFETHANRIVGFFSKIIG   ELPN  IEAD
Artemia T1   FVE  FRNIP  ASELAS    SERLLYHGGRVLSSIDEAIA   GIATPDRAVKT
Artemia T2   FRQ  FRDVD  LDKLGE    SNSFVAHVFRVVAAFDGIIH   ELDNNQFIVST
Artemia T3   FRF  SQGMP  LASLVE    SPKFAAHTQRVVSALDQTLL   ALNRPSDFVYM
Artemia T4   FVNKGFADVP   LVSLRE   DERFISHMANVLGGFDTLLQ   NLDESSYFIYS
Artemia T5   FKK  FEQYS  VEELPS    TDAFHYHISLVMNRFSSIGK   VIDDNVSFVYL
Artemia T6   FTGR FGN    VDNINE    NAPFKAHLHRVLSAFDIVIS   TLDDSEHLIRQ
Artemia T7   FPK  FSDVP  LDKLED    TPAVGKHSISVTTKLDELIQ   TLDEPANLALL
Artemia T8   FTRTGFGDTP   LTKLDD   NPAFGTHIIKVMRAFDHVIQ   ILGKPKTLMAY
Artemia T9   FSN  LAQYE  ITGLEG    SPELNTHARNVMAQLDTLVG   SLQNSIELGQS


             F   4   8   FG      G        16         GH  5
Myoglobin    LKPLAQSHAT   KHKI   PIKYLEFISEAIIHVLHSR   HPGDF
Chironomus   VNTFVASHKP   RGV    THDQLNNFRAGFVSYMKAH   T D F
Artemia T1   LLALGERHIS   RGT    VRRHFEAFSYAFIDELKQR   G    V
Artemia T2   LKKLGEQHIA   RGT    DISHFQNFRVTLLEYLKEN   G    M
Artemia T3   IKELGLDHIN   RGT    DRSHFENYQVVFIEYLKET   LGDSL
Artemia T4   LRNLGDAHIQ   RKA    GTQHFRSFEAILIPILQES   Q G L
Artemia T5   LKKLGREHIK   RGL    SRKQFDQFVELYIAEISSE   L  S
Artemia T6   LKDLGLFHTR   LGM    TRSHFDNFATAFLSVAQDI   APNQL
Artemia T7   ARQLGEDHIV   LRV    NKPMFKSFGKVLVRLLEND   LGQRF
Artemia T8   LRSVGADHIA   TNV    ERRHFQAFSNALIPVMQHD   LKAQL
Artemia T9   LAQLGKDHVP   RKV    NRVHFKDFAEHFIPLMKAD   LGDEF


                             HA1  5    10    16
             H    10                 22                A1
Myoglobin    GADAQGAMNKALELFRKDIAAKYKELGYQG
Chironomus   A GAEAAWGATLDTFFGMIFSKM
Artemia T1   ESADLAAWRRGWDNIVNVLEAGL   LRRQIDLEVTGL   SCVDVANIQESW
Artemia T2   NGAQKASWNKAFDAFEKYISMGL   SSLKRVDPITGL   SGLEKNAILSTW
Artemia T3   DEFTVKSFNHVFEVIISFLNEGL   RQADIVDPVTHL   TGRQKEMIKASW
Artemia T4   DAASVEAWKKFFDVSIGVIAQGLKVATSEEADPVTGL   YGKEIVALRQAF
Artemia T5   DT GRNGLEKVLTFATGVIEQGL   FQLGQVDSNT L   TALEKQSIQDIW
Artemia T6   TVLGRESLNKGFKLMHGVIEEGL   LQLERINPITGL   SAREVAVVKQTW
Artemia T7   SSFASRSWHKAYDVIVEYIEEGL   QQSYKQDPVTGI   TDAEKALVQESW
Artemia T8   RPDAVAAWRKGLDRIIGIIDQGL   IGLKEVNPQNAF   SAYDIQAVQRTW
Artemia T9   TPLAESAWKRAFDVMIATIEQGQ   EGSSHALSSFLT   NPLA*
Petromyzon                            PIVDTGSVAPL   SAAEKTKIRSAW
```

The A Helix. The Trp at *A*12 is almost diagnostic; a Phe occupies this position in *T*5. Bashford recorded only one instance of *A*12 being neither Trp nor Phe. Although the Leu appearing at *A*14 in *T*7, *T*8 and *T*9 incurs a high penalty of 1.0 against the template, the *Artemia* sequences appear to be reliable at *A*14 since the Leu occurs in three successive domains as part of a consensus: -Thr-Trp-X-Leu-X-Lys-Pro-Asp-Leu-Met-Gly-X-Gly-. The Val at *A*6 in *T*5 is out of character but is sterically comparable with the Thr often observed.

AB Turn. In domain *T*6 the strength of the alignment in the *A*, *B* and *C* helices places -Ser-Thr- (followed by Gly) at *AB*. Although a single Ala or Gly *AB* turn predominates in globins, longer *AB* turns are encountered, such as the comparable -Ser-Val-Gly- in *Tylorrhynchus* Hb.

B Helix. Ile replaces Phe at *B*14 in *T*2. It is interesting that *B*10 is invariably Phe in all *Artemia* domains, since in the database 204 Leu and only five Phe are found. Although not included in the template, a Gly at *B*6 is highly conserved in globins. In *T*6, *B*6 is an Ala; this will be discussed when the *E* helix is considered.

Figure 1 (Left). A region of the *Chironomus* HbIII heme environment (solid lines). The upper fragment shows *C*1–*C*5, progressing from left to right; the lower left fragment is the *G* helix receding clockwise from the viewer. Dashed lines: replacement of the *C*4 Ile with a Tyr to simulate *Artemia* domains. The Tyr ring has been rotated to clear the *B*14 Phe (not shown). (Right). Same as on the *left*, except mutated further to replace the *C*2 Pro and *C*3 Ser with a single Gly as is present in domain *T*9.

C *Helix and early* CD *Turn.*   The alignment with other globins of the *Artemia C* and early *CD* sequences in the critical heme environment is unmistakable. Recognizable are a consensus (-His-Pro-Glu-) at C1-3, identical to many Mbs, and the invariant *CD*1 Phe. As found in *T*7, a Phe at C1 is common. The Tyr at C4, not found in any other globin in the database, is conserved completely throughout the *Artemia* domains. Figure 1 shows a reconstruction of the C4 Tyr region based on *Chironomus* globin III, which itself has a bulky and hydrophobic Ile at C4. In common with *Chironomus*, eight of the nine *Artemia* domains have a Phe at B14, and it was possible to adjust the rotation of the substituent C4 Tyr ring to clear B14.

Though a Pro is usually at C2, in this case, T9 alone was an exception as the C helix sequence fell short of the expected length. The alignment of T9 throughout B13–14, C1, C4–5, C7 and CD1 leaves no scope for major adjustment, and the inference is that C2–3, typically -Pro-Glu-, has been replaced by a single Gly. The region constitutes an exposed surface which in the mammalian equivalent would be more involved with allosteric switching than with accommodation of the heme. Despite the limited predictive value of a reconstruction of the mainchain, the previously unknown deletion of C2 or C3 was sufficiently intriguing to justify some exploratory modeling. It was found that in *Chironomus* the amide nitrogen atoms of residues C2 and C4 could be positioned approximately 2.94 Å apart and could well be connected by a single Gly.

CD *Turn and* D *Helix.*   Working back from the almost mandatory E7 His, an unbroken alignment convincingly positions hydrophobic residues coincident with the buried D5 Ile and D2 Leu of *Chironomus*. This confines most length discrepancies to the CD turn. Why is the hydrophobic environment associated with CD4, nearly always Phe, and usually CD7? Both sites are represented in *Artemia* provided T6 has an extra residue inserted in the first half of the CD turn, and that T4 and T8 have two extra residues, accommodated at the exposed external corner. Domain T3 does not have a suitable CD4 candidate but this domain corresponds to the amino acid sequence known as E1 determined directly by Moens *et al.* (7), which indeed showed a Phe at CD4. The cDNA sequence, however, is equally unequivocal in this region and the discrepancy is attributed to polymorphism.

E *Helix.*   The allocation of a His to position E7 in all the *Artemia* domains and the choice between alternative His candidates in T5 and T6 were influenced by the resultant Val at E11 and the identical linear separation of 32 residues between E7 and the F8 His. The hydrophobic requirements at E4, E12 and E15 were consistently met.

This strong alignment does not prove that E7 His in *Artemia* fulfills the distal His role since in *Chironomus* the E11 Ile sidechain takes its place, the E7 His sidechain being rotated towards the start of the helix. In *Artemia*, on the one hand the E7-His-E11 Val motif is conserved and is consistent with a conventional distal His. On the other hand, a Val, invariably found at *E*11 in *Artemia*, is a good substitute for Ile in a distal role, particularly since one of the γ carbons of the Ile in *Chironomus* is closest to the iron. The *Artemia* sequences could support either scenario.

The *E* helix alignment contained the highest density of deviants from Bashford's Template II. Eleven of the discrepancies are hydrophobic residues at normal surface sites, namely *E*5, *E*6, *E*9, *E*10, *E*17 and *E*20. These are not hydrophobic in *Chironomus*, except that *E*6 is polymorphically either Thr or Ile (12). A net increase in hydrophobicity at the *E* helix surface may support interactions between domains and between the multi-domain subunits.

The usual Gly at *B*6 and *E*8 in higher globins permits close proximity of the helix axes at the *B/E* crossing. In contrast, *Chironomus* has Pro-Val at *B*6–*B*7, fitting between Ala at *E*8 and Val at *E*12. Although the absence of a central Gly prevents *Chironomus* from meeting the Richmond and Richards (13) class 1 crossing criteria with mainchains in contact, its dimensions are not very different from Mb. The departure from a Gly/Gly crossing is unlikely to be of any greater consequence in *Artemia* than in *Chironomus*. Eight of the *Artemia* domains in fact have a Gly at *B*6.

EF *Turn and* F *Helix.* Alignment of the *F* helix sequence is dictated by the obligatory *F*8 proximal His. This alignment is strongly supported through the residues flanking the *F*8 His, which are a uniform Gly at *F*5 and the well conserved *F*4 Leu in the heme environment. The strength of alignment in the *E* and *F* helices consistently directs 10 residues into the *EF* turn.

FG *Turn and* G *Helix.* Around the *FG* turn, the *Artemia* sequences match *Chironomus* better than do higher globins. Domains *T*1–*T*5 and *T*9 have an Arg aligned with the Arg in *Chironomus* that hydrogen bonds with the heme propionate like the *FG*2 His of Mb; thus we number it *FG*2 in *Artemia*. Domains *T*6–*T*8 are less obliging, with Leu or Thr at *FG*2. Although *T*6 has an Arg at *F*10, the mainchain would need to be adjusted about 1 Å closer to restore the hydrogen bonds. The problem is greater for *T*7 and *T*8 where hydrogen bond donors appear in *FG*3, which is too distant (12.3 Å from its $C-\alpha$ to the nearer heme oxygen). The scope for reconstruction is limited by the 'severe' size constraint at *FG*4 (6) and it is evident that *T*6, *T*7 and *T*8 follow different pathways through the *FG* turn.

Key features in identification of the *G* helix are the hydrophobic residues at *G*5 and *G*8, having 'severe' and 'medium' size constraints (6). Seventeen of these 18 positions in *Artemia* are filled by Phe. Discrepancies at interior positions *G*12, *G*15 and *G*16 are few, and in seven out of nine domains *G*4 is a His, otherwise rare in the database (4/226).

GH *Turn to End of Domain.* The *Artemia* sequences are variously nine to 14 residues longer than *Chironomus*, between the end of the *G* helix and the start of the following *A* helix, *via* inter-domain linkages of unknown lengths.

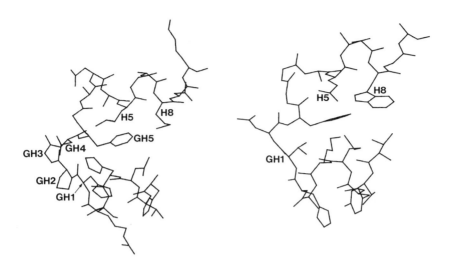

Figure 2. Comparison of the *GH* turn and associated *G* and *H* helices in sperm whale Mb (left) and *Chironomus* HbIII (right). Distances between *GH*1 and *H*5 α carbons (or between *GH*1 and *H*8) are indistinguishable between species and the position of the Phe (*GH*5 in Mb) is similar. Modeling of the *GH* turns of the *Artemia* domains is discussed in the text.

Nevertheless a powerful alignment through this variable region was achieved by working backwards. The alignment was maintained for 30 positions from *A*1 back to *H*6, disturbed only by a single deletion in *T*5 and by a two residue insertion in *T*4.

A consensus Trp appearing at *H*8 was a key to the alignment since *H*8 is one of the few positions (*A*5, *A*12, *C*3, *H*8) where its large sidechain is commonly found. Comparison of Mb and *Chironomus* revealed almost identical spacing between the *GH*1 and *H*8 α carbons (12.9 Å) or between *GH*1 and *H*5 (10.9 Å), although their sequences differ in length. Variations in length thus appear to be accommodated around the *GH* turn and the variants are back in register by *H*5. Figure 2 illustrates how *GH*1 to *H*5 is spanned with a different number of residues in Mb and *Chironomus*; both retain the well conserved Phe (*GH*5 in Mb) buried between the *G* and *H* helices.

In *Artemia*, *T*3 and *T*6–*T*9 are the same length as Mb through this region and a Leu or Phe appears at *GH*5 satisfactorily. *T*1, *T*2, *T*4 and *T*5 have variously four to six residues to fit between *GH*1 and *H*5 and the structure is unpredictable. Tolerable *GH*5 residues are implied in Table 1; *T*1, *T*3, *T*4 and *T*8 have an *H*5 sidechain that alternatively could be rotated to occupy the *GH*5 space.

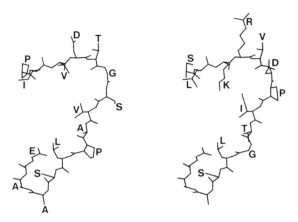

Figure 3. First 15 residues of *Petromyzon* globin V, ending with residue *A*4, viewed from the surface side (left); the same structure mutated to the sequence linking *Artemia* domains *T*2 and *T*3, aligned as in Table 1 (right).

Inter-Domain Linker. The excess of about 12 residues between each *H* helix and the following *A* helix implies that the linker is not an adaptation of the *H* or the *A* helix. We have given the designation *HA* to linker sequence, starting with *HA*1 at the equivalent of *H*22. Interpretation of the *HA* sequence is a different problem from the rest of the molecule since there is no known structural precedent. The *HA* sequence has, however, over-representation (compared with proteins in general (14)) of Thr, Ser, Asp, and especially Gly and Gln, all residues associated with natural protein linkers (15).

The *HA* block, being among the most conserved within the *Artemia* globin alignment, indicates that the linkers are of uniform structure. This in turn is compatible with a regular arrangement of domains such as the cylindrical morphology inferred from electron micrographs of the isolated molecule (16). A potential model for the linker may exist in globin V of *Petromyzon*, for which the sequence (17) and the crystal structure (11) are known. Preceding its *A* helix *Petromyzon* has a sequence of 11 residues (Table 1) of remarkably similar character to those found in the *Artemia* linker sequences. There are steric or character similarities at *HA*7, *HA*9, *HA*10, *HA*13 and *HA*16, with turn-provoking residues at *HA*12 and *HA*15 that are, to within a residue, a transposition of the Gly-X-X-X-Pro sequence of *Petromyzon*. Figure 3 shows a pair of models in which the *Petromyzon* structure has been mutated in accordance with the *Artemia* sequence. If the *Artemia* sequence does conform to a similar template, the inference would be that the major part of the *HA* sequence is applied to the surface of the molecule with the inter-domain bridge confined to the proximal part.

Acknowledgements

We thank the Trustees of the E.G. Johnstone Trust, New Zealand, for conference support and the Trustees of the J.D.S. Roberts Trust (New Zealand) for computing equipment. Grants from the Medical Research Council (New Zealand) and the University Grants Committee are gratefully acknowledged. We thank the numerous people we have consulted over computing problems including Drs. Peter Stockwell, Donald Bashford, Geoff Wyvill, Brendan Murray and Andrew Trotman.

References

1. Moens, L. and Kondo, M. (1978) *Eur. J. Biochem* **82**: 65–72.
2. Waring, G., Poon, M.C. and Bowen, S.T. (1970) *Int. J. Biochem.* **1**: 537–543.
3. Manning, A.M., Ting, G.S., Mansfield, B.C., Trotman, C.N.A. and Tate, W.P. (1986) *Biochem. Int.* **12**: 715–724.
4. Bowen, S.T., Moise, H., Waring, G. and Poon, M.C. (1976) *Comp. Biochem. Physiol.* **55**B: 99–103.
5. Manning, A.M., Trotman, C.N.A. and Tate, W.P. (1990) *Nature* **348**: 653–656.
6. Bashford, D., Chothia, C. and Lesk, A.M. (1987) *J. Mol. Biol.* **196**: 199–216.
7. Moens, L., Van Hauwaert, M.-L., De Smet, K., Geelen, D., Verpooten, G., van Beeumen, J., Wodak, S., Alard, P. and Trotman, C.N.A. (1988) *J. Biol. Chem.* **263**: 4679–4685.
8. Moens, L., Van Hauwaert, M.-L., De Smet, K., Ver Donck, K., Van de Peer, Y., van Beeumen, J., Wodak, S., Alard, P. and Trotman, C.N.A. (1990) *J. Biol. Chem.* **265**: 14285–14295.
9. Steigemann, W. and Weber, E. (1979) *J. Mol. Biol.* **127**: 309–338.
10. Philips, S.E.V. (1980) *J. Mol. Biol.* **142**: 531–554.
11. Honzatko, R.B., Hendrickson, W.A. and Love, W.E. (1985) *J. Mol. Biol.* **184**: 147–164.
12. Osmulski, P.A. and Leyko, W. (1986) *Comp. Biochem. Physiol.* **85**B: 701–722.
13. Richmond, T.J. and Richards, F.M. (1978) *J. Mol. Biol.* **119**: 537–555.
14. McCaldon, P. and Argos, P. (1988) *Proteins* **4**: 99–122.
15. Argos, P. (1990) *J. Mol. Biol.* **211**: 943–958.

16. Wood, E., Barker, C., Moens, L., Jacob, W., Heip, J. and Kondo, M. (1981) *Biochem. J.* **193**: 353–360.
17. Hombrados, I., Rodewald, K., Neuzil, E. and Braunitzer, G. (1983) *Biochimie* **65**: 247–257.

28
Structure of the Extracellular Hemoglobin of *Tylorrhynchus heterochaetus*

Toshio Gotoh,[a] Tomohiko Suzuki[b] and Takashi Takagi[c]

[a]Department of Biology, College of General Education, University of Tokushima, Tokushima 770, Japan
[b]Department of Biology, Faculty of Science, Kochi University, Kochi 780, Japan
[c]Biological Institute, Faculty of Science, Tohoku University, Sendai 980, Japan

Introduction

We have attempted to construct a model of the molecular architecture of the giant Hb of the polychaete *Tylorrhynchus heterochaetus* mainly by determining the amino acid sequence of each constituent chain (1). The giant *Tylorrhynchus* Hb consists of about 200 polypeptide chains of several distinct subunits and has an M_w of 3–4 x 10^3 kDa (2). Our strategy was primarily to isolate each constituent chain, determine its precise M_w from its sequence, and calculate the molar ratios of the constituent polypeptide chains (3). To construct the model of the multisubunit Hb, we have to consider the heme content, electron microscopic image and M_w of the whole molecule as well as the symmetry of the distribution of its constituents (3, 4). *Tylorrhynchus* Hb, like other annelid Hbs (5), consists of double-layered hexagonal submultiples (4). Scanning transmission electron microscopy (STEM) reveals that there is a filamentous structure in the central cavity of the molecule that appears to be anchored to the surfaces of all 12 submultiples (4). This Hb contains four major and some minor components. The major components exist as a "monomer" and a disulfide-bonded "trimer" in the molecule. We have already reported the sequences of the four major components (2, 4, 6–8). The sequences of four polypeptide chains of Hb of the oligochaete *Lumbricus terrestris* have also been reported recently (9, 10).

In the present paper, we first briefly review our sequencing studies on the major components of *Tylorrhynchus* Hb in comparison with those of the corresponding chains of *Lumbricus* Hb (9, 10), and propose a new nomenclature for the polypeptide chains common to the polychaete and oligochaete Hbs. Then, we report preliminary information on the sequences of two minor components and propose a new model for the molecular assembly of the polychaete Hb. The findings reported are not consistent with the "192-chain" model (3), but instead support the "bracelet" model proposed by Vinogradov *et al.* (11).

Materials and Methods

The brackish-water worm, *Tylorrhynchus heterochaetus* (the so-called "Japanese Palolo"), was collected during the time of reproductive swarming from the basin of the River Yoshino, Tokushima, Shikoku Island.   When swarming occurred vigorously, more than 10 kg of the mature worms could easily be scooped out of channels of the river with a net within one hour.

The monomeric chain of *Tylorrhynchus* Hb was isolated by gel filtration on Ultrogel AcA-44 (LKB) in the presence of 1 mM EDTA at alkaline pH (6). The four major components, a monomeric chain and trimeric chains could be isolated simultaneously in the presence of a reducing agent by chromatofocusing on a PBE94 (Pharmacia) column (7). Two other minor components were separated by the reverse-phase column chromatography on Cosmosil $5C_{18}$-300 (Nacalai Tesque). The amino acid sequences of proteins and peptides were determined with an automated sequencer (Applied BioSystems Model 477A).

Results and Discussion

Nomenclature of Constituent Chains. Since we intend to eventually construct a common model for the molecular architecture of the giant annelid Hbs, we need to establish a standard nomenclature of the constituent chains common to all Hbs. Since annelid Hbs are apparently quite similar in size and shape (1, 5, 12), the multisubunit proteins in different species appear to be composed of homologous constituents. SDS-PAGE is the most convenient method for identifying the constituent chains and estimating their molecular sizes, but annelid Hbs exhibit a variety of patterns on SDS-PAGE, and so it is usually hard to identify corresponding chains from the SDS-PAGE patterns of

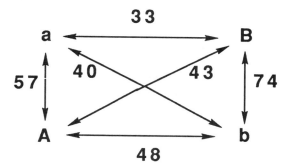

Figure 1. Sequence homologies (number of identical residues) between the heme-containing chains *a, A, b* and *B* of *Tylorrhynchus* Hb.

oligochaete and polychaete Hbs. For instance, the trimeric chains of *Tylorrhynchus* Hb could not be separated by SDS-PAGE in the presence of a reducing agent (7), whereas the trimeric chains of *Lumbricus* Hb could be separated into three bands in this way (13).

When we finally identified four distinct chains of *Tylorrhynchus* Hb in terms of amino acid sequence, we found that phylogenetically there are two groups of constituent chains (2). Considerable homology is observed between chains *a* and *A* (40%), and also between chains *b* and *B* (50%), as illustrated in Figure 1. This finding has been extended further to the idea (14) that there are, in general, two distinct groups, strains *A* and *B*, of heme-containing chains in the Hbs of both the polychaete *Tylorrhynchus* and the oligochaete *Lumbricus*. Furthermore, Fushitani *et al.* have shown two very clear subfamilies in the phylogenetic tree of *Lumbricus* and *Tylorrhynchus* Hbs (10). Thus, it is now possible to propose the common nomenclature shown in Table 1 for sequenced chains in order to facilitate the comparison between "monomeric" and "trimeric" chains of different species. The phylogenetic nomenclature is based on the findings of two strains (or subfamilies) and characteristic numbers of half-cystine residues (2, 4, 9, 10, 14). The smaller, monomeric chain with two half-cystine residues is called chain *a*. Chain *A* has three half-cystine residues. Strain *A* has both chains *a* and *A*. Similarly, chains *b* and *B* in strain *B* may be defined as those having four and three half-cystine residues, respectively. All the chains have one intrachain-disulfide bond and no free *SH* group. Therefore, chain *a* cannot form an interchain-disulfide bond and exists as a monomer. Chain *b* was experimentally shown to be situated in the center of the trimer *AbB* (4, 10). The distributions of half-cystine residues in *Tylorrhynchus* and *Lumbricus* Hbs are almost the same (2, 10). As the polychaete *Tylorrhynchus* and oligochaete

Lumbricus are very far apart in the phylum Annelida, the proposed nomenclature may be extended to many other annelid Hbs. A slight modification of the proposed nomenclature will, however, be necessary because some annelid Hbs contain disulfide-bonded tetramers or dimers of heme-containing chains instead of trimers (1). We will refer to the nomenclature of other minor components later.

"192-Chain" Model or "Bracelet" Model ?   In a previous paper we proposed a "192-chain" model for the formation of *Tylorrhynchus* Hb (3) that may be represented as $(a \cdot AbB)_{48}$ according to the new nomenclature proposed above. The tetramer, $a \cdot AbB$, is the minimum structural entity, which may correspond to one unit in the tetrahedral structure of the submultiple observed by STEM (4). Thus, the submultiple consists of 16 chains. According to the "192 chain" model, this protein consists of 27,935 amino acid residues and 192 heme groups, and is linked by 288 disulfide bonds. The M_w is 3,275,808 and the minimum M_w per heme is 17,062. Although the "192-chain" model is in good agreement with electron microscopic observations of the tetrahedral structure of a submultiple (4), it does not explain the filamentous structure in the central cavity of the molecule. There are three other defects in this model. First, the measured value of the minimum M_w per heme is much higher than the value calculated from the model (15). Each chain in this model contains one heme. On the other hand, Vinogradov and his colleagues consider that not all chains contain a heme group (11, 13, 16). Second, the SDS-PAGE pattern of *Tylorrhynchus* Hb varies and often shows one or two extra minor components with M_ms approximately twice those of chains containing heme (15). Third, there are some indications that the M_ms of annelid giant Hbs are clustered in the second half of 3,000 kDa (17).

Recently, Vinogradov and his colleagues proposed a novel "bracelet" model for the molecular assembly of *Lumbricus* Hb (11). This model overcomes all the defects in the "192-chain" model mentioned above. They noted fifth and sixth components that can always be observed as faint bands V and VI on the SDS-PAGE of *Lumbricus* Hb preparations (13). In the "bracelet" model these non-heme chains have the remarkable role of linking 12 submultiples composed of heme-binding subunits *a* and *AbB*. We would name these components "linkers" because of this possible role. Although there is no direct evidence that the linkers are components of the giant Hbs, these chains may correspond to the filamentous structure in the central cavity of the molecules (11, 18, 19). However, it still seems uncertain whether the linkers are actually common components of all annelid Hbs (1), because the only linker chain that has been isolated and well characterized is from the Hb of a deep-sea tube worm *Lamellibrachia sp.* (20) which is assigned to the phylum Vestimentifera (21). The discrepancy between the "192-chain" model and "bracelet" model may be explained in large part by sequencing the linker chains from at least two species of Hbs. Accordingly, we have attempted to isolate and sequence the postulated "linkers" of *Tylorrhynchus* Hb.

Sequences and Electrophoretic Mobilities of Tylorrhynchus *Hb "Linkers."* Figure 2 shows the amino acid sequences of the two linker chains of *Tylorrhynchus* Hb; *T*1 and *T*2 consist of 253 and 236 amino acid residues and have M_ws of 28,200 and 26,316, respectively. These chains were characterized by their high contents of half-cystine residues. They show 23–27%

```
        1                  10                  20                      30
LV                                                      A A V Q P L S V S D
T1   G A L G D R N G D C A C D R P S P R G Y W G G G M T G R S A F A D P H
T2             D D C V C                                 P G G R E W A

            40                  50                  60                  70
LV   A M G A R V D A Q A W R V D R L T K Q F Q A I S D A A D T S I G A A K
T1   I A E G R L A N Q D A R L G T L E E E V D K L Q H K Y D D F I A G K T
T2   S V A S K A D S Q E A R V N R L A G R V E A L A E N L R A G G D R L S
                     *           *               *

            80                  90                  100
LV   S G G D I A R H M L N S     H L D D H W C P S K Y H R C G N S P   Q C
T1   A R R E R F A Q L K D R V W G L E A H H C D D D H L S C K D V A F T C
T2   H Y K S E F R E L E F R V D E L E G N G C E P R H F Q C G G S A M E C
                                   *               *           *       *

            110                 120                 130           #   140
LV   M S N M A F C D G V N D C K N H F D E D E N R C V V P V T A N S T W I
T1   I G H N L V C D G H K D C L N G H D E D E E T C S I A A S V G S S F E
T2   I S D L L T C D G S P D C A N G A D E D E D V C H I P P I A G T L L V
             * * *       * *   *       * * * *       *

            150                 160                 170
LV   G Y P     A Y D H C T Q R R P Y E M V I S I T S A P S D I V Y K V H Q
T1   G Q I R   Q L D T C T K R K P S A F R F I I T N V D V P K Y F P Q E P
T2   G H L N T D H D F C T K R K P N E F D L F I S S V Q R S S Y F Q S R L
     *               *   * *   *   *               *

            180                 190                 200                 210
LV   P L K V Q V D L F S K K G G L K Q S A S L H G D A V Y C K G S Q R L I
T1   H V K A T I L M T S S K D G H E T Q S S L A V D G V Y D F T H R K V I
T2   K V K G N L Q I K Y T A E G R D Q E D V L Q V K G Y Y N F G T H Q L V
         *                       *               *           *

            220                 230                 240
LV   V A P P E D D R L E I I G Q F D G V S N D R F K G Y I V R E M S G D K
T1   L Y S P D K D N L I F E C T F P R H D N N H C K G   V M K H S G G D V
T2   I L P P E D D R L G I V C N F R A G N D D R C R A H I V H E A S L E H
             *   * *   *                   *

            250
LV   C A E F R F F K Q
T1   C L T F T L E R I D
T2   C G D D F V F V K E D D H H
     *
```

Figure 2. Alignment of the amino acid sequences of *Tylorrhynchus* linker chains *L*1 (*T*1), *L*2 (*T*2) and *Lamellibrachia* linker chain *A*V (*L*V) (20). The residues conserved in the three chains are marked with asterisks.

Table 1. Proposed common nomenclature for the chains of multisubunit annelid Hbs, and the corresponding names for *Tylorrhynchus* and *Lumbricus* Hbs used hitherto.

Hb	Chain name				Reference
Tylorrhynchus	I	IIA	IIC	IIB	Suzuki *et al.* (7)
Lumbricus	I	II	III	IV	Shlom & Vinogradov (13)
Lumbricus	*d*	*b*	*a*	*c*	Fushitani *et al.* (10)
Common name	*a*	*A*	*b*	*B*	This paper
Strain	A	A	B	B	Gotoh *et al.* (14)
No. of Cys	2	3	4	3	Suzuki *et al.* (2)
					Shishikura *et al.* (9)
					Fushitani *et al.* (10)

homology with the linker of tube-worm Hb (20, 22): eight of 13 half-cystine residues are identical in the three linker chains. Since the existence of a linker chain in the tubeworm Hb has already been well defined (20, 23), we have no doubt that the sequenced "linkers" are actual components of *Tylorrhynchus* Hb. Thus, their identification disproves the "192-chain" model (3) and supports the "bracelet" model proposed by Vinogradov *et al.* (11). The *Tylorrhynchus* linker chains are significantly homologous with other heme-containing chains and should represent a distinct branch in the molecular evolution of globin chains. Details of the phylogenetic and biochemical properties of *Tylorrhynchus* linker chains will be reported elsewhere (22). Here we will limit our discussion to the subunit assembly.

Figure 3 is a schematic representation of the SDS-PAGE patterns of the Hbs of *Tylorrhynchus* and oligochaetes (13, 24–26). In the presence of SDS, *Tylorrhynchus* Hb dissociates into four subunits: subunits $L1$ and $L2$ (55 kDa), subunit T (trimer AbB, 41 kDa) and subunit M (monomer a, 14 kDa). In the absence of a reducing agent, *Tylorrhynchus* subunits $L1$ and $L2$ are not clearly separated and move more slowly than does the trimer on SDS-PAGE, in contrast to those of oligochaete linkers. Upon reduction, *Tylorrhynchus* linkers dissociate into chains with apparent M_ms of 34–36 kDa. Thus, subunits $L1$ and $L2$ appear to be disulfide-bonded homodimers.

Subunit Assembly of Tylorrhynchus *Hb.* *Tylorrhynchus* linkers 1 and 2 were isolated by reverse-phase chromatography. The ratios of the peak areas of $a \cdot AbB : L1 : L2$ were estimated to be 81 : 6 : 13. A new model for the molecular assembly of *Tylorrhynchus* Hb is proposed, based on the precise M_ms of subunits a, AbB, $L1$ and $L2$, and their molar ratios. In the model, the unit of one twelfth of *Tylorrhynchus* Hb consists of four tetramers of $a \cdot AbB$ and two linker chains, having M_ms of 273.0 kDa (82%) and 53.8 kDa (18%), respectively. Consequently, the whole molecule is composed of 192-heme containing chains (12 submultiples) and 24 non-heme linker chains, giving a calculated M_m of 3,921.6 kDa. Therefore, the minimum M_m per mole of heme is calculated to be 20,425 g, in good agreement with the value of 20,680 g obtained experimentally (22). Although we have discarded the "192-chain" model and support the "bracelet" model (11), we still consider that each submultiple is composed of 16 heme-containing chains. In the novel bracelet model (11), linker chains are proposed to form a closed circular collar or bracelet decorated with 12 complexes of three copies each of monomer *a* and trimer *AbB*, giving the appearance by electron microscopy of a symmetrical hexagonal bilayer. An alternative explanation is that the linker chains do not form a continuous bracelet structure, but may act as linkers between submultiples (11).

Table 2 summarizes the recently proposed models of multisubunit annelid Hbs that include linker or non-heme chains. These models depend greatly upon the estimated value of the heme content, which varied significantly in different laboratories (1). There is no direct evidence for the absence of heme groups in linker chains, but chains *a, A, b* and *B* have been shown to contain a heme group in the Hbs of *Tylorrhynchus* (7) and *Lumbricus* (10, 14). Our model is in fairly good agreement with the one proposed for *Lumbricus* Hb (27) in which it was reported that linker chains constitute 11.2% of the total. On the other hand, Vinogradov *et al.* (11) reported that the relative proportions of linker chains constitute 35% of the total protein in *Lumbricus* Hb, and Kapp *et al.* (19) reported that they constitute 29% in the leech *Macrobdella decora* Hb. In *Tylorrhynchus* Hb, the two linker chains were estimated to constitute 19% of the total protein. These values reflect differences in the numbers of not only linker

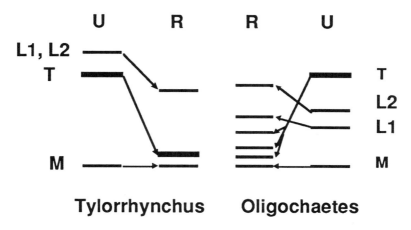

Figure 3. Schematic representation of SDS-PAGE of *Tylorrhynchus* Hb in comparison with those of some oligochaete Hbs (13, 24–26) under unreduced (*U*) and reduced (*R*) conditions.

Table 2. Models of annelid Hbs with linker or non-heme chains.

Hb	No. of chains	No. of hemes	M_r (kDa)	Reference
Lumbricus (28)	204	144	3,770	Vinogradov *et al.*
Lumbricus (11)	ca. 200	156	3,800	Vinogradov *et al.*
Lumbricus (27)	204	192	3,770	Fushitani & Riggs
Macrobdella	180	144	3,528	Kapp *et al.* (19)
Tylorrhynchus	216	192	3,922	This paper

chains but also of heme-containing chains in these models. In the models listed in Table 2, the numbers of linker chains are thought to be 12 in *Lumbricus* Hb (27), 24 in *Tylorrhynchus* Hb and 36 in *Macrobdella* Hb (19). In estimating the number of linker chains, the accuracy of the value for the M_m of each chain must also be considered. In fact, the M_ms of *Tylorrhynchus* linkers one and two, respectively, were determined to be 28.2 Da and 26.3 kDa from their sequences, and 34–36 kDa from their mobilities in reduced conditions on SDS-PAGE. By SDS-PAGE, the M_ms of linker chains were estimated to be 31 kDa and 37 kDa for *Lumbricus* Hb (11, 27) and 30 kDa for *Macrobdella* Hb (19). These values are probably over-estimates.

It should also be noted that *Lumbricus* (27) and *Tylorrhynchus* Hbs were both concluded to contain 192 heme-binding chains, whereas *Macrobdella* Hb was reported to contain 144 (19). This discrepancy appears to reflect a difference in the subunit assemblies of submultiples; that is, in the tetrahedral arrangement of the four minimum entities, *a·AbB*, in *Lumbricus* (27) and *Tylorrhynchus* Hbs, and in the triad arrangement of tetramers in *Macrobdella* Hb (19). Using two-dimensional reconstructions of electron microscopic images of *Ophelia bicornis* Hb, it has been determined that the submultiple contains three globular units in both dorsal and lateral projections (29). Thus it was concluded that the simplest way of representing this submultiple is as a tetrahedron—which in both projections shows only three of its components, the fourth being masked by the others. This was also confirmed in the case of *Tylorrhynchus* Hb (4). No disulfide bonding was found between the basic building blocks, *a·AbB*, or between these blocks and the linker chains in the Hbs of *Tylorrhynchus*, *Lumbricus* (15) and *Macrobdella* (19). In these cases, the tetrahedral structure of the submultiple appears to be more stable than does the triad structure (30).

Svedberg also wondered about the discrepancy in reports of 144 to 192 chains when he realized that annelid Hbs have a high M_w (12, 31). Although there is as yet no consensus on details, there have been many fruitful studies on the stimulating and challenging problem of constructing a model of the molecular architecture common to annelid multisubunit Hbs. In particular, the proposal of the "bracelet" model (11) has stimulated much interest in the biology and chemistry of invertebrate Hbs.

References

1. Gotoh, T. and Suzuki, T. (1990) *Zool. Sci.* **7**: 1–16.
2. Suzuki, T. and Gotoh, T. (1986) *J. Biol. Chem.* **264**: 9257–9267.
3. Suzuki, T. and Gotoh, T. (1986) *J. Mol. Biol.* **190**: 119–123.
4. Suzuki, T., Kapp, O.H. and Gotoh, T. (1988) *J. Biol. Chem.* **263**: 18524–18529.
5. Roche, J. (1965) In *Studies in Comparative Biochemistry*, ed. K.D. Munday, 62–80. Oxford: Pergamon Press.
6. Suzuki, T., Takagi, T. and Gotoh, T. (1982) *Biochim. Biophys. Acta* **708**: 253–258.
7. Suzuki, T., Furukohri, T. and Gotoh, T. (1985) *J. Biol. Chem.* **260**: 3145–3154.
8. Suzuki, T., Yasunaga, H., Furukohri, T., Nakamura, K. and Gotoh, T. (1985) *J. Biol. Chem.* **260**: 11481–11487.
9. Shishikura, F., Snow, J.W., Gotoh, T., Vinogradov, S.N. and Walz, D.A. (1987) *J. Biol. Chem.* **262**: 3123–3131.
10. Fushitani, K., Matsuura, M.S.A. and Riggs, A.F. (1988) *J. Biol. Chem.* **263**: 6502–6517.
11. Vinogradov, S.N., Lugo, S.D., Mainwaring, M.G., Kapp, O.H. and Crewe, A.V. (1986) *Proc. Natl. Acad. Sci. U.S.A.* **83**: 8034–8038.
12. Svedberg, T. (1937) *Nature* **139**: 1051–1062.
13. Shlom, J.M. and Vinogradov, S.N. (1973) *J. Biol. Chem.* **248**: 7904–7912.
14. Gotoh, T., Shishikura, F., Snow, J.W., Ereifej, K.I., Vinogradov, S.N. and Walz, D.A. (1987) *Biochem. J.* **241**: 441–445.
15. Gotoh, T. and Kamada, S. (1980) *J. Biochem.* **87**: 557–562.
16. Vinogradov, S.N. (1985) *Comp. Biochem. Physiol.* **82**B: 1–15.
17. Vinogradov, S.N. and Kolodziej, P. (1988) *Comp. Biochem. Physiol.* **91**B: 577–579.
18. Kapp, O.H., Mainwaring, M.G., Vinogradov, S.N. and Crewe, A.V. (1987) *Proc. Natl. Acad. Sci. U.S.A.* **84**: 7532–7536.
19. Kapp, O.H., Qabar, A.N., Bonner, M.C., Stern, M.S., Walz, D.A., Schmuck, M., Pilz, I., Wall, J.S. and Vinogradov, S.N. (1990) *J. Mol. Biol.* **213**: 141–158.
20. Suzuki, T., Takagi, T. and Ohta, S. (1990) *J. Biol. Chem.* **265**: 1551–1555.
21. Jones, M.L. (1985) *Biol. Soc. Wash. Bull.* **6**: 117–158.
22. Suzuki. T., Takagi, T. and Gotoh, T. (1990) *J. Biol. Chem.* **265**: 12168–12177.
23. Suzuki, T., Takagi, T. and Ohta, S. (1988) *Biochem. J.* **255**: 541–545.
24. Ochiai, T. and Enoki, Y. (1981) *Comp. Biochem. Physiol.* **68**B: 275–279.
25. Chiancone, E., Vecchini, P. and Verzili, D. (1984) *J. Mol. Biol.* **172**: 545–55.
26. Suzuki, T. (1989) *Eur. J. Biochem.* **185**: 127–134.

27. Fushitani, K. and Riggs, A.F.   (1988)   *Proc. Natl. Acad. Sci. U.S.A.* **85**: 9461–9463.
28. Vinogradov, S.N., Shlom, J.M., Hall. B.C., Kapp, O.H. and Mizukami, H. (1977) *Biochim. Biophys. Acta* **492**: 136–155.
29. Ghiretti-Magaldi, A., Ghiretti, F., Tognon, G. and Zanotti, G.   (1986)   In *Invertebrate Oxygen Carriers*, ed. B. Linzen, 45–55. Berlin: Springer-Verlag.
30. Klotz, I.M.   (1967)   *Science* **155**: 697–698.
31. Svedberg, T.   (1933)   *J. Biol. Chem.* **103**: 311–325.

29
cDNA Cloning, Sequencing and Expressing the cDNA for *Glycera dibranchiata* Monomer Hemoglobin Component IV

Peter C. Simons[a] and James D. Satterlee[b]

[a]Department of Biochemistry, University of New Mexico, Albuquerque, NM 87131, USA
[b]Department of Chemistry, Washington State University, Pullman, WA 99164, USA

Introduction

Glycera dibranchiata is a polychaetous marine annelid commonly referred to as the "blood worm." This descriptive name reflects the fact that it contains Hb in nucleated erythrocytes. Vinogradov *et al.* (1) first demonstrated that the Hb content of these cells was separable by Sephadex chromatography into polymeric and monomeric fractions. We subsequently demonstrated that the application of high-resolution bioanalytical techniques permitted further resolution of the monomer Hb fraction into three major monomer Hbs (2–4)

These monomer Hbs are of interest to those studying heme protein structure-function relationships because original x-ray crystallography (5) and primary sequencing (6) revealed that at least one of the components of the monomer fraction displayed an exceptional amino acid substitution: $E7$ His to Leu. His $E7$, the so-called distal His, is an amino acid that lies in closest proximity to the heme ligand-binding site in the vast majority of normal vertebrate Hbs and Mbs. In this position, it is uniquely capable of influencing ligand binding. Whereas other species, such as elephant and opossum, also possess fully functional Mbs and Hbs in which the distal His is replaced, one unique characteristic of the *G. dibranchiata* monomer Hbs is that His *E*7 is replaced by an amino acid (Leu) with a completely non-polar side chain.

The expected implication that this substitution would have on ligand binding dynamics seems to have been widely recognized. However, comparing ligand binding studies from different laboratories is difficult due to the diverse isolation and purification methods employed. The sequencing and molecular biology work described herein contribute to the idea that the monomer Hb fraction actually contains three distinct major components. This conclusion has been independently corroborated (7).

Our goal in this work is two-fold. First, we intend to use the modern techniques of molecular biology to prove whether or not the three major monomer Hb components are truly individual proteins. Second, we wish to express the genes for one or more of the monomer Hbs and for specific site-directed mutants that will be used in subsequent biophysical studies. This article describes some of the background rationale for our effort and the progress we have made to date.

Materials and Methods

Monomer Hb component IV was the first protein examined. The details of creating the cDNA library, plasmid purification, cloning and cDNA sequencing have been previously published (8). Briefly, amino terminal sequencing of the three isolated wild-type monomer Hb components allowed the identification of a short segment of homology common to the three components (*WKDIA*), for which a 12-fold degenerate oligonucleotide probe was created. RNA was isolated from freshly prepared erythrocytes. The mRNA (poly (A+) RNA) was used to prepare double-stranded cDNA as described in the BRL cDNA cloning manual. This mRNA was dC-tailed and mixed with an optimal amount of PstI-cut, dG-tailed *pBR*322 (*BRL*). This was subsequently used to transform *Escherichia coli DH5*, which were plated onto *L*+ tetracycline plates and grown overnight. The cells were rinsed off of the plates with *L* broth and the resulting library was stored in 20% glycerol at $-70°C$. Colonies were screened using the labeled oligonucleotide probe. Positive colonies were grown, plasmid mini-preparations were made, the DNA was digested and the longest inserts were ligated to the PstI cut site of *pIB*I76. The ligation mixtures were used to transform *E. coli NM*522, and, ultimately, the cDNA was sequenced using Sanger's dideoxy chain termination method. The first complete cDNA sequence corresponded to *G. dibranchiata* monomer globin component IV (*GMG*4).

Expression has been carried out in *E. coli* strain *BL*21(*DE*3) using pET3d, an expression system developed and generously given to us by Dr. F. William Studier. The *GMG*4 gene was cloned into the NcoI to BamHI region in plasmid pET3d, as shown in Figure 1. Cells were grown to $A_{600} = 0.7$ and induced to

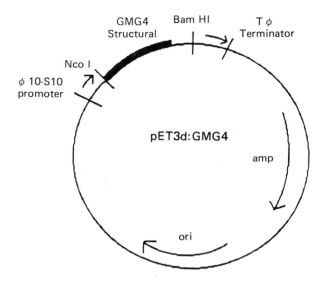

Figure 1. Plasmid construct used for expression in these studies.

produce *GMG*4 by the addition of isopropyl thio-β-*D*-galactoside (IPTG) to 0.4 mM. After 2h the cells were harvested by centrifugation, resuspended in 1/10 volume of 50 mM Tris, pH 7.6, 1 mM EDTA, and frozen. The thawed suspensions were lysed by sonication and spun for five minutes in a microcentrifuge. The supernatants were removed and mixed with an equal volume of 2X SDS loading buffer. For SDS-PAGE, the pellets were dissolved in an identical volume of this same buffer. For holoprotein formation, the pellet was dissolved in 1/5 volume of the Tris buffer (described above) containing 8M urea.

Holoprotein was prepared by carefully adding protohemin IX (equine, Sigma Chem. Co., St. Louis, MO, USA) that had been dissolved in a minimal amount of 0.1 M NaOH, then diluted with 0.2 M potassium phosphate, pH = 8, to the stirred, expressed apoglobin solution.

Results and Discussion

Uniqueness of the Monomer Component Hbs. The question of the extent of protein heterogeneity (and its source) in the monomer Hb fraction has persistently plagued our work. Reviewers were so skeptical of our results that it took two years for the initial results (3) to be published. However, several comparative studies from this laboratory (4, 8–12) and another (7, 13, 14)

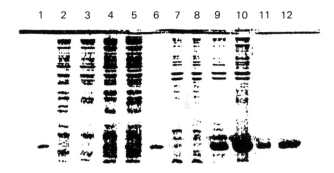

Figure 2. SDS-PAGE results of an assay of the expression system shown in Figure 1. Lanes 11 and 12 confirm the production of the *GMG*4 apo-protein.

provide compelling evidence in support of the fact that the monomer Hb fraction contains three distinct—but related—fully functional components, which we label as components II, III and IV. The amino terminal sequencing of the individual wild-type monomer components' globins provides additional support for this idea. These partial sequences are as follows (starting at the amino terminal *G* labeled as 1):

Comp II: GLSAAQRQV I AATWKDIAGN DNGAGV - - D
Comp III: GLSAAQRQVV ASTWKDIAGA DNGAGVGKEC LIKFIHA⁻P
Comp IV: GLSAAQRQVV ASTWKDIAGS DNGAGV.

The differences in these three primary sequences at positions 10, 12, 20 and 29 further emphasize that these are distinct proteins and not merely degradation products of a common precursor globin. Although limited in scope, it is instructive to compare these partial sequences with a previously published partial sequence (15) and the published complete amino acid sequence (6). The partial and complete amino acid sequences (6, 15) were performed on protein crystallized from the unseparated monomer fraction without further purification. Two different primary sequences were obtained. The partial sequence (15) apparently corresponds to our component III globin, as indicated by sequence correspondences between the two at positions 10 (Val) 20 (Ala), 29 (Glu), 32 (Ile), 34 (Phe), 35 (Ile) and 39 (Pro) (8). The complete sequence published by

Imamura *et al.* (6), evidently corresponds to our component II, as indicated by sequence correspondences at positions 10 (Ile), 12 (Ala), 20 (Asn) and 29 (Asp).

These results clearly indicate that sequences of three different monomer globins have appeared in the literature. The two protein sequences, partial and complete, apparently originated from identical preparations from the same laboratory. The fact that these two sequences correlate with the amino terminal sequences of two of our components also provides strong evidence for the existence of three different monomer Hbs, as do the physical studies described elsewhere (2, 3, 7, 9–14).

GMGA *Expression.* Using Studier's pET3d plasmid expression system we have successfully expressed the gene for *G. dibranchiata* globin IV in the *E. coli* strain *BL*21 (*DE*3). We also included Studier's plasmid *p*Lys*S*, yielding *T*7 lysozyme which helps cell lysis during the procedure implemented to purify the expressed globin. It has the added value of inhibiting *T*7 polymerase which may be produced in the absence of IPTG stimulation. The IPTG stimulates the promoter upstream from the *GMG*4 insert (Figure 1). Monomer globin IV is produced in inclusion bodies which are subsequently solubilized, reconstituted with heme and purified. UV-visible spectra are identical to the wild type component IV monomer metHb (4, 10, 11).

Figure 2 shows an SDS-PAGE gel, stained with Coomassie blue, and used to assay the expression system. Lanes 1, 6, 11 and 12 correspond to the wild-type component IV globin, loaded at successively high concentrations. Lanes 2–5 report the contents of the supernatant following cell lysis and pelleting. In this case, lanes 2 and 3 were obtained from cells not induced by IPTG, whereas lanes 4 and 5 were from cells induced for two hours. Lanes 7–10 were obtained from the resolubilized pellet that occurred upon cell lysis. Again, only lanes 9 and 10 correspond to cells induced with IPTG for two hours.

Lanes 9 and 10 clearly reveal a substantial amount of protein at the position corresponding to wild-type globin IV, indicating its inducible production in this system. Combined with the optical spectroscopy described above we conclude that we have successfully expressed the wild-type *GMG*4 gene.

The results of this line of research indicate that the nucleated erythrocytes of *G. dibranchiata* are transcriptionally active. Furthermore, our success in isolating, sequencing, cloning and expressing the *GMG*4 gene places us in a position to begin physical studies of structure-function relationships. This effort is already underway, initiated by the creation of two additional pET3d plasmids in which the *E*7 position has been mutated first to His, and then to Asn. The first of these mutants is designed to mimic normal vertebrate Hbs and Mbs. The second is the beginning of a systematic study designed to elucidate factors regulating ligand binding, heme association and heme isomerization within the holoprotein. Asn presents a side chain of virtually identical size and geometry as Leu, but one of increased polarity. This mutation will test the influence of polarity changes at the heme ligand binding location while maintaining constant steric effects.

Acknowledgements

This work was supported by a research grant from the National Institutes of Health (DK30912) and by a National Institutes of Health Research Career Development Award to J.D.S. (HL01758).

References

1. Vinogradov, S.N., Machlik, C.A. and Chao, L.L.   (1970)   *J. Biol. Chem.* **245**: 6533–6538.
2. Kandler, R.L. and Satterlee, J.D.   (1983)   *Comp. Biochem. Physiol.* **75B**: 499–503.
3. Kandler, R.L., Constantinidis, I.C. and Satterlee, J.D.   (1985)   *Biochem. J.* **226**: 131–138.
4. Constantinidis, I.C. and Satterlee, J.D.   (1988)   *Biochemistry* **27**: 3069–3076.
5. Padlan, E.A. and Love, W.E.   (1974)   *J. Biol. Chem.* **249**: 4067–4078.
6. Imamura, T., Baldwin, T.O. and Riggs, A.F.   (1972)   *J. Biol. Chem.* **247**: 2785–2797.
7. Cooke, R.M. and Wright, P.E.   (1985)   *Biochim. Biophys. Acta* **832**: 357–364.
8. Simons, P.C. and Satterlee, J.D.   (1989)   *Biochemistry* **28**: 8525–8530.
9. Constantinidis, I.C., Satterlee, J.D., Pandey, R.K., Leung, H.-K. and Smith, K.M.   (1988)   *Biochemistry* **27**: 3069–3076.
10. Mintorovitch, J. and Satterlee, J.D.   (1988)   *Biochemistry* **27**: 8045–8050.
11. Mintorovitch, J., Van Pelt, D. and Satterlee, J.D.   (1989)   *Biochemistry* **28**: 6099–60104.
12. Satterlee, J.D.   (1984)   *Biochim. Biophys. Acta* **791**: 384–394.
13. Cooke, R.M. and Wright, P.E.   (1985)   *FEBS Lett.* **187**: 219–223.
14. Cooke, R.M., Dalvit, C., Narula, S.S. and Wright, P.E.   (1987)   *Eur. J. Biochem.* **166**: 399–408.
15. Li, S.L. and Riggs, A.F.   (1971)   *Biochim. Biophys. Acta* **236**: 208–210.

30
Identification of the cDNA for Some of the Polymeric Globins of *Glycera dibranchiata*

Li-Hui Chow, Rasheeda S. Zafar and Daniel A. Walz

Department of Physiology, Wayne State University School of Medicine, Detroit, MI 48201, USA

Introduction

The polymeric fraction of the intracellular Hb of the marine polychaete *Glycera dibranchiata* consists of at least six components which are different in their amino acid sequences (1, 2). Among these sequence differences, two major heterogeneous sequences *AMEEKVP* and *AMNSKVP* have been found in residues 117 to 123. To distinguish these two kinds of transcripts, two 17-base oligonucleotide probes (one 32-fold degenerate (*MEEKVP*) and the other 96-fold degenerate (*AMNSKV*)) were used to screen the cDNA library. Since the cDNA library was constructed from poly(A+) mRNA of *Glycera* erythrocytes in *pBR*322, this screening provided the necessary information to determine if the heterogeneity exists at the mRNA level, and, if it does, what would be its distribution ratio.

Materials and Methods

The cDNA library of poly(A+) mRNA of *Glycera* erythrocytes in *pBR*322 was kindly donated by Drs. P.C. Simons and J.D. Satterlee. The oligonucleotide probes (Figure 1), synthesized by Synthetic Genetics (San

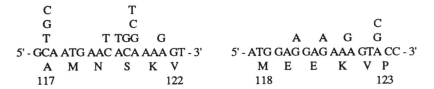

Figure 1. Sequences of two 17-base oligodeoxynucleotide probes with 32-fold and 96-fold redundancy. Partial amino acid sequencing of the "polymeric" globin peptide revealed that sequences *MEEKVP* in residues 118–123 and *AMNSKV* in residues 117–122 were different from the "monomeric" globins.

Diego, USA), were end-labeled by γ-^{32}P-ATP using polynucleotide kinase and were purified with a cartridge column (NENSORB 20) according to the manufacturer's instructions (Pharmacia, Piscateway, NJ, USA). Aliquots of the frozen library were diluted and plated to produce about 3000 colonies on six nitrocellulose filters. Two replica filters were made from each master filter. The colonies on the replica filters were lysed, denatured, neutralized and baked at 80°C in a vacuum oven. The filters were prehybridized in 5X SSC, 50 mM sodium phosphate, 5X Denhardt's solution, 0.2% SDS, 200 mg/ml sonicated salmon testis DNA at 37°C for four hours, and were then hybridized overnight under the same conditions with a 5'-end-labeled probe. The washing stringency for the hybridizing colonies was 5X SSC/0.05% SDS or 2X SSC/0.05% SDS at temperatures between 37 and 50°C. The filters were exposed to x-ray film at –70°C before completely dry. Plasmids of selected positive clones were amplified in 5 ml liquid culture. DNA was isolated by a modification of the method of Birnboim and Doly (3) and was digested with Pst 1. The size of the insert cDNA was determined by electrophoresis of the digested plasmid DNA in 1.2% agarose gels. After being transferred to nitrocellulose filters, the cDNA inserts were hybridized with the former probes in a Southern blot to determine if they were real.

Results and Discussion

Figure 2 reveals many positive clones from the screening of the cDNA library with the *MEEKVP* probe. Three positive clones were chosen from each plate for making stock cultures and further characterizations. Since some of these may have been false signals, hybridization with the same probe in a Southern blot of

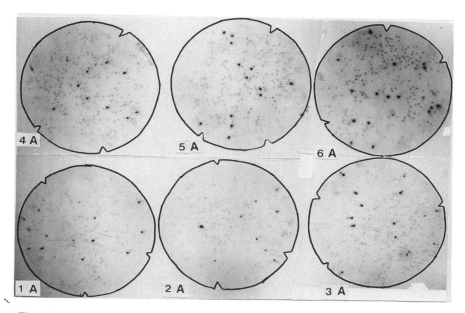

Figure 2. Screening of the cDNA library in *pBR*322 by hybridization with the 32*P*-end-labeled *MEEKVP* probe. Autoradiography was carried out for 18 hours at –70°C.

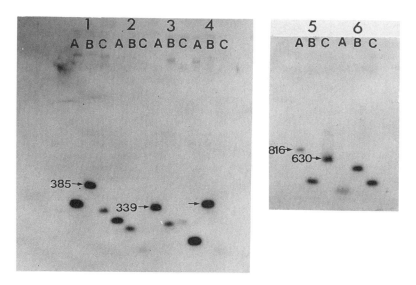

Figure 3. Southern blot analysis of cDNA inserts from a Pst 1 digestion of 18 positive clones with *MEEKVP* probe. Arrows indicate the size (base pairs) of five larger cDNAs that were selected for sequencing.

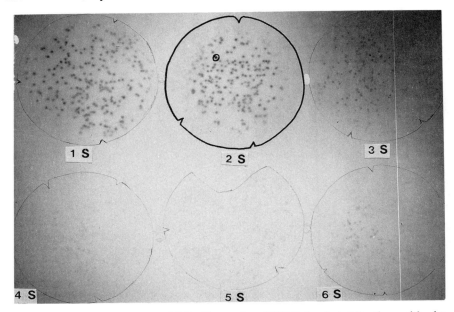

Figure 4.   Screening of the cDNA library in *pBR*322 by hybridization with the ^{32}P-end-labeled *AMNSKV* probe.   Autoradiography was carried out as in the *MEEKVP* screening.

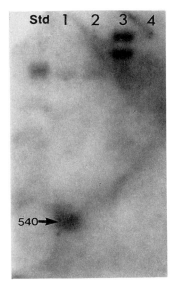

Figure 5.   Southern blot of the cDNA insert from the only positive clone with *AMNSKV* probe. Std: Hind III fragments of DNA. Lane 1: Pst 1 digested DNA of the positive clone. Lane 2: Pst 1 digested DNA of the negative control. Lane 3: uncut DNA of the positive clone. Lane 4: uncut DNA of the negative control.

the cDNA inserts (Figure 3) was an important check to perform. Based on the size shown, the five larger clones were selected for cDNA sequence determination. Only one positive clone was found in the screening with the *AMNSKV* probe, as shown in Figure 4. The cDNA insert of this clone was shown to be a real positive through hybridization with the same *AMNSKV* probe in a Southern blot (Figure 5). The sequences of the six cDNA inserts were determined and are provided elsewhere (4). The ratio of *MEEKVP* screening to *AMNSKV* screening is 18 : 1. This ratio possibly reflects the fact that, of the six identified clones, five contain *MEEKVP* while only one contains *AMNSKV*.

Acknowledgements

This work was supported in part by National Institutes of Health grant DK 30382 (D.A.W.). R.S.Z. was supported by National Institutes of Health Training Grant T32HL 07602.

References

1. Zafar, R.S., Chow, L.H., Stern, M.S., Vinogradov, S.N. and Walz, D.A. (1990) *Biochim. Biophys. Acta* **1041**: 117–122.
2. Zafar, R.S., Chow, L.H., Stern, M.S., Scully, J.S., Sharma, P.R., Vinogradov, S.N. and Walz, D.A. (1990) *J. Biol. Chem.* **265**: 21843–21851.
3. Birnboim, H.C. and Doly, J. (1979) *Nucleic Acids Res.* **7**: 1513–1516.
4. Zafar, R.S., Chow, L.H., Stern, M.S. and Walz, D.A. (1991) This volume.

31
The cDNA Sequences Encoding the Polymeric Globins of *Glycera dibranchiata*

Rasheeda S. Zafar, Li-Hui Chow, Mary S. Stern and Daniel A. Walz

Department of Physiology, Wayne State University School of Medicine, Detroit, MI 48201, USA

Introduction

The nucleated erythrocytes of the marine polychaete *Glycera dibranchiata*, the bloodworm, contain several globin chains of *ca.* 17 kDa, separable into two Hb fractions (1–3): a "monomeric" fraction and a "polymeric" fraction consisting of self-associated globin chains. Both fractions are present in individual erythrocytes (4). The amino acid sequences of two "monomeric" *Glycera* globins have been determined, *M*-II (5) and *M*-IV (6), and the crystal structure of *M*-II (7) has been refined recently to a 1.5 Å resolution (8). In the "monomeric" globins, the distal His (*E*7) is replaced by a Leu residue. Furthermore, the two fractions exhibit extensive heterogeneity with at least five or six distinct components in the "monomeric" fraction (9–15) and a similar number of components in the "polymeric" fraction (16, 17).

In this work, we report the cDNA sequences of several components of the "polymeric" fraction of *Glycera* Hb and compare them with the two known "monomeric" sequences.

Table 1. Amino acid sequences determined by Edman degradation of cyanogen bromide (C) and tryptic (T) peptides obtained from the "polymeric" fraction of *Glycera* Hb (E = exact match; R = partial match; N = not found).

Position Number		Amino Acid Sequence	P 1	P 2	P 3	P 4	P 5	P 6
1–10	(C)	MKLTADGIAT	R	R	R	N	N	N
1–15	(C)	MHLTADQVAALKASW	E	R	E	N	N	N
13–27	(T)	ASWPEVSAGDGGAQL	E	R	E	R	N	N
13–34	(T)	ASWPEVSAGDGGQLGLEMFTK	R	R	R	R	N	N
35–44	(T)	YFDENPQMMF	R	R	R	R	N	N
44–58	(C)	FVFGYSGRTQALKSN	R	R	R	R	R	N
44–82	(C)	FIFGYSGRTDALKHNAKLQNHGKLIV NQIGQAVAELCDA	R	R	R	E	R	R
45–51	(C)	VFGYSGR	R	R	R	R	E	N
46–59	(C)	FFYSGRTQALKHNP	R	R	R	R	R	N
61–66	(T)	LQNHGK	R	E	E	E	E	E
69–83	(C)	GDAKQVAGTLHALGV	R	R	R	R	R	E
74–83	(T)	QAAGTLHALGV	R	R	R	E	E	R
98–102	(T)	GFGDI	E	E	E	E	E	E
104–117	(T)	AEYFPALGNALLEAA	R	R	R	R	R	R
104–121	(T)	ADFFPALGMCLLDAMEEK	R	E	R	E	E	R
119–145	(C)	EEKVPGLNRTLWAAAYREISDALIAGL	E	R	E	E	E	E
122–126	(T)	VPGLN	E	E	E	E	E	E
128–135	(T)	TLWAAAYR	E	E	E	E	E	E
128–138	(T)	TLWAAGYAALK	R	R	R	R	R	R
128–138	(T)	VLWAAQYAALK	R	R	R	R	R	R
136–146	(T)	VISDALIAGLE	R	R	E	R	R	R
136–146	(T)	VISEALIAGLE	R	R	R	R	R	R

Materials and Methods

The preparation of the "polymeric" Hb fraction, the removal of heme, the globin separation and the protein sequencing are described elsewhere (16, 17). The cDNA library of poly(A+)mRNA of *Glycera* erythrocytes in *pBR*322 was a generous gift of Drs. P. C. Simons and J. D. Satterlee. Eighteen positive clones were selected with the *MEEKVP* probe. Only one positive clone was revealed by the *AMNSKV* probe, the details of which are given elsewhere (18). The sequences of both strands of the cDNA were determined using the dideoxy chain termination reaction (19).

Results and Discussion

The partial amino acid sequences obtained by Edman degradation of cyanogen bromide and tryptic peptides of the *Glycera* "polymeric" globin fraction are shown in Table 1. The table also lists the relationships of the peptide sequences with the corresponding cDNA sequences found in the three complete *P1*, *P2* and *P3* clones and the three partial clones, *P4*, *P5* and *P6*. Exact agreement is denoted by *E*, partial agreement is denoted by *R* and the absence of correspondence is marked by *N*. At least four of the isolated peptides have sequences which do not exactly match any of the cDNA sequences; suggesting that the family of "polymeric" globins may contain more components than the six already identified (16, 17).

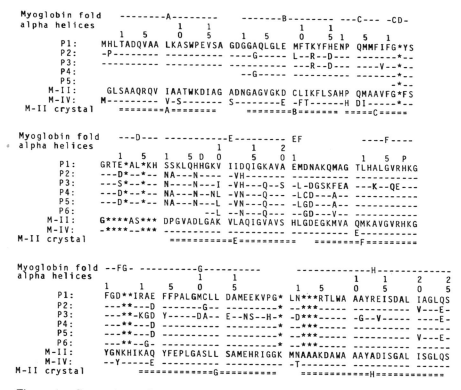

Figure 1. Comparison of the amino acid sequences of the "polymeric" globins *P1* through *P6* (16, 17) and the "monomeric" globins *M*-II (5), corrected as the results of crystal structure determination (8), and *M*-IV (6) using the alignment of Bashford *et al.* (20). The deletions are marked by an asterisk. *D* and *P* mark the locations of the distal (*E*7) and the proximal (*F*8) residues, respectively. The helical segments observed in the crystal structure of *M*-II (8) are shown in the bottom line.

Figure 1 depicts the three complete (*P*1–*P*3) and the three partial (*P*4–*P*6) "polymeric" globin sequences inferred from the cDNA sequences (16, 17), which are aligned with the two known "monomeric" *Glycera* sequences *M*-II (5) and *M*-IV (6) using the method of Bashford *et al.* (20) between the *M*-II sequence and the vertebrate globin sequences. The "monomeric" *M*-II sequence has been corrected as the result of the recent high-resolution structure (8). The actual α helical segments observed in the latter are shown in the bottom line of Figure 1. In addition, the α helical segments of the Mb fold, based on the alignment of Bashford *et al.* (20), are shown in the top line. We identify the positions in the *Glycera* globin sequences by their location in the Mb fold helices. Although all *Glycera* globins retain the two invariant residues, Phe at *CD*1 and the proximal His at *F*8, the "polymeric" globins retain the distal His which is altered to a Leu in the "monomeric" globins. In addition, comparison of the two types of sequences indicates that the number of identical amino acids ranges from 54 to 59, corresponding to 36–40% identity.

The crystal structure of the "monomeric" globin *M*-II (7, 8) shows the absence of helix *D* and thus resembles human α chains more than it does human β chains. The several deletions in the two "monomeric" sequences relative to the "polymeric" sequences occur in this region. Although the structure of "polymeric" globins is unknown, it may be possible for them to have a *D* helix. The deletions in the "polymeric" relative to the "monomeric" sequence occur at the *FG* corner, at the end of the *G* helix and at the beginning of the *H* helix.

Table 2. List of positions where the identical/similar surface residues in the *Glycera* "monomeric" globins *M*-II and *M*-IV are altered from polar to nonpolar or from small to large hydrophobic residues in the "polymeric" globins.

Position[a]	Amino Acids							
	*P*1	*P*2	*P*3	*P*4	*P*5	*P*6	*M*-II	*M*-IV
A5	V	V	V	-	-	-	R	R
A6	A	A	A	-	-	-	Q	Q
A13	P	P	P	-	-	-	K	K
A17	A	A	A	-	-	-	G	G
C5	M	M	M	M	-	-	A	A
C6	F	F	F	F	-	-	A	A
D4	L	L	L	L	L	-	S	S
E10	V	V	I	L	L	L	K	K
FG1	F	F	F	F	F	F	Y	Y
G6	P	P	P	P	P	P	E	E
H21	A	A	A	A	A	A	S	S

[a]Named using the Mb fold α helices according to the alignment of Bashford *et al.* (20) shown in Figure 1.

At the 44 separate positions at which the residues are identical or very similar in the six "polymeric" globins, they differ from the pair of corresponding identical/similar residues in M-II and M-IV (17). This list should include the amino acid residues that are possibly involved in the self-association of the "polymeric" globins. The likelihood of the involvement of a specific residue in transition from the "monomeric" to the "polymeric" sequences was considered to depend on the following two criteria: (1) location at the surface of the globin and (2), alteration from a polar to a nonpolar residue or from a small hydrophobic residue to a larger one. Table 2 lists the 11 positions at which an alteration satisfies the foregoing restrictions. These positions are the following: $A5$ (Arg→Val), $A6$ (Gln→Ala), $A13$ (Lys→Pro), $A17$ (Gly→Ala), $C5$ (Ala→Met), $C6$ (Ala→Phe), $D4$ (Ser→Leu), $E10$ (Lys→Val/Ile), $FG1$ (Tyr→Phe), $G6$ (Glu→Pro) and $H21$ (Ser→Ala). Except for residues at $A5$, $C5$ and $D4$, the altered residues also possess surface accessibility areas calculated from three-dimensional modeling of the $P3$ sequence with the sperm whale metMb crystal structure as template (17), which are about 60 Å^2 or greater. In particular, the Phe residues at $C6$ and at $FG1$ have accessibility areas of about 160 and 190 Å^2, respectively. Thus, self-association of the "polymeric" globins mediated by the removal of one or both of these residues from exposure to solvent water would appear to be rather likely.

Acknowledgements

We thank Drs. J.D. Satterlee and P.C. Simons for the generous gift of their cDNA library of poly ($A+$) mRNA from *Glycera* erythrocytes. This work was supported in part by National Institutes of Health grant DK 30382 (D.A.W.). R.S.Z. and M.S.S. were supported by National Institutes of Health Training Grant T32 HL 07602.

References

1. Hoffman, R.J. and Mangum, C.P. (1970) *Comp. Biochem. Physiol.* **36**: 211–228.
2. Vinogradov, S.N., Machlik, C.A. and Chao, L.L. (1970) *J. Biol. Chem.* **245**: 6533–6538.
3. Seamonds, B., Forster, R.E. and George, P. (1971) *J. Biol. Chem.* **246**: 5391–5397.
4. Mangum, C.P., Colacino, J.M. and Vandergon, T.L. (1989) *J. Exp. Zool.* **249**: 144–149.
5. Imamura, T., Baldwin, T.O. and Riggs, A. (1972) *J. Biol. Chem.* **247**: 2785–2797.
6. Simons, P.C. and Satterlee, J.D. (1989) *Biochemistry* **28**: 8525–8530.
7. Padlan, E.A. and Love, W.E. (1974) *J. Biol. Chem.* **249**: 4067–4078.
8. Arents, G. and Love, W.E. (1989) *J. Mol. Biol.* **210**: 149–161.

9. Kandler, R.L. and Satterlee, J.D.   (1983)   *Comp. Biochem. Physiol.* **75**B: 499–503.

10. Kandler, R.L., Constantinidis, I.C. and Satterlee, J.D.   (1984)   *Biochem. J.* **226**: 131–138.

11. Cooke, R.M. and Wright, P.E.   (1985)   *Biochim. Biophys. Acta* **832**: 365–372.

12. Cooke, R.M. and Wright, P.E.   (1985)   *Biochim. Biophys. Acta* **832**: 357–364.

13. Constantinidis, I.C. and Satterlee, J.D.   (1987)   *Biochemistry* **26**: 7779–7786.

14. Constantinidis, I.C., Satterlee, J.D., Pandrey, R.K., Leung, H.K. and Smith, K.M. (1988) *Biochemistry* **27**: 3061–3076.

15. DiFeo, T.J. and Addison, A.W. (1989) *Biochem. J.* **260**: 863–871.

16. Zafar, R.S., Chow, L.H., Stern, M.S., Vinogradov, S.N. and Walz, D.A. (1990) *Biochim. Biophys. Acta.* **1041**: 117–122.

17. Zafar, R.S., Chow, L.H., Stern, M.S., Scully, J., Sharma, P., Vinogradov, S.N. and Walz, D.A. (1990) *J. Biol. Chem.* **265**: 21843–21851.

18. Chow, L.H., Zafar, R.S. and Walz, D.A. (1991) This volume.

19. Sanger, F., Nicklen, S. and Coulson, A.R. (1977) *Proc. Natl. Acad. Sci. U.S.A.* **74**: 5463–5467.

20. Bashford, D., Chothia, C. and Lesk, A.M.   (1987)   *J. Mol. Biol.* **196**: 199–216.

32
Primary Structure of 440 kDa Hemoglobin from the Deep-Sea Tube Worm *Lamellibrachia*

Takashi Takagi,[a] Hisashi Iwaasa,[a] Suguru Ohta[b] and Tomohiko Suzuki[c]

[a]Biological Institute, Faculty of Science, Tohoku University, Sendai 980, Japan
[b]Ocean Research Institute, University of Tokyo, Tokyo 164, Japan
[c]Department of Biology, Faculty of Science, Kochi Univeristy, Kochi 780, Japan

Introduction

The deep-sea tube worms *Riftia* and *Lamellibrachia* belong to the phylum Vestimentifera (1) and are found in hydrothermal vents or cold seeps at depths of 600–2500 m (2, 3). These invertebrate animals are sustained by mutual symbiosis with sulfide-oxidizing bacteria (4), and their Hb-containing blood transports hydrogen sulfide, which is extremely toxic to heme-containing proteins and to internal symbiont bacteria (5), and which facilitates O_2 transport (6).

The tube worm *Lamellibrachia* lives in the cold-seep area of Sagami Bay, Japan, and has two giant extracelullar Hbs, a 3000 kDa Hb and a 440 kDa Hb (6). The former consists of four heme-containing chains (AI–AIV) and two linker chains (AV–AVI) for the assembly of the heme-containing chains. The latter has four heme-containing chains (BI–BIV) and no linker chain. In this paper, we describe the complete amino acid sequence of the four heme-containing chains of 440 kDa Hb, and of the hydrogen sulfide binding site.

Results and Discussion

Four heme-containing *BI–BIV* chains were isolated by HPLC (7), and the amino acid sequences of these chains were determined with an automated sequencer using intact protein and peptides derived from enzymatic and chemical cleavages as described elsewhere (8). The sequence of the *B*III chain was identical with that of the *A*III chain (8). The amino acid sequences of *BI–BIV* were compared with the heme-containing chains of annelid Hbs. Thus far, the amino acid sequences have been reported for the four chains of both *Tylorrhynchus* (9) and *Lumbricus* (10, 11), and one chain of *Pheretima* (12), as shown in Figure 1. The sequences of the *Lamellibrachia* chains show a significant homology (30-50% identical) with those of annelid Hbs. In thirteen Hb sequences in Figure 1, 14 residues, including Cys-7, Phe-54 (invariant-Phe at *CD*1), His-70 (distal-His at *E*7), His-97 (heme-binding proximal-His at *F*8) and Cys-141, appear to be invariant.

In annelid Hbs, the Cys residues play an important role in subunit assembly. In fact, the Cys residues are restricted to positions 6, 7, 132 and 141 in the alignment of Figure 1, and they are all involved in either intra- or inter-chain disulfide bonds (10, 11, 13). Thus, there is no free Cys residue in annelid Hbs. The subunit composition of annelid heme-containing chains is one monomer and a disulfide-bonded trimer: the Cys residues at positions 7 and 141 participate in intra-molecular disulfide bonding, and those at positions 6 and 132 are involved in the formation of a trimer (11, 13). On the other hand, *Lamellibrachia* 440 kDa Hb is composed of two monomers (*B*III and *B*IV) and a disulfide-bonded dimer (*BI–BII*) (14): all chains have Cys residues at positions 7 and 141 and *BI* and *BII* chains have a Cys residue at position 132. Judging from the structural homology, it is very likely that the former two Cys residues form an interchain disulfide bond, as in annelid Hbs. If so, the remaining Cys residues in the *B*II and *B*III chains at positions 71 and 81 must be free Cys residues, since the *B*III chain exists as a monomer and the BII chain makes a dimer with the BI chain via the Cys at position 132.

Arp *et al.* (5, 15) have demonstrated that both 3000 kDa and 440 kDa Hbs of the tube worm *Riftia*, a phylogenetically closely related species to *Lamellibrachia*, have special site(s) for the binding of hydrogen sulfide in order to supply it to internal bacterial symbionts. Therefore, *Riftia* Hbs have two important functions: the transport of O_2 and of hydrogen sulfide. Although the subunit structures, chain compositions and amino acid sequences of *Riftia* Hbs are not available, data for amino acid composition, SDS-PAGE pattern and electron-microscopic observation strongly suggest a structural similarity between *Riftia* and *Lamellibrachia* Hbs (16, 17). In addition, both *Lamellibrachia* Hbs also bind hydrogen sulfide (14). Arp *et al.* have proposed that hydrogen sulfide is bound via thiol disulfide exchange at the disulfide bonds in *Riftia* Hbs (14). However, this is unlikely because *Lumbricus* Hb has no ability to bind hydrogen sulfide (15), and because the tube worm Hbs share many characteristics with annelid Hbs, including the disulfide-bridge configuration discussed above.

Alternatively, we propose that the free Cys residues at positions 71 and 81, which are not present in annelid Hbs, are responsible for hydrogen sulfide binding ability. In fact, both the 3000 kDa and the 440 kDa Hbs of *Lamellibrachia* contain free thiol groups that are detectable by fluorescent

```
           1        10        20          30        40          50        60
Phe.I         DcNTLKRFKVKHQWQQVFSGE    HHRTEFSLHFWKEFLHDHPDLVSLFKRVQGE
TW.BI         DcNILQRLKVKMQWAKAYGFG    AERAKFGNSLWTSIFNYAPDARELFDSVKSK
TW.BIII       YEcGPLQRLKVKRQWAEAYGSG   NDREEFGHFIWTHVFKDAPSARDLFKRVRGD
Lum.d         ECLVTEGLKVKLQWASAFGHA    HQRVAFGLELWKGILREHPEIKAPFSRVRGD
Lum.b         KKQcGVLEGLKVKSEWGRAYGSG   HDREAFSQAIWRATFAQVPESRSLFKRVHGD
Tyl.I         TDcGILQRIKVKQQWAQVYSVG    ESRTDFAIDVFNNFFRTNPD RSLFNRVNGD
Tyl.IIA       SSDHcGPLQRLKVKQQWAKAYGVG  HERVELGIALWKSMFAQDNDARDLFKRVHGE
TW.BII        SSNScTTEDRREMQLMWANVWSAQF TGRRLAIAQAVFKDLFAHVPAAIGLFDRVHGT
TW.BIV        SKFcSEGDATIVIKQWNQIYNAGI  SAGSRLTMGNKIFSTLFKLKPESEALFSNVNVA
Lum.c         DEHEHccSEEDHRIVQKQWDILWRDTESSKIKIGFGRLLLTKLAKDIPDVNDLFKRVDIE
Lum.a         ADDEDccSYEDRREIRHIWDDVWSSSF TDRRVAIVRAVFDDLFKHYPTSKALFERVKID
Tyl.IIB       DDccSAADRHEVLDNWKGIWSAEF  TGRRVAIGQAIFQELFALDPNAKGVFGRVNVD
Tyl.IIC       DTccSIEDRREVQALWRSIWSAED  TGRRTLIGRLLFEELFEIDGATKGLFKRVNVD
              *         *                                         *  *

           70        80        90          100       110         120
Phe.I         NIYSPEFQAHGIRVLAGLDSVIGVLDEDDTFTVQLAHLKAQHTER GTKPEYFDLFGTQL
TW.BI         EMQSPQFKAHVARVIGGLDRVISMLDNAEALNADLEHLKSQHDPR GLDALNFAVFGKAL
TW.BIII       NIHTPAFRAHATRVLGGLDMCIALLDDEGVLNTQLAHLASQHSSR GVSAAQYDVVEHSV
Lum.d         NIYSPQFGAHSQRVLSGLDITISMLDTPDMLAAQLAHLKVQHVER NLKPEFFDIFLKHL
Lum.b         DTSHPAFIAHAERVLGGLDIAISTLDQPATLKEELDHLQVQHEGR KIPDNYFDAFKTAI
Tyl.I         NVYSPEFKAHMVRVFAGFDILISVLDDKPVLDQALAHYAAFHKQF GTIP   FKAFGQTM
Tyl.IIA       DVHSPAFEAHMARVFNGLDRVISSLTDEPVLNAQLEHLRQQHIKL GITGHMFNLMRTGL
TW.BII        DVNSNEFKAHCIRVVNGLDSAIGLLSDPSTLNDQLLHLATQHQERAGVTKGGFSAIAQSF
TW.BIV        NMSSGAFHAHTVRVLSGLDMGINYLNDAATLTSLTSHLATQHVARTGLKAVYFDAMGKVL
Lum.c         HAEGPKFSAHALRILNGLDLAINLLDDPPALDAALDHLAHQHEVREGVQKAHFKKFGEIL
Lum.a         EPESGEFKSHLVRVANGLDLLINLLDDTLVLQSHLGHLADQHIQRKGVTKEYFRGIGEAF
Tyl.IIB       KPSEADWKAHVIRVINGLDLAVNLLEDPKALQEELKHLARQHRERSGVKAVYFDEMEKAL
Tyl.IIC       DTHSPEEFAHVLRVVNGLDTLIGVLGDSDTLNSLIDHLAEQHKARAGFKTVYFKEFGKAL
              *    *    * *     *         *          *      *

           130       140       150
Phe.I         FDILGDKLGTH FDQAAWRDc YAVIAAGIKP
TW.BI         FATVGGKFGV cFDLPAWESc YKVIAKGITGNDMFN
TW.BIII       MMGVEHEIGQNVFDKDAWQAc LDVITGGIQGN
Lum.d         LHVLGDRLGTH FDFGAWHDc VDQIIDGIKDI
Lum.b         LHVVAAQLGR cYDREAWDAc IDHIEDGIKGHH
Tyl.I         FQTIAEHIHG  ADIGAWRAcYAEQIVTGITA
Tyl.IIA       AYVLPAQLGR cFDKEAWAAcWDEVIYPGIKHD
TW.BII        LRVMPQVAS  cFNPDAWSRc FNRITNGMTEGLA
TW.BIV        MTVLPALID  NFNPDAWRNc LLPLKSAIAEGLP
Lum.c         ATGLPQVLD  DYDALAWKSc LKGILTKISSRLNA
Lum.a         ARVLPQVLS  cFNVDAWNRc FHRLVARIAKDLP
Tyl.IIB       LKVLPQVSS  HFNSGAWDRc FTRIADVIKAELP
Tyl.IIC       NHVLPEVAS  cFNPEAWNHc FDGLVDVISHRIDG
              **   *
```

Figure 1. Alignment of the amino acid sequence of *Lamellibrachia* B1–BIV chains with those of annelid heme-containing chains. The sequence was aligned by the program ALIGN (18). Asterisks indicate the 14 invariant residues in 13 Hbs. Cys residues involved in the disulfide bond formation and free thiol groups are indicated in small letters and bold, capital, underlined letters, respectively. Key: TW., tube worm, *Lamellibrachia* (this work); Tyl., Tylorrhychus (9); Lum., *Lumbricus* (10, 11); Phe., *Pheretima* (12).

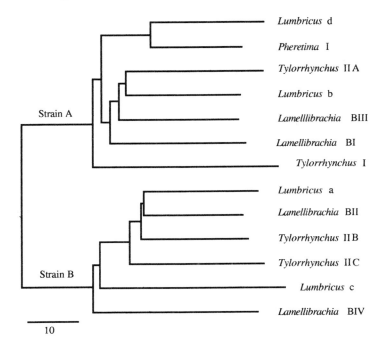

Figure 2. Phylogenetic tree for the 13 heme-containing chains from annelid and tube worm Hbs. This figure was constructed with the program TREE (18).

labeling (8). Such Cys residues in deep-sea tube worm Hbs may have been acquired after the divergence of the Hb from polychaetes and oligochaetes as a high molecular adaptation for symbiosis with sulfide-oxidizing bacteria.

Recently, the deep-sea tube worms *Riftia* and *Lamellibrachia* were placed in a new phylum, Vestimentifera, on the basis of their unique outward appearance, such as the very long trunk region and absence of a mouth, gut and anus (1). However, all biochemical data on tube worm Hbs suggest that they are probably members of the phylum Annelida (7, 14, 16, 17). Now that we have obtained the complete amino acid sequence of *Lamellibrachia* 440 kDa Hb, we have constructed a phylogenetic tree for annelid and tube worm Hbs by using the program TREE (18), as shown in Figure 2. The results clearly show that there are two globin strains *A* and *B*, as suggested by Gotoh *et al.* (19) for annelid and tube worm Hb chains. None of the *Lamellibrachia* chains separated from the annelid chains: all chains diverged almost evenly in the phylogenetic tree. These results clearly show that *Lamellibrachia*, *Tylorrhynchus* and *Lumbricus* diverged from their ancestors at approximately the same time. On the basis of these results, we conclude that deep-sea tube worms *Lamellibrachia* and *Riftia* should be placed in the phylum Annelida instead of in the Vestimentifera, but a new class may be needed to accommodate them.

Acknowlegements

We thank Dr. R.F. Doolittle, University of California, San Diego, for supplying us with the computer programs. We also thank Drs. K. Konishi and T. Furukohri for their encouragement during this work.

References

1. Jones, M.L. (1985) *Biol. Soc. Wash. Bull.* **6**: 117–158.
2. Corlis, J.B., Dymond, J., Gordon, L.I., Edmond, J.M., Von Herzen, R.P., Ballard, R.D., Green, K., Williams, D., Bainbridge, A., Crane, K. and Van Adel, T.H. (1979) *Science* **203**: 1073–1083.
3. Kenneticutt, M.C., Brooks, J.M., Bidigare, R.R., Fay, R.R., Wade, T.L. and McDonald, T.J. (1985) *Nature* **317**: 351–353.
4. Childress, J.J., Felback, H. and Somero, G.N. (1987) *Sci. Am.* **256**: 106–158.
5. Arp, A.J. and Childress, J.J. (1983) *Science* **219**: 259–297.
6. Arp, A.J. and Childress, J.J. (1981) *Science* **213**: 342–344.
7. Suzuki, T., Takagi, T. and Ohta, S. (1988) *Biochem. J.* **255**: 541–544.
8. Suzuki, T., Takagi, T. and Ohta, S. (1990) *Biochem. J.* **266**: 221–225.
9. Suzuki, T. and Gotoh, T. (1986) *J. Biol. Chem.* **261**: 9257–9267.
10. Shishikura, F., Snow, J.W., Gotoh, T., Vinogradov, S.N. and Walz, D.A. (1987) *J. Biol. Chem.* **262**: 3123–3131.
11. Fushitani, K.M., Matsuura, S.A. and Riggs, A.F. (1988) *J. Biol. Chem.* **263**: 6502–6517.
12. Suzuki, T. (1989) *Eur. J. Biochem.* **185**: 127–134.
13. Suzuki, T., Kapp, O.H. and Gotoh, T. (1988) *J. Biol. Chem.* **263**: 18524–18529.
14. Suzuki, T., Takagi, T., Okuda, K., Furukohri, T. and Ohta, S. (1989) *Zool. Sci.* **6**: 915–926.
15. Arp, A.J., Childress, J.J. and Vetter, R.D. (1987) *J. Exp. Biol.* **128**: 139–158.
16. Terwilliger, R.C., Terwilliger, N.B. and Schabtach, E. (1980) *Comp. Biochem. Physiol.* B **65**: 531–535.
17. Terwilliger, R.C., Terwilliger, N.B., Bonaventura, C., Bonaventura, J. and Schabtach, E. (1985) *Biochim. Biophys. Acta* **829**: 227–233.
18. Feng, D.F., Johnson, M.S. and Doolittle, R.F. (1985) *J. Mol. Evol.* **21**: 1122–1125.
19. Gotoh, T., Shishikura, F., Snow, J.W., Ereifej, K., Vinogradov, S.N. and Walz, D.A. (1987) *Biochem. J.* **241**: 441–445.

33
Protozoan Hemoglobins: Their Unusual Amino Acid Sequences

Hisashi Iwaasa, Takashi Takagi and
Keiji Shikama

Biological Institute, Faculty of Science, Tohoku University,
Sendai 980, Japan

Introduction

Wakabayashi *et al.* (1) recently reported a Hb-like protein from *Vitreoscilla* which combines with molecular oxygen and exhibits maximum sequence homology (24%) with yellow lupine legHb. This finding of a dimeric bacterial Hb has now upset the common belief that Hb is a protein that is found only in eukaryotes. Despite such a rare occurrence, however, Hb or Mb is found mostly in higher animals. Thus, Hbs found in organisms of the lowest phyla of the animal kingdom, would provide an important molecular basis for the understanding of the evolution of the globin family from prokaryotes to eukaryotes, as well as from protozoa to higher animals.

The occurrence of a Hb-like protein in protozoa was first reported by Sato and Tamiya on the basis of spectroscopic observations on cell suspensions of *Paramecium caudatum* (2). Subsequently, Keilin and Ryley also observed a heme protein in *Tetrahymena pyriformis* (3), a ciliated species belonging to the same class as *Paramecium*. Although the isolation and characterization of Hb from several species of *Paramecium* has been performed, until recently the primary structure of protozoan Hb remained undetermined.

Very recently, we have succeeded in isolating monomeric Hbs from both *P. caudatum* (4) and *T. pyriformis* (5), and have determined their amino acid sequences. Here, we compare them with the other known proteins of the globin family.

Materials and Methods

The packed cells of *P. caudatum* (syngen 3, stock StG1) and of *T. pyriformis*, harvested from the mass culture, were homogenized with a Teflon-glass homogenizer and the extract was fractionated with ammonium sulfate. The

Table 1. Amino acid compositions of Hbs and Mbs.

	Number of Residues				
Amino Acid	*Paramecium* Hb	*Tetrahymena* Hb	Sperm whale Mb[a]	*Aplysia* Mb[b]	*Vitreoscilla* Hb[c]
Asp	3	7	7	9	8
Asn	8	9	1	10	4
Thr	12	9	5	1	8
Ser	3	0	6	15	1
Glu	4	8	14	5	9
Gln	8	5	5	2	9
Pro	2	3	4	6	7
Gly	11	8	11	9	8
Ala	22	11	17	26	23
Cys	1	0	0	0	1
Val	14	6	8	9	14
Met	2	7	2	3	3
Ile	3	5	9	4	12
Leu	8	12	18	12	14
Tyr	1	3	3	2	4
Phe	6	5	6	13	4
Lys	2	15	19	11	10
His	2	6	12	1	4
Arg	3	2	4	4	2
Trp	1	0	2	2	1
Total	116	121	153	144	146

[a]Taken from Romero-Herrera and Lehmann (6).
[b]Taken from Suzuki *et al*. (7).
[c]Taken from Wakabayashi *et al*. (1).

protein was then applied to gel filtration followed by DEAE- or CM-cellulose chromatography. After the removal of heme, each apoHb was reduced and carboxymethylated, and was then subjected to sequence determination (4, 5).

Results and Discussion

Table 1 summarizes the amino acid compositions of the two protozoan Hbs, of sperm whale Mb, of the Mb from *Aplysia kurodai*, a common gastropod sea mollusc and of the bacterial Hb from *Vitreoscilla*, in terms of the monomer unit. There are marked differences in the total number of residues, and also in the contents of some specific residues.

Figure 1 displays the complete amino acid sequences of protozoan Hbs from *P. caudatum* and *T. pyriformis*, both blocked at the N-terminus. Although the blocking group was not yet identified directly for *T. pyriformis* Hb, judging from the elution time of its N-terminal peptide on HPLC, it seems also to be an acetyl group.

The Hb from *P. caudatum* is composed of 116 amino acid residues with an M_m of 12,565 daltons, including a heme group. On the other hand, *T. pyriformis* Hb consists of 121 amino acid residues and its M_w was calculated to be 14,342, if one supposes that the N-terminus was also blocked by an acetyl group. At any rate, these values are much smaller than the other monomeric globins sequenced so far. *P. caudatum* Hb contains two His residues at positions 68 and 84 with a single Cys at position 50, whereas *T. pyriformis* Hb has six His residues at positions 36, 45, 63, 73, 82 and 113, and lacks Ser, Cys and Trp.

We have carried out a computer search for the sequence homology of the two protozoan Hbs, using the program GENAS (Kyushu University) with the National Biomedical Research Foundation protein sequence database, but failed

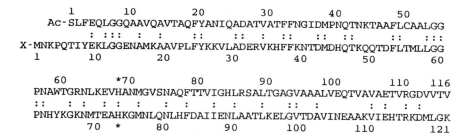

Figure 1. The amino acid sequences of protozoan Hbs from *P. caudatum* (upper) and *T. pyriformis* (lower). Identical residues between the two globins are shown by the colon (:). The His residue which is thought to be heme-binding is marked by an asterisk. Ac, acetyl group; X, unidentified blocking group.

```
Helix        AAAAAAAAAAAAAAAA BBBBBBBBBBBBBBBBBCCCCCCC      DDDDDDDEE
position     1234567891111111 12345678911111111234567      123456712
                    0123456            0123456
                                                   *
HUHBA        VLSPADKTNVKAAWGKVGAHAGEYGAEALERMFLSFPTTKTYFPHF DLSH   GSA
HUHBB        VHLTPEEKSAVTALWGKV NVDEVGGEALGRLLVVYPWTQRFFESFGDLSTPDAVMGNP
SWMB         VLSEGEWQLVLHVWAKVEADVAGHGQDILIRLFKSHPETLEKFDRFKHLKTEAEMKASE
AKMB         SLSAAEADLVGKSWAPVYANKDADGANFLLSLFEKFPNNANYFADFKG KSIADIKASP
CTHB         LSADQISTVQAS   FDKVKGDPVGILYAVFVKADPSIMAKFTQFAG KDLESIKGTA
LLLB         GALTESQAALVKSSWEEFNANIPKHTHRFFILVLEIAPAAKDLFSFLKG TSEVP QNNP
PCHB         SLFEQLG    GQAAVQAVTAQFYANIQADATVATFF            NGI
TPHB         MNKPQTIYEKLG    GENAMKAAVPLFYKKVLADERVKHFF          KNT
```

```
Helix        EEEEEEEEEEEEEEEEEE          FFFFFFFFFF    GGGGGGGGGGGGGGGG
position     34567891111111112          1234567891    123456789111111
                    01234567890                  0           012345
                                             *
HUHBA        QVKGHGKKVADALTNAVAHVDDM    PNALSALSDLHAHKLRVDPVNFKLLSHCLLVT
HUHBB        KVKAHGKKVLGAFSDGLAHLDNL    KGTFATLSELHCDKLHVDPENFRLLGNVLVCV
SWMB         DLKKHGVTVLTALGAILKKKGHH    EAELKPLAQSHATKHKIPIKYLEFISEAIIHV
AKMB         KLRDVSSRIFTRLNEFVNNAADAGKM SAMLSQFASEHVGF GVGSAQFENVR SMFPA
CTHB         PFETHANRIVGFFSKIIGELPNI    EADVNTFVASHKPR GVTHDQLNNFRAGFVSY
LLLB         ELQAHAGKVFKLVYEAAIQLEVTGVVVSDATLKNLGSVHVSK GVADAHFPVVKEAILKT
PCHB         DMPNQTNKTAAFLCAALGGPNAW    TGRNLKEVHANM GVSNAQFTTVIGHLRSA
TPHB         DMDHQTKQQTDFLTMLLGGPNHY    KGKNMTEAHKGM NLQNLHFDAIIENLAAT
```

```
Helix        GGGG      HHHHHHHHHHHHHHHHHHHHHHHHHHH
position     1111      12345678911111111112222222
             6789               01234567890123456
HUHBA        LAAHLPAEFTPAVHASLDKFLASVSTVLTSKYR
HUHBB        LAHHFGKEFTPPVQAAYQKVVAGVANALAHKYH
SWMB         LHSRHPGDFGADAQGAMNKALELFRKDIAAKYKELGYQG
AKMB         FVASLSAPPADD    AWNKLFG LIVAALKAAGK
CTHB         MKAHTDFAGAEA    AWGATLDTFFGMIFSKM
LLLB         IKEVVGAKWSEELNSAWTIAYDELAIVIKKEMDDAA
PCHB         LT    GAGVAAALVEQTVAVAETVRGDVVTV
TPHB         LK    ELGVTDAVINEAAKVIEHTRKDMLGK
```

Figure 2. Multiple sequence alignment of protozoan Hbs with other proteins of the globin family. The globins examined are: HUHBA, human Hb α chain (10); HUHBB, human Hb β chain (10); SWMB, sperm whale Mb (6); AKMB, *A. kurodai* Mb (7); CTHB, *Chironomus thummi thummi* Hb CTT-III (11); LLLB, *Lupinus luteus* legHbII (12); PCHB, *P. caudatum* Hb; TPHB, *T. pyriformis* Hb. The helix position refers to that of sperm whale Mb (13). Asterisks denote the residues which are identical among all eight globins.

to find any notable degree of similarity with other proteins of the globin family. However, it should be noted that *T. pyriformis* Hb shows the maximum sequence homology (33.9 % identical) with *P. caudatum* Hb (Figure 1). As mentioned already, the two Hbs differ in their His contents, but one His is conserved at position 68 for *P. caudatum* and at position 73 for *T. pyriformis*. Therefore, these His seem to be the proximal heme-binding residues that is essential for the function of all Hbs and Mbs.

At this point, it is of interest to note that the circular dichroism measurements of *P. caudatum* Hb in terms of the mean residue molar ellipticity show a value of $-20,600$ deg cm^2 dmol^{-1} at 222 nm in 0.01 M phosphate buffer at pH 7.4 (4). This value seems rather similar to the value of $-24,000$ ($\pm$ 500) deg cm^2 dmol^{-1} observed for sperm whale Mb (8).

The sequence comparison of the two protozoan Hbs with other known globins is shown in Figure 2. The alignments were made with the program ALIGN of Feng and Doolittle (9). In these positions, Phe-33 and His-68 of *P. caudatum* Hb, as well as Phe-38 and His-73 of *T. pyriformis* Hb, are placed as the invariant Phe at *CD*1 and as the proximal His at *F*8, respectively. If these are correct, both protozoan Hbs will lack the usual distal His at *E*7, its position being replaced by another kind of residue, such as Gln-41 for *P. caudatum* and Gln-46 for *T. pyriformis*. These alignments also suggest the possibility that protozoan Hbs lack the *D* helix and have short *A* helices.

According to our results, the protozoan Hbs are about 20–30 residues smaller than the other known proteins of the globin family, and their amino acid sequences also show no notable degree of similarity with those of other Hbs. Therefore, the study of the crystal structure of these protozoan Hbs and an analysis of their gene structure will no doubt deepen our understanding of the chemistry and evolution of Hbs and Mbs.

Acknowledgements

We thank Drs. K. Hiwatashi and T. Watanabe of this University, and Drs. Y. Watanabe and T. Takemasa, University of Tsukuba, for supplying the seed animals of *P. caudatum* and *T. pyriformis*, respectively. We also thank Dr. R.F. Doolittle, University of California at San Diego, for supplying the computer programs.

References

1. Wakabayashi, S., Matsubara, H. and Webster, D.A. (1986) *Nature* **322**: 481–483.
2. Sato, T. and Tamiya, H. (1937) Cytologia (Tokyo), Fujii Jubilee Volume, 1133–1138.
3. Keilin, D. and Ryley, J.F. (1953) *Nature* **172**: 451–452.
4. Iwaasa, H., Takagi, T. and Shikama, K. (1989) *J. Mol. Biol.* **208**: 355–358.
5. Iwaasa, H., Takagi, T. and Shikama, K. (1990) *J. Biol. Chem.* **265**: 8603–8609.
6. Romero-Herrera, A.E. and Lehmann, H. (1974) *Biochim. Biophys. Acta* **336**: 318–323.

7. Suzuki, T., Takagi, T. and Shikama, K. (1981) *Biochim. Biophys. Acta* **669**: 79–83.
8. Shikama, K., Suzuki, T., Sugawara, Y., Katagiri, T., Takagi, T. and Hatano, M. (1982) *Biochim. Biophys. Acta* **701**: 138–141.
9. Feng, D.-F. and Doolittle, R.F. (1987) *J. Mol. Evol.* **25**: 351–360.
10. Braunitzer, G., Gehring-Muller, R., Hilschmann, N., Hilse, K., Hobom, G., Rudloff, V. and Wittmann-Liebolb, B. (1961) *Hoppe-Seyler's Z. Physiol. Chem.* **325**: 283–286.
11. Buse, G., Steffens, G.J., Braunitzer, G. and Steer, W. (1979) *Hoppe-Seyler's Z. Physiol. Chem.* **360**: 89–97.
12. Egorov, T.A., Kazakov, V.K., Shakhparonov, M.I., Feigina, M.Y. and Kostetsky, P.V. (1978) *Bioorg. Khim.* **4**: 476–480.
13. Takano, T. (1977) *J. Mol. Biol.* **110**: 537–568.

34
Primary Structure of the Beta Chain of the Tetrameric Hemoglobin from *Anadara broughtonii*

Hiroaki Furuta* and Akihiko Kajita

Department of Biochemistry, Dokkyo University School of Medicine, Mibu, Tochigi 321-02, Japan

Introduction

Respiratory proteins are widely distributed throughout most of the animal kingdom. They consist of a broad range of molecules that are adapted to the environment the organisms live. Compared to mammals, a great deal of variation in the structure and function of these proteins is found in invertebrates. These respiratory proteins include the Hbs, Hcs, Hrs and Chls.

In the Arcid family (e.g., *Anadara broughtonii*, *Anadara trapezia*, and *Scapharca inaequivalvis*), dimeric and tetrameric Hbs having an $\alpha_2\beta_2$ subunit structures are packed in the red cells (1–3). These structural features are rarely found among invertebrate O_2 carriers. *A. broughtonii* tetrameric Hb binds O_2 cooperatively, but the binding profiles are independent from effectors such as organic phosphates and protons. This tetramer is of great interest as an example of the parallel evolution of Hb molecules.

This report describes the primary structure of the β chain of the tetrameric HbII from *A. broughtonii* and compares it to that of other Hbs.

---

* Present address: Department of Biochemistry, Vanderbilt University School of Medicine, Nashville, TN 37232.

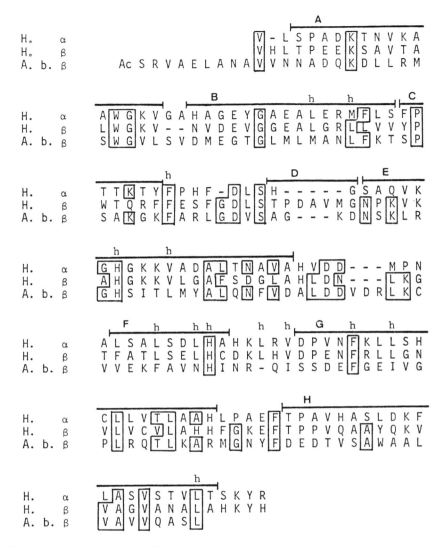

Figure 1. Sequences of the β chain of *A. broughtonii* and the human α and β chains. A.b., *A. broughtonii*; H., human. Bars denote helical sections; and *h*, the amino acid residues in contact with the heme group in the human HbA.

Results and Discussion

Because the amino terminus of the β chain is masked by an acetyl group (4), no sequences are detectable on Edman degradation of the whole chain. The amino acid sequence of the β chain was elucidated by aligning the sequences of the peptides derived from enzymatic digestions and chemical cleavages. The sequence is shown in Figure 1. The β chain is composed of 151 residues from amino terminal acetyl-serine to carboxyl terminal Leu. The total number of the residues of the whole β chain is two residues longer than that of the corresponding α chain. The β chain is also the longest chain of the three distinct chains in the homodimer and the heterotetramer Hbs of *A. broughtonii*. The extent of homology between the β chain and the α chain is 62% and between it and the globin of the homodimer, 51%.

Compared to the human α and β chains, the clam β chain has nine extra residues in the N-terminal region, corresponding to an additional pre-helix forming section. In contrast, the C-terminal region of the clam β chain is five residues shorter. The C-terminal Leu is located at a particular position which is mostly invariant in the vertebrate and invertebrate Hbs. As seen in Figure 1, some residues seem to be lacking in the *D* helical segment as in the human α chain. It was recently reported in an analysis of 2.4 Å x-ray data that the *Scapharca* homodimer Hb lacks a *D* helix (6).

The two His residues in the *A. broughtonii* β chain are located at the same proximal and distal positions as in the corresponding α chain and in the globin I from the homodimer. The heme contact positions, including the His above, are very likely to have been well-preserved from an early point in the evolution of Hb since eleven out of thirteen residues are identical or homologous, and maintain hydrophobicity (Figure 1).

The hydrophobicity profiles of the three globin chains in the *E* helix region are shown in Figure 2. This region is adjacent to ligand-binding sites, an especially important part of Hb function. The profiles in this narrow region, *E*

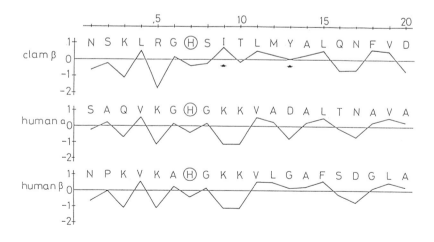

Figure 2. Hydrophobicity profiles of *A. broughtonii* HbII β chain and human Hb α and β chains in the *E* helix region.

helix, closely resemble each other. However, the homologies of the clam β chain to the human α and β chains in the 20 residues are only 30% and 20%, respectively. This is also seen in the *Scapharca* globin chain (7). The two residues marked with arrows in the clam β chain seem to be rather different in polarity or in the side chain volume from those of the human Hb. However, these residues are also evolutionary leftovers in the Arcid Hbs and are likely to have an important effect on subunit interfaces in building a basic dimer. In *Scapharca* homodimer Hb, Ile-71 and Tyr-75 are included in the dimer interface, and associate with *E–E* and *E–F* helical interactions (6). The entire sequence provides some information about invariant and key residues used to maintain the unique quaternary assembly of the Arcid Hb family.

Acknowledgements

We thank Dr. Masato Ohe for his advice and Ms. Yuriko Nozawa and Ms. Naomi Takayama-Hoshino for their technical assistance.

References

1. Furuta, H., Ohe, M. and Kajita, A. (1977) *J. Biochem.* **82**: 1723–1730.
2. Gilbert, A.T. and Thompson, E.O.P. (1985) *Aust. J. Biol. Sci.* **38**: 221–236.
3. Chiancone, E., Vecchini, P. Verzili, D., Ascoli, F. and Antonini, E. (1985) *J. Mol. Biol.* **152**: 577–592.
4. Furuta, H. and Kajita, A. (1986) In *Invertebrate Oxygen Carriers*, ed. B. Linzen, 117–121. Berlin: Springer-Verlag.
5. Furuta, H. and Kajita, A. (1983) *Biochemistry* **22**: 907–914.
6. Royer, W.E., Hendrickson, W.A. and Chiancone, E. (1989) *J. Biol. Chem.* **264**: 21052–21061.
7. Petruzzelli, R., Goffredo, B.M., Barra, D., Bossa, F., Boffi, A., Verzili, D., Ascoli, F. and Chiancone, E. (1985) *FEBS Lett.* **184**: 328–332.

35

Partial Amino Acid Sequence of Hemoglobin II from *Lucina pectinata*

Jerrolynn Hockenhull-Johnson,[a]
Mary S. Stern,[a] Daniel A. Walz,[a]
David W. Kraus[b] and
Jonathan B. Wittenberg[b]

[a]Department of Physiology, Wayne State University School of Medicine, Detroit, MI 48201, USA
[b]Department of Physiology and Biophysics, Albert Einstein College of Medicine, Bronx, NY 10461, USA

Introduction

The Puerto Rican clam *Lucina pectinata*, a bivalve mollusc, has three cytoplasmic Hbs associated with its gill: HbI, HbII and HbIII. The clam utilizes these three Hbs in symbiosis with intracellular chemoautotrophic bacteria (1). These bacteria reside in the gill of the clam where its requirement for O_2 and hydrogen sulfide is provided by the clam *via* the Hbs (2).

In the presence of O_2, HbI reacts with hydrogen sulfide to form ferric Hb sulfide, which helps to convert CO_2 to hexoses (1). HbI is called sulfide-reactive Hb. HbII and HbIII remain oxygenated in the presence of H_2S and are called O_2-reactive Hbs. It is believed that HbII and HbIII are involved in O_2 transport, which benefits both the clam and the bacteria, possibly functioning as a terminal oxidase (1). Oxygen kinetic studies of the three Hbs suggest that another residue replaces the distal His in all three Hbs (2).

Table 1. HbII sequence data from CNBr hydrolysis.

Fragment[a]	Sequence	No. of Amino Acids
1	EDHNHM	6
2	DNGVSNGQGFYM	12
3	SSFVDHLDDNM	11
4	KAQSLVFCNGM	11
5	LVVLIQKM	8
6	SSFVDGLDDNM	11
7	Blocked N-terminal	19
8	AKLHNNRGIRASDLRTAYDILIHYM	25
9	KEL	3
10	DLFKAHPETLTPFKSLFGGLTLAQLQDNPKM	31
11	VGGAKDAWEVFVGFICKTLGDYM	23

[a]Fragment assignment is based on the relative elution position for a C-18 column developed with a gradient of 0.1% TFA and 0.1T% TFA in 90% acetonitrile.

Table 2. HbII sequence data following tryptic digestion.

Peptide[a]	Sequence	No. of Amino Acids
1	SSWSK	5
2	Blocked N-terminal	8
3	TLGDYMK	7
4	AHPETLTPFK	10
5	AGPETLTPFK	10
6	SLFGGLTLAQLQDNPK	16
7	FMDNGVSNGQGFYMDLFK	18
8	TAYDILIHYMEDHNHMVGGAK	21
9	DAWEVFVGFICK	12
10	AQSLVFCNGMSSFVDHLDDNM	21

[a]Peptide assignment is based on the relative elution position of peptides from a C-18 column developed with 90% acetonitrile in 0.1% trifluoroacetic acid.

The focus of our current work is on HbII. Our goal is to ascertain the complete amino acid sequence of HbII and, therefore, to elucidate which amino acid replaces the distal His; it has been proposed that Tyr is the residue replacing His (2).

Results and Discussion

HbII, prepared as described elsewhere (2), was heme-stripped, purified on a C8 reversed-phase column, and any disulfide bonds were reduced and carboxymethylated. HbII was then fragmented with cyanogen bromide, which hydrolyzes proteins at methionine residues, and also fragmented with trypsin. The cyanogen bromide (CNBr) hydrolysis yielded 11 peptides and the tryptic digest yielded 10 peptides. The sequence data for the CNBr peptides is provided in Table 1 and that for the tryptic peptides is given in Table 2.

Based on amino acid composition data (not shown), HbII has approximately 150 residues. The amino-terminus of HbII is blocked, and thus prevents direct sequence determination. The results for the CNBr sequence (Table 1) account for 160 residues. A careful examination of the 11 fragments examined suggests that fragments 3 (*SSFVDHLDDNM*) and 6 (*SSFVDGLDDNM*) are related, since they are identical in 10 of the 11 positions and differ only in *G* versus *H* at position 6. These two fragments were well-resolved on C18 columns (data not shown), although the His-containing fragment (fragment 3) is recovered in greater quantities. Trypsin proteolysis yielded 10 distinct peptides (Table 2) which could account for 128 amino acids. All peptides, as well as fragments following CNBr hydrolysis, were analyzed for amino acid content in addition to sequence analysis. Thus, CNBr fragment 7 (Table 1) and tryptic peptide 2 (Table 2) were blocked with respect to sequence analysis, which placed them as the NH2-terminal pieces, and their approximate length was established by amino acid composition. One more pair of near-identical peptides, peptides four and five (Table 2) were resolved; again they differed by a Gly and a His. This peptide was previously identified as a component of CNBr fragment 10, where the questioned residue was identified as a His.

We have recently completed the analyses of a series of peptides derived from separate chymotrypsin and endopeptidase Asp-N hydrolyses and are able to overlap virtually all of the tryptic and cyanogen peptides. Based upon these analyses, coupled with the knowledge that the amino-terminus of this globin is blocked, we propose the following alignment of the CNBr fragments as presented in Table 1:

7-2-10-4-3-5-8-1-11-9

Similarly, the alignment of the tryptic data from Table 2 would have the following sequence of peptides:

2-1-7-4-6-10-8-9-3

The tentative sequence of *Lucina* HbII structure, beginning with residue 20 and terminating with residue 148, is as follows:

(residues 1–17 unidentified)-FMDNGVSNGQGFYMDLFKAHPETLTPFKSL FGGLTLAQLQDNPKMKAQSLVFCNGMSSFVDHLDDNMLVVLIQKMAKL HNNRGIRASDLRTAYDILIHYMEDHNHMVGGAKDAWEVFVGFICKTLGD YMK-(residues 148–150 unidentified).

With such information and on the basis of tentative alignments with known globin sequences, we are able to make the additional observation that this globin probably lacks a His in the distal position and appears to have a Gln (Gln65, underlined), instead. Furthermore, the *Lucina* globin contains 2 Cys residues at positions 70 and 139 which are disulfide linked, perhaps placing unusual conformational constraints on the folding of the heme binding site.

Acknowledgements

This work was supported by a grant from the U.S. Public Health Service, DK 30382. Training grants T32 HL07624 and T32 HL07602 supported J.H.J. and M.S.S., respectively.

References

1. Wittenberg, J.B. and Wittenberg, B.A. (1990) *Annu. Rev. Biophys. Chem.* **19**: 217–241.
2. Kraus, D.W. and Wittenberg, J.B. (1990) *J. Biol. Chem.* **265**: 16043–16053.

36
The Globin Composition of *Daphnia pulex* Hemoglobin

Kris Peeters and Luc Moens

Department of Biochemistry, University of Antwerp, Universiteitsplein 1, 2610 Wilrijk, Belgium

Introduction

Although the majority of Crustacea have Hc as an O_2 carrier, extracellular Hbs are mainly found in the hemolymph of the Branchiopoda. Basing our analysis on globin structure and using Vinogradov's classification, we may distinguish two groups: the multi-domain–multisubunit Hbs found in Anostraca, and the two domain–multisubunit Hbs found in Notostraca, Conchostraca and Cladocera (1). The Hb of *Daphnia pulex* is a representative of the latter. It has an $S_{20°,w}$ of 16.4–16.9, an M_r of 420,000–470,000, contains one mole heme per $M_r = 18,000–20,000$, and is composed of different globin chains with $M_r = 31,000–37,000$. Limited proteolysis suggests that one globin chain is comprised of two heme-containing O_2 binding domains with $M_r = 17,000$ (2). In this work we describe the globin composition of the *D. pulex* Hb.

Results and Discussion

The globin composition of *D. pulex* Hb, purified as described (3), was analyzed by two-dimensional electrophoresis. Using the O'Farrell technique with SDS-PAGE ($T = 12.5\%$, $C = 2.7\%$) in the second dimension, three globin chains with very similar M_r and p*I*, α, β and γ, and material having an M_r around 70,000, may be distinguished (Table 1). The amount of this higher M_r material is highly increased when 2-mercaptoethanol is omitted from the denaturation buffer. When run on a 7.5% SDS gel without reduction, three bands (I, II and III) are observed (Table 1). When reduced in 5% 2-

Table 1. Physicochemical characteristics of the globin chains and their dimers of *D. pulex.*

Globin Chain or Protein Fraction	M_r (SDS-PAGE)	Apparent pI (in 8 M urea)	Globin-Chain Composition
α	35,500 ± 500	7.5	–
$\beta1$	34,100 ± 400	7.6	–
$\beta2$	33,700 ± 400	7.6	–
γ	32,600 ± 700	7.7	–
I	71,600	–	α - α
II	69,200	–	α - βx
III	68,000	–	βx - βx

mercaptoethanol and rerun on a 12.5% gel, each band dissociates into α- and β-globin chains, demonstrating that they are disulfide linked dimers. Disulfide linked globin chains are also observed in the Hb of *Lepidurus apus* (4) and *Caenestheria inopinata* (5), and many other invertebrates, particularly the Annelida (1). When an SDS-PAGE with lower bisacrylamide concentrations (C = 0.7%) is used, the β-globin band may be separated into two subspecies, $\beta1$ and $\beta2$ (Table 1).

The single Hb phenotype of *D. pulex* thus contains four globin chains, α, $\beta1$, $\beta2$ and γ. A similar chain heterogeneity is observed in all vertebrate Hbs and in most invertebrate Hbs, as it is essential in establishing cooperative O_2 binding (6). Accepting an apparent M_r of 420,000–460,000 for the Hb and an M_r of 32,000–35,000 for the globin chains without heme, the presence of 12 globin chains in the native Hb is suggested. Considering that a certain percentage of dimers are identified, it is not clear if the globin chains first form dimeric subunits before aggregating into the functional molecule (as in many annelid Hbs) or if these dimers are the artificial products of the isolation procedure.

In order to separate the globin chains, Hb was denatured in 6M guanidinium chloride, 2% 2-mercaptoethanol, 100°C, and loaded onto a wide-pore Bakerbond C4-column (4.3 mm x 25 cm). Protein material was eluted at 1 ml/min with a gradient of acetonitrile in 0.1% trifluoroacetic acid (Figure 1A). In contrast with many vertebrates and invertebrates (7, 8), no clear-cut separation of all globin chains was obtained. Only the α-chain elutes as a single peak, whereas the other globins co-elute (β, γ) and/or run as aggregates ($\beta - \gamma$, $\alpha - \beta - \gamma$) (Figure 1B). Carboxymethylation of the samples or modification of the separation conditions (gradient, pH and ionic strength) do not result in a dissociation of the aggregate or in a better separation.

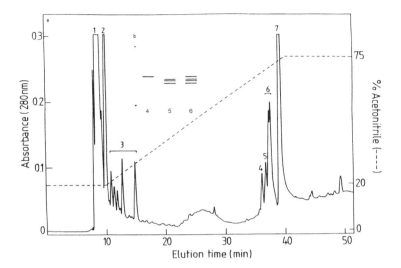

Figure 1. (*a*) Reversed-phase chromatography of 100 µg denatured *D. pulex* Hb. Conditions as described in text; (*b*) 15% SDS-PAGE of protein fractions 4, 5 and 6 after reversed-phase chromatography. Fraction 7 contains heme.

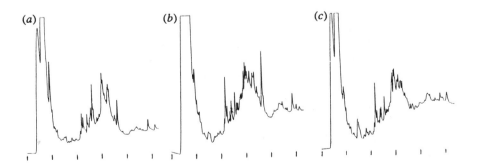

Figure 2. Peptide mapping of globin chains of *D. pulex*. Method as described in text; (*a*) α chain, (*b*) β chain, (*c*) γ chain.

Table 2. Amino-acid composition of the Hb and constituent globin chains of *D. pulex*. Hb was purified by column chromatography, globin chains by semi-preparative SDS-PAGE. Acid hydrolysis was performed in 6 N HCl vapour in the presence of 1% phenol and 4% thioglycolic acid for 24 hours. Values for Ser, Thr, Ile, and Val were corrected according to Reeck (9). Amino-acid analysis was carried out by RPLC according to the PicoTag method (10). Values are expressed as mole %. Trp was not determined.

	Hb	α chain	β chain	γ chain
Lys	6.4	2.8	4.1	5.2
His	1.9	2.3	1.7	1.5
Arg	3.6	6.6	5.6	5.8
Asx	10.1	9.4	7.5	7.4
Thr	7.5	9.2	8.8	8.3
Ser	6.6	9.1	7.8	9.4
Glx	12.5	12.4	8.3	9.1
Pro	5.1	8.4	11.3	8.6
Gly	5.9	9.6	11.4	11.8
Ala	12.3	11.1	9.7	9.7
Cys[a]	0.4	0.1	0.4	0.5
Val[b]	6.1	6.1	6.1	6.1
Ile	5.2	3.3	3.8	4.0
Leu	9.5	5.0	6.9	6.8
Tyr	1.7	1.5	2.4	2.1
Phe	4.0	1.7	2.2	2.6
Met	0.9	1.2	1.6	1.5

[a]Val content of globin chains not reliable and taken as a constant.
[b]Determined as carboxymethyl Cys in Hb and not analyzed separately in globin chains.

As an alternative, the α-, β – (β1 + β2) and γ-chains were isolated on a small scale by preparative SDS-PAGE. The purity of the a-chain isolated by SDS-PAGE is confirmed by vapor phase sequencing. Twenty-six of the amino terminal residues could be identified unambiguously. The amino acid composition of the *D. pulex* Hb is typical for invertebrate globins, showing high Glx, Asx and Ala values (Table 2).

Due to the method of isolating the globin chains (SDS-PAGE), their amino acid compositions must be interpreted with care. The determination of Val and Pro are unreliable because impurity peaks, presumably derived from the PAGE system, coelute with these PTC-amino acids. Furthermore, glycine from the electrophoresis buffer might not be completely removed by Poly *F* chromatography after eluting the gel pieces (11). However, the amino acid compositions of the three globin chains are very similar to each other, as might be expected from their very similar physicochemical characteristics. The β- and γ-chains possess almost the same amino acid composition.

This conclusion is confirmed by peptide mapping (Figure 2a–c). Globin chains separated by SDS-PAGE were digested with trypsin in the gel matrix and the resulting peptides were analyzed by RPLC, according to Eckerskorn and Lottspeich (12). Peptide patterns of β- and γ-chains are very similar, with the exception of one major peak in the β-chain that is lacking in the γ-chain. The pattern of the α-chain differs considerably from that of the other chains.

Acknowledgements

This work was supported by the Fund for Joint Basic Research, Belgium (Program No. 3.0103.90) and by the Belgian Institute for the Encouragement of Scientific Research in Industry and Agriculture.

References

1. Vinogradov, S.N. (1985) *Comp. Biochem. Physiol.* **82**B: 1–15.
2. Dangott, L.J. and Terwilliger, R.C. (1980) *Comp. Biochem. Physiol.* **67**B: 301–306.
3. Peeters, K., Mertens, J., Hebert, P. and Moens, L. (1990) *Comp. Biochem. Physiol.* **97**B: 369–381.
4. Dangott, L.J. and Terwilliger, R.C. (1979) *Biochim. Biophys. Acta* **579**: 452–461.
5. Ilan, E., David, M.M. and Daniel, E. (1981) *Biochem.* **20**: 6190–6194.
6. Wood, E.J. (1980) *Essays Biochem.* **16**: 1–47.
7. Fushitani, K., Matsuura, M.S.A. and Riggs, A.F. (1988) *J. Biol. Chem.* **263**: 6502–6517.
8. Moens, L., Ver Donck, K., De Smet, K., Van Hauwaert, M.-L., van Beeumen, J., Allard, P., Wodak, S. and Trotman, C.N.A. (1989) In *Cell and Molecular Biology of Artemia Development*, eds. A.H. Warner, T.H. MacRae and J.C. Bagshaw, 429–438. New York: Plenum Press.
9. Reeck, G. (1970) In *Handbook of Biochemistry*, ed. H.A. Sober, C121. Cleveland: CRC-Press.
10. Bidlingmeyer, B.A., Cohen, S.A. and Tarvin, T.L. (1984) *J. Chromat.* **336**: 93–104.
11. Van Fleteren, J.R. and Peeters, K. (1990) *J. Biochem. Biophys. Methods* **20**: 227–235.
12. Eckerskorn, C. and Lottspeich, F. (1989) *Chromatographia* **28**: 92–94.

37

The Hemoglobin of *Ascaris suum*: Structure and Partial Sequence of Domain 1

Ivo de Baere,[a] Lu Liu,[a] Jozef van Beeumen[b] and Luc Moens[a]

[a]Department of Biochemistry, University of Antwerp, Universiteitsplein 1, 2610 Wilrijk, Belgium
[b]Laboratory of Microbiology, State University of Ghent, Ledeganckstraat 35, 9000 Ghent, Belgium

Introduction

Contrary to what is expected for an intestinal parasite, the common roundworm, *Ascaris suum*, is a normal aerobic animal that uses O_2 as its terminal electron acceptor (1, 2). Consequently, O_2 carriers are present in this species. Heme-containing respiratory pigments, intracellular in the body wall (3) and extracellular in the perienteric fluid (4), have been identified, and because of their locale, labeled Mb and Hb, respectively. Both types of molecules have been partially characterized. Their physicochemical and functional characteristics are summarized in Table 1.

Both molecules exhibit exceptionally high O_2 affinity, although that of the Hb is 100 times higher than that of the Mb. This suggests that, contrary to all other systems, O_2 is transferred from the Mb to the Hb, which raises questions about the functional role of the latter (9).

Based on the M_r of the native Hb and of its constituent globin chains, the molecule must be classified as a two-domain, multi-subunit Hb (10). Electron microscopic observations show rectangular particles with dimensions of 70 x

Table 1. Physicochemical characteristics of *Ascaris* Hb and Mb.

Native Molecule	Hb	Mb
$S_{20°,w}$	11.7–11.8 (5)	3.1 (3)
M_r equil. sed.	328,000 (5,6)	37,000 (3)
gel filtration	425,000 (7)	–
pI	5.0 (4)	–
n (g/ml)	0.725 (4)	–
f/f_0	1.51 (5)	–
heme/M_r	1/21,6000 (6)	1/37,000 (3)
	1/40,000 (4)	–
E. M. dimensions (Å)	70 x 200 (7)	–
no. of phenotypes	1 (4)	2 (3)

Ligand Binding	O_2	CO	O_2
P_{50} (mm Hg)	0.001–0.004	0.1 (8)	0.11 (9)
k' (mM^{-1}.s^{-1})	1.5	0.21 (8)	1.2 (9)
$t^{1/2}$ (msec)	342	3.3 (8)	–
$k\ aff$ (s)	0.004	0.018 (8)	0.23 (9)
$t_{1/2}$ (s)	173	139 (8)	–

Globin Chains	Hb	Mb
M_r equil. sed.	40,600 (5, 6)	37,000 (3)
SDS-PAGE	43,000 ± 2,000 (6)	–
	38,500 ± 500 (7)	–
pI	6.6 (5, 7)	–
Sugar Content (%w/w)	3 (7)	–
Number of Chains	1 (7)	–

200 Å. Assuming an M_r of 328,000 for the native molecule and a two-domain structure for each globin chain of 38,500, an overall model of the Hb molecule is presented in Figure 1.

In contrast with Svedberg's hypothesis (1 heme/M_r = 16,000), heme analysis of the *Ascaris* Hb originally suggested a heme/M_r = 40,600 (4). Recently, Darawshe et al. (6) concluded from heme titration experiments that each globin chain is essentially able to bind two heme groups.

The exceptional O_2 affinity and heme binding must be the result of structural changes in the heme environment of the globin-folded domains. Perutz (9) suggested that the high O_2 affinity may result from an additional hydrogen bond between the O_2 ligand and a more highly charged residue, such as Asp, at position $E11$. Similarly, the introduction of more highly charged residues in the critical heme environment might weaken the heme binding. Only with sequence data may we confirm or disprove this hypothesis.

Results and Discussion

Ascaris Hb was purified from the perienteric fluid essentially as described (4). The preparation was finally purified by FPLC Superose 6 gel filtration. Analysis of the native Hb by different electrophoretic techniques (carrier and pH) clearly shows the presence of a single phenotype. Similarly, two dimensional electrophoresis (11) only shows a single globin chain. However, the globin spot is rather broad, suggesting the presence of glycosidic groups. Glycosylation of the *Ascaris* Hb was already suggested by the presence of glycosamine in the amino acid composition (4), confirmed by glycoprotein staining on SDS-PAGE and quantified with phenol sulfuric acid staining (7).

Sequence analysis on a vapor phase sequencer (Applied Biosystems 477 A) confirms the presence of a single globin chain (7). The primary structure of the first domain up to residue 121 is reconstructed from peptides obtained by cleavage with trypsin, chymotrypsin, *Staphylococcus aureus* V8 protease and cyanogen bromide and limited acid hydrolysis (Figure 2).

Analysis of this sequence, using Template II of Bashford *et al.* (12), indicates the globin-like nature of the first domain. Further partial sequence information (not shown) indicates that this is true for domain 2 as well. Helices *A* to *G* are clearly present in the first domain, although the *A* helix is rather short and lacks the highly conserved residues Leu and Trp at *NA2* and *A*12. Major deviations of the template are recognized in the *E, F* and *G* helices (Table 2) for both surface and interior locations. The latter suggests structural changes in the heme environment.

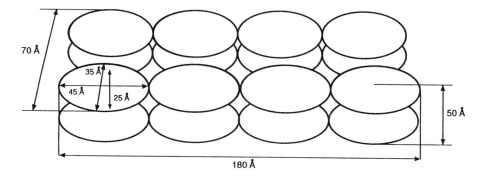

Figure 1. Hypothetical model of *Ascaris* Hb.

```
              NA      A                AB     B                 C        CD    D
              1 2     1   5   0   1    1      1   5   0   1 5   1 5 0 0 0 1 1   1 1 5 0
HAHU          ------V-L  SPADKTNVKAAWGKVG  A----  HAGEYGAEALERMFLS  FPTTKTY  FPHF  -DLS----HG
HBHU          ---VHL     TPEEKSAVTALWGKV-  -----  NVDEVGGEALGRLLVV  YPWTQRF  FESF  GDLSTPDAVMG
MYWHP         ------V-L  SEGEWQLVLHVWAKVE  A----  DVAGHGQDILIRLFKS  HPETLEK  FDRF  KHLKTEAEMKA
GGICE3        ------V-L  SADQISTVQASFDKVK  G----  ---DPVGILYAVFKA   DPSIMAK  FTQF  AG KDLESIKG
GGLMS         PIVDTGSVAPL SAAEKTKIRSAWAPVY S----  TYETSGVDILVKFFTS  TPAAQEF  FPKF  KGLTTADQLKK
GGGAA         ------SL   SAAEADLAGKSWAPVF  A----  NKNANGADFLVALFEK  FPDSANF  FADF  KGKSVAD-IKA
GGNKID        PSVYDAAAQ-L TADVKKDLRDSWKVIG S----  DKKGNGVALMTTLFAD  NQETIGY  FKRL  GNV--SQGMA
GPYL2         ------GAL  TESQAALVKSSWEEFN  A----  NIPKHTHRFFILVLEI  APAAKDL  FS-F  LK-GTSEVPQN
GGNW1B        ------GL   SAAQRQVIAATWKDIA  GN--D  NGAGVGKDCLIKHLSA  HPQMAAV  FG-F  SG----AS---

ASCDOM1       ------     -NKTRELEMKSLEHAK  VDTSN  EARQDGIDLYKHMFEN  YPPLRKY  FKNR  EEYTAEDV-QN
penalty score            0 0 * 000000          0 00 000           0 0 0    00


              E                     EF            F          FG       G
              1 5   1 2 0           1             1 5        1 4      1 5
HAHU          SAQVKGHGKKVADALTNAVA  HV---D--DMPNA  LSALSDLHA  HKL--RV  DPVNFKLLSHCLLVTLAAH  :
HBHU          NPKVKAHGKKVLGAFSDGLA  HL---D--NLKGT  FATLSELHC  DKL--HV  DPENFRLLGNVLVCVLAHH  :
MYWHP         SEDLKKHGVTVLTALGAILK  K---K-GHHEAE   LKPLAQSHA  TKH--KI  PIKYLEFISEAIIHVLHSR  :
GGICE3        TAPFETHANRIVGFFSKIIG  EL--P--NIEAD   VNTFVASHK  PRG---V  THDQLNNFRAGFVSYMKAH  :
GGLMS         SADVRWHAERIINAVNDAVA  SM--DDTEKMSMK  LRDLSGKHA  KSF--QV  DPQYFKVLAAVIADTVAAG  :
GGGAA         SPKLRDVSSRIFTRLNEFVN  DA--ANAGKMSAM  LSQFAKEHV  GFG---V  GSAQFENVRSMFPGFVASV  :
GGNKID        NDKLRGHSITLMYALQNEID  QL--DNPDDLVCV  VEKFAVNHI  TRK---I  SAAEFGKINGPIKKVLASK  :
GPYL2         NPELQAHAGKVFKLVYEAAI  QLEVTGVVVSDAT  LKNLGSVHV  SKG---V  ADAHFPVVKEAILKTIKEV  :
GGNW1B        DPAVADLGAKVLAZIGVAVS  HL--GDZGKMVAQ  MKAVGVRHK  GYGNKHI  KGQYFEPLGASLLSAMEHR  :

ASCDOM1       DPFFAKQGQKILLADHVLEA  TY DDRETFNAY  TRELLDRHA  RDLV-QM  PPEVWTDFWKLFEEYLGKK  :
penalty score 00*000700000*0* 50*0    000700       0007 0    2      70 0*0 0  000
```

Figure 2. Alignment of the partial sequence of domain 1 of *Ascaris* Hb (ASCDOM1) with globin sequences of known crystallographic structure. Penalty scores (without decimal point) of ASCDOM1 with Bashford's Template II (12) are placed under the sequence. Amino acids indicated with an asterisk are not listed in the template. Abbreviations are those of the *NBRF* databank: *HAHU*: α chain human Hb; *HBHU*: β chain human Hb; *MYWHP*: sperm whale Mb; *GGICE3*: *Chironomus* globin III; *GGLMS*: sea lamprey globin V; *GGGAA*: *Aplysia limacina* globin; *GGNKID*: *Scapharca inaequivalvis* globin I; *GPYL2*: *Lupinus* legHbII; *GGNW1B*: *Glycera dibranchiata* globin I.

Although the overall hydrophobicity of the *Ascaris* Hb sequence is quite similar to that of other globins, more hydrophobic regions may be observed in the *A*, *B* and *G* helices (Figure 3).

Inspection of the residues involved in heme binding (14) (Table 2) clearly indicates that, with the exceptions of *E*15 (Asp), *F*7 (Arg) and *G*5 (Trp), their physicochemical character (hydrophobicity/volume) is conserved.

The combined presence of Gln (*E*7) and Ile (*E*11) is rare but not uncommon. This is also observed in *Chironomus* globin IIIA and *Myxine glutinosa* globin. Exceptional O_2 affinity was not observed in either case. The Ile at *E*11 does not fit into Perutz's hypothesis.

From the available sequence information, it is inconclusive whether or not the first domain is responsible for the strong O_2 binding or for the incomplete heme binding, or if both phenomena may occur in the same domain. In our opinion,

Table 2. Summary of amino acids differing from Template II and amino acids involved in heme binding.

Position	Residue	Penalty Score (12)	Location	Amino Acid Observed in 226 Sequences (12)
A8	E	1	interior	V(154) I(65)
C4	L	0.2	heme contact	T(194)
C7	Y	no data	heme contact	Y(92) F(70) K(36)
CD1	F	0	heme contact	F(226)
CD3	N	no data	heme contact	no data
E3	F	1	interior	K(78) Q(62) D(43)
E7	Q	0.7	int./heme cont.	H(219)
E10	K	0	heme contact	K(156) T(42)
E11	I	0	heme contact	V(198) I(22)
E13	L	1	surface	T(62) A(29)D(42)G(34)
E14	A	0	heme contact	A(138) S(54)
E15	D	1	int./heme cont.	L(118) F(73)
E17	V	0.5	surface	D(51) E(62)N(51)G(24)
E19	E	1	interior	V(90) L(75)
F4	L	0	heme contact	I(215)
F5	L	0.7	interior	F(162)
F7	R	no data	heme contact	L(142)
F8	H	0	heme contact	H(226)
FG2	V	0	heme contact	L(155) H(39) G(19)
FG4	M	0.2	int./heme cont.	V(174) I(50)
G4	V	no data	heme contact	N(152) W(36)
G5	W	0.7	int./heme cont.	F(186)
G8	F	0	heme contact	L(161) I(40) F(13)
G9	W	1	surface	G(88) S(81)

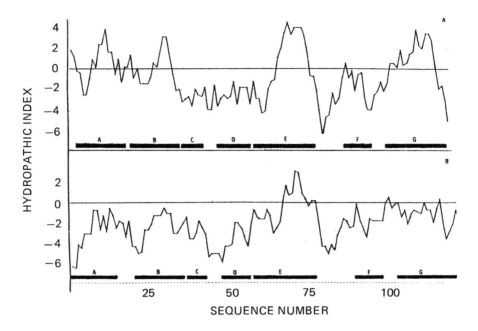

Figure 3. Comparison of the hydrophobicity profiles of sperm whale Mb (*A*) and domain 1 of *Ascaris* Hb (*B*). The profiles were calculated according to Kyte and Doolittle (15) using a window length of seven residues.

the exceptional characteristics, heme binding and O_2 affinity, of the *Ascaris* Hb may not be attributed to the substitution of a single residue. Most likely they will be the result of several substitutions scattered over the sequence, and to the interaction of the domains in the native molecule.

Acknowledgements

This work was supported by the Belgian Institute for Encouragement of Scientific Research in Industry and Agriculture.

References

1. Smith, M.H. (1969) *Nature* **223**: 1129–1132.
2. Lee, D.L. and Smith, M.H. (1965) *Exp. Parasitol.* **16**: 392–424.

3. Okazaki, T., Wittenberg, B.A., Briehl, R.W. and Wittenberg, J.B. (1967) *Biochim. Biophys. Acta* **140**: 258–265.
4. Wittenberg, B.A., Okazaki, T. and Wittenberg, J.B. (1965) *Biochim. Biophys. Acta* **111**: 485–495.
5. Okazaki, T., Briehl, R.W., Wittenberg, J.B. and Wittenberg, B.A. (1965) *Biochim. Biophys. Acta* **111**: 496–502.
6. Darawshe, S., Tsafadyah, B. and Daniel, E. (1987) *Biochem. J.* **242**: 689–694.
7. De Baere, I., Liu, L., Leurentop, L., van Beeumen, J. and Moens, L. (1991) In *Invertebrate Dioxygen Carriers*, ed G. Préaux. Leuven: Leuven University Press. In press.
8. Okazaki, T. and Wittenberg, J.B. (1965) *Biochim. Biophys. Acta* **111**: 503–511.
9. Perutz, M. (1987) Personal Communication.
10. Vinogradov, S.N. (1985) *Comp. Physiol. Biochem.* **82B**: 1–15.
11. O'Farrell, P.H. (1975) *J. Biol. Chem.* **250**: 4007–4021.
12. Bashford, D., Clothia, C. and Lesk, A.M. (1987) *J. Mol. Biol.* **196**: 199–216.
13. Lesk, A.M. and Clothia, C. (1980) *J. Mol. Biol.* **136**: 225–270.
14. Kyte, J. and Doolittle, R.F. (1982) *J. Mol. Biol.* **157**: 105–132.

38
Nomenclature of the Major Constituent Chains Common to Annelid Extracellular Hemoglobins

Toshio Gotoh,[a] Tomohiko Suzuki[b] and
Takashi Takagi[c]

[a]Department of Biology, College of General Education, University of
Tokushima, Tokushima 770, Japan
[b]Department of Biology, Faculty of Science, Kochi University, Kochi
780, Japan
[c]Biological Institute, Faculty of Science, Tohoku University, Sendai
980, Japan

We propose a new nomenclature for the similar constituent chains in the Hbs
of the polychaete *Tylorrhynchus heterochaetus* and the oligochaete *Lumbricus
terrestris*, and suggest that, since these two species are far apart in the phylum
Annelida, the proposed nomenclature may be extended to other extracellular
annelid Hbs, as well (1).

We have proposed a model for the subunit assembly of *Tylorrhynchus* Hb that
consists of 216 chains of six different polypeptide chains (2). To construct a
common model for the molecular architecture of annelid extracellular Hbs, we
must establish a nomenclature of the constituent chains that is common to these
molecules. If the structure of an annelid Hb were represented by common letters
for the constituent chains, such as $\alpha_2\beta_2$ used for vertebrate Hbs, discussions
about the molecular assembly of multi-subunit Hbs (see reference 3 for review)

would be much clearer and simpler for both biologists and protein chemists to understand. Different names have been used even for the respective constituent chains of *Lumbricus* Hb in different laboratories. Vinogradov's group named the heme-binding chains of *Lumbricus* HbI, II, III and IV (4); whereas Riggs' group employed the alphabetical names *a* (corresponding to chain IV), *b* (II), *c* (III) and *d* (I) for the globin chains (5). These arbitrary names were given according to either the order of mobility on SDS-PAGE (4) or the elution order on column chromatography (5). An individual species, whether of organisms or of bio-macromolecules, should have only one name to avoid confusion. Animals are named according to the *International Rules of Zoological Nomenclature* (6). The best way to name polypeptide chains might be to construct a system based on evolutionary relationships. In fact, when Garlick and Riggs found that one of the constituent chains of *Lumbricus* "erythrocruorin" is homologous with vertebrate Hbs (7), they recommended discarding the old name "erythrocruorin" and using "hemoglobin" instead.

Considering the homology of amino acid sequences, we would like to propose a common nomenclature for the constituent chains of *Tylorrhynchus* and *Lumbricus* Hbs. Alignment of the amino acid sequences of the four major globin chains of *Tylorrhynchus* Hb (8) with those of the four chains of *Lumbricus* Hb (5) showed that the eight chains could be clearly distinguished as two strains (5, 9): 'A', consisting of *Tylorrhynchus* chains I and IIA and *Lumbricus* chains I (*d*) and II (*b*), and 'B', consisting of *Tylorrhynchus* chains IIB and IIC and *Lumbricus* chains III (*c*) and IV (*a*). Moreover, the two chains of each strain in both species may be distinguished by their characteristic numbers of half-cystine residues (5, 8). For instance, *Tylorrhynchus* chain I and *Lumbricus* chain I both have two half-cystine residues. Thus, these chains are named '*a*' (read "little *a*"). The other chain of strain *A* in both species has three half-cystine residues, and is named '*A*' (read "big *A*"). Similarly, chains '*b*' (*Tylorrhynchus* chain IIC and *Lumbricus* chain IV) and '*B*' (*Tylorrhynchus* chain IIB and *Lumbricus* chain III) in strain *B* may be defined as those having four and three half-cystine residues, respectively. Chain *a* is characterized as the smallest polypeptide of the four heme-binding chains *a*, *A*, *b*, *B* (5, 10). It has two half-cystine residues and an intra-chain disulfide bond, and thus exists as a "monomer" in the Hbs of both *Tylorrhynchus* and *Lumbricus* (5, 10). On the other hand, chains *A*, *b* and *B* form a disulfide-bonded trimer "A–b–B," or "AbB" (5, 10). Note that *little b* is situated between *big A* and *big B*. Biologists might be familiar with the symbolic term "AaBb" used to designate a heterozygote of genetic factors by G. Mendel. Therefore, we do not think it is unsuitable to use the proposed name "$a \cdot AbB$" for the minimum entity of a submultiple of the giant Hbs. Each submultiple of *Tylorrhynchus* Hb may be represented as " $(a \cdot AbB)_4$ " (11). the 12 submultiples, $(a \cdot AbB)_{48}$, might be stabilized by 12 homodimers of linker chains to form an Hb molecule with 216 polypeptide chains (2). Table 1 shows the recent models proposed for the subunit assembly of giant annelid Hbs represented by the proposed nomenclature. Differences between the three models may be clearly seen.

Table 1. Recent models of the subunit assembly of *Tylorrhynchus* and *Lumbricus* Hbs using the proposed nomenclature.

Hb	Model	Reference
Tylorrhynchus	$(a \cdot AbB)_{48} \cdot L_{24}$	Suzuki *et al.* (3)
Lumbricus	$(a \cdot AbB)_{48} \cdot L_{12}$	Fushitani & Riggs (12)
Lumbricus	$(a \cdot AbB)_{36} \cdot L_{36}$	Vinogradov *et al.* (13); Kapp *et al.* (14)

The proposed nomenclature is not yet perfect. As already mentioned (1), it must be slightly modified because leech Hbs contain disulfide-bonded dimers of heme-containing chains instead of trimers (14) and the chlorocruorins consist of tetramers and dimers (15). Furthermore, we have to devise some way to indicate possible heterogeneity of components. Fortunately, the chains of *Tylorrhynchus* and *Lumbricus* appear to show little, if any, heterogeneity (5, 8), perhaps because of steric limitations of subunit assembly of these multi-subunit Hbs.

The current use of different names for the same polypeptide chains of *Lumbricus* Hb in different laboratories results in unnecessary confusion. We should devise a suitable nomenclature for the polypeptide chains based on the *International Rules of Zoological Nomenclature* (6). The international code has developed some concepts such as synonym, homonym and *nomen conservandum* to avoid confusion. We hope that our proposal will be the first step towards the establishment of an ideal nomenclature for the polypeptide chains of invertebrate giant Hbs.

Comments

Chiancone, E. (University of Rome, Rome): It is indeed most desirable to agree on a common nomenclature for annelid Hb chains. Each classification in use has merits; for example, the one proposed in your paper brings out the evolutionary relationship among the chains. Furthermore, it stresses the dichotomy of *A* and *B* type polypeptide chains, which is reminiscent of mammalian Hbs and may prove to be of importance in the understanding of cooperative phenomena in annelid Hbs.

However, I think that the nomenclature proposed is unsuitable for lectures, since the difference between capital and small characters is difficult to appreciate in the spoken language. Moreover, if type *A* (or *B*) chains were present in three (or more) homologous polypeptides in any Hb or Chl yet to be characterized, one would run into problems.

Therefore, I still favor a classification with distinct letters for each polypeptide chain, like that used by Fushitani *et al.* (5), perhaps reserving the first letters of the alphabet for one phylogenetic kind of chain and the last ones for the other.

Riggs, A.F. (University of Texas, Austin): The primary purpose of a name, whether of organisms or molecules, is unambiguous identification. The name given to a polypeptide that has been fully sequenced should allow one to go immediately to the paper describing the sequence. The most important attribute of a name is that it stays attached to the thing it designates. A secondary purpose of a name is to show relationships, but taxonomic relationships are frequently too complex to permit a simple designation to describe the relationship. Amino acid sequence papers provide designations for each heme-binding chain for *Tylorrhynchus, Lumbricus* and *Lammellibrachia*. I think the least confusing course is to retain these names. Gotoh *et al.* have proposed a new nomenclature for the chains *a, A, b* and *B*. This does have the advantage of showing relationships, but may only be meaningful to someone who already knows the relationships and the meanings to be attributed to the symbols. Chains *a* and *b* of the proposed nomenclature correspond to chains *d* and *a* of Fushitani *et al.* and must be clearly distinguished from chains *a* and *b* (5). Furthermore, oral presentations would have to distinguish *a* and *A* (*little a, big A*). Consider the convenience of describing a disulfide-linked trimer: IIA–IIC–IIB trimer, II–III–IV, *AbB* trimer or *abc* trimer. For all of these reasons, I do not think we should introduce a new nomenclature. It would, however, be helpful to provide cross-referencing to avoid confusion.

Walz, D.A. (Wayne State University, Detroit): The most difficult aspect of unifying diverse, published identification terms into a single nomenclature is the sense of loyalty the original authors have towards their terminology coupled with the limitation such a nomenclature might place on new discoveries. Nonetheless, I do believe some effort should be made at a consensus for organizing these globin chains. The proposal to refer to chains as *a, A, b* or *B* was coolly received at the Satellite Symposium, perhaps because of the cumbersome nature of oral referrence to them (do you say *big A* and *little a*?) as well as the possibility that another class of chains will be discovered upon analysis of tetramer complexes or dimer complexes of globins. Perhaps an alternative could arise from the molecular biology examples, which refer to restriction enzymes as EcoR1, and so forth. Hence, *Lumbricus* globins would then be *Lte* I, *Lte* II, and so forth; *Tylorrhynchus* would be the I, and so forth. Variations of each globin could assume lower case letters (*lte* I1, etc.).

Ghiretti-Magaldi, A. (University of Padova, Padova): The suggestion of a new nomenclature is quite useful. Naturally, I would have liked to have seen some Chl included. I hope that you may soon extend some sequence work also to Chls.

Vinogradov, S.N. (Wayne State University, Detroit): In 1973, Shlom and Vinogradov first assigned numbers to the subunits of *Lumbricus* Hb on the basis of their mobilities on SDS polyacrylamide gels, using numbers starting with 1 for native (*i.e.*, unreduced) material, and I for the reduced material. On this basis, the main subunits of the unreduced molecule were the monomer (subunit 1), several species of nonreducible subunits 2–5 and the disulfide-bonded trimer (subunit 6). The main subunits of the reduced material were chain I corresponding to subunit 1, chains II, III and IV corresponding to subunit 6, and chains V*A*, V*B* and VI to correspond to the nonreducible subunits. Fushitani and Riggs chose in 1988 to provide a new nomenclature for the four globin chains on the basis of chromatographic mobility. The nomenclature proposed by Gotoh *et al.* is based on the known sequence homologies between the four globin chains of *Lumbricus* and the corresponding chains of *Tylorrhynchus* Hb. Although I agree that their nomenclature is desirable for most annelid globin chains, it is also necessary to consider the nomenclature of the *ca.* 30 kDa linker chains and of the globin chains of leech Hbs and polychaete Chls.

Acknowledgements

We are very grateful to E. Chiancone of Rome University, A.F. Riggs of Texas University, A. Ghiretti-Magaldi of Padova University, and S.N. Vinogradov and D.A. Walz of Wayne State University for invaluable comments on the first version of this paper.

References

1. Gotoh, T., Suzuki, T. and Takagi, T. (1991) This volume.
2. Suzuki, T., Takagi, T. and Gotoh, T. (1990) *J. Biol. Chem.* **265**: 12168–12177.
3. Gotoh, T. and Suzuki, T. (1990) *Zool. Sci.* **7**: 1–16.
4. Shlom, J.M. and Vinogradov, S.N. (1973) *J. Biol. Chem.* **248**: 7904–7914.
5. Fushitani, K., Matsuura, S.M.A. and Riggs, A.F. (1988) *J. Biol. Chem.* **263**: 6502–6517.
6. *International Rules of Zoological Nomenclature.* (1948) Paris.
7. Garlick, R. L. and Riggs, A.F. (1982) *J. Biol. Chem.* **257**: 9005–9015.
8. Suzuki, T. and Gotoh, T. (1986) *J. Biol. Chem.* **264**: 9257–9267.
9. Gotoh, T., Shishikura, F., Snow, J.W. Ereifej, K.I., Vinogradov, S.N. and Walz, D.A. (1987) *Biochem. J.* **241**: 441–445.
10. Suzuki, T., Kapp, O.H. and Gotoh, T. (1988) *J. Biol. Chem.* **263**: 18524–18529.
11. Suzuki, T. and Gotoh, T. (1986) *J. Mol. Biol.* **190**: 119–123.
12. Fushitani, K. and Riggs, A.F. (1988) *Proc. Natl. Acad. Sci. U.S.A.* **85**: 9461–9463.
13. Vinogradov, S.N., Lugo, S.D., Mainwaring, M.G., Kapp, O.H. and Crewe, A.V. (1986) *Proc. Natl. Acad. Sci. U.S.A.* **83**: 8034–8038.
14. Kapp, O.H., Qabar, A.N., Bonner, M.C., Stern, M.S., Walz, D.A., Schmuck, M., Pilz, I., Wall, J.S. and Vinogradov, S.N. (1990) *J. Mol. Biol.* **213**: 1551–1555.

Part IV
Gene Structure and Physiological Role

39
Complete Nucleotide Sequence of a Hemoglobin Gene Cluster from the Midge *Chironomus thummi piger*

T. Hankeln,[a] C. Luther,[b] P. Rozynek[a] and E.R. Schmidt[a]

[a]Institute of Genetics, Johannes Gutenberg University, Becherweg 32, 6500 Mainz, Federal Republic of Germany
[b]Institute of Genetics, Ruhr University, Universitätsstrasse 150, 4630 Bochum 1, Federal Republic of Germany

Introduction

The aquatic larvae of non-biting midges (Chironomidae, Diptera) contain a variety of Hb proteins in their hemolymph that enable them to survive in an anoxic environment (1). In *Chironomus thummi thummi*, 12 different Hb variants have been identified and their amino acid sequences determined (2). Based on these primary structures, the evolutionary relationships between the five monomeric and the seven dimeric Hb proteins have been deduced (2). The two groups are thought to have evolved in two different lineages which separated more than 255 million years ago.

In 1984, Antoine and Niessing reported the nucleotide sequence of the genes coding for the monomeric HbIII and IV of *C. th. thummi* (3). Surprisingly, these Hb genes did not contain introns, in contrast to the Hb genes from all other organisms characterized so far (4). This has subsequently been confirmed by the sequencing of genes coding for the dimeric HbVIIB variants of *C. th.*

thummi (5) and the closely related subspecies *Chironomus thummi piger* (6, 7). Furthermore, these studies have shown that the Hb genes in *Chironomus* are organized in large clusters. Two such clusters have been mapped on the chromosomes of *Chironomus* (6, 8).

Here we report the organization and complete nucleotide sequence of an Hb gene cluster from *C. th. piger*. Nine different genes could be identified in 13 kb of cloned DNA. The deduced amino acid sequences allow a classification of these genes and an analysis of their phylogenetic relationships.

Materials and Methods

Clone λpiHb1 was isolated from a genomic library of *C. th. piger* constructed in λEMBL3 (9), using an HbIV gene containing a restriction fragment of *C. th. thummi* as an heterologous probe (3). After construction of the λpiHb1 restriction map by standard procedures, overlapping fragments were subcloned into the plasmid vector pUC 18 (10). Sequencing was done by the dideoxy-method (11) using double-stranded DNA as template (12). Templates were prepared by a modified "boiling" method (13), and the T7 Polymerase sequencing kit of Pharmacia was employed according to the instructions of the manufacturer. Approximately 85% of the sequences were determined on both DNA strands. The other parts were sequenced at least two to three times on the same strand. Multiple alignments of sequences and constructions of dendrograms were performed on an IBM PC-AT 386-compatible computer, using the "Clustal" program of Higgins and Sharp (14).

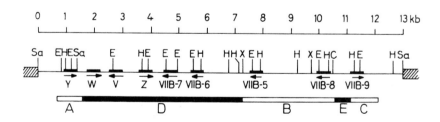

Figure 1. Restriction map of clone λpiHb1. Gene regions are marked by boxes, their transcriptional orientations are indicated by arrows. The sequenced region (see Figure 2) is represented below by the box: the nucleotide sequences of subregions A, B and C have been published previously (6, 7). Figure 2 shows the complete sequence of 11 kb, including parts D and E. (C = ClaI; E = EcoRI; H = HindIII; Sa = SalI; X = XbaI).

Results and Discussion

The clone λpiHb1 contains an insert of 13 kb of *C. th. piger* DNA (Figure 1), cross-hybridizing with an HbIV gene probe (3). This DNA was analyzed in detail by restriction mapping and nucleotide sequencing. Parts of the sequences of the subregions A, B and C (see Figure 1) have been published (6, 7). Now we would like to present the contiguous, complete primary sequence of this region including parts *D* and *E* (Figure 2), lacking only short stretches of DNA at the 5' and 3' ends of the clone.

Within 11 kb of DNA a total number of nine open reading frames, potentially coding for Hb proteins, were identified by comparing the inferred amino acid sequences with known sequences of *C. th. thummi* Hb proteins (2). These genes are organized within λpiHb1 in a tightly clustered manner, with different orientation of the various coding regions (Figure 1). After denaturation and partial renaturation of DNA molecules, this arrangement of Hb genes leads to the formation of short palindromic structures, thus facilitating the identification and mapping of Hb gene-containing regions by electron microscope analysis (8).

Although the functionality of the gene regions in λpiHb1 has not been proven, they all contain the regulatory signals necessary for gene function such as TATA-boxes and the cap- and polyadenylation-signals (Figure 2, Table 1). Furthermore, all open reading frames encode proteins of nearly the same length (160–163 amino acids), and these proteins all contain a 16 amino acid signal peptide, which is typical for chironomid Hbs (3, 15). No Hb pseudogenes have been found within the sequence.

The inferred amino acid sequences of the *C. th. piger* Hb genes were compared with the 12 Hb protein variants in *C. th. thummi* described by Goodman *et al.* (2) (Figure 3). The multiple alignment of amino acid sequences shows that five *C. th. piger* Hb genes encode proteins which are closely related, but are not identical, to the dimeric HbVIIB component of *C. th. thummi*. These genes are therefore called Ctp HbVIIB-5, -6, -7, -8 and -9 (the numbers were chosen to account for the finding of possibly orthologous HbVIIB genes from *C. th. thummi* (5)). None of the inferred amino acid sequences of these Ctp HbVIIB genes is in full agreement with the one published for the *C. th. thummi* HbVIIB variant (2). This might be due to allelic differences between the two subspecies, but it is more probable that several gene variants exist, which code for different HbVIIB variants. Several such different HbVIIB genes have in fact been reported in *C. th. thummi* (5). Since there exist at least five slightly different genes for only one protein variant, the multiplicity of *C. thummi* Hb genes is probably much higher than originally anticipated (7).

This conclusion is supported by the fact that we have found four additional Hb genes in λpiHb1 for which corresponding proteins are not yet known (Ctp HbY, Ctp HbW, Ctp HbV, Ctp HbZ). Antoine *et al.* (16) have also reported a new Hb gene. This gene *E* is possibly expressed only in minor amounts and thus escapes detection by protein analysis.

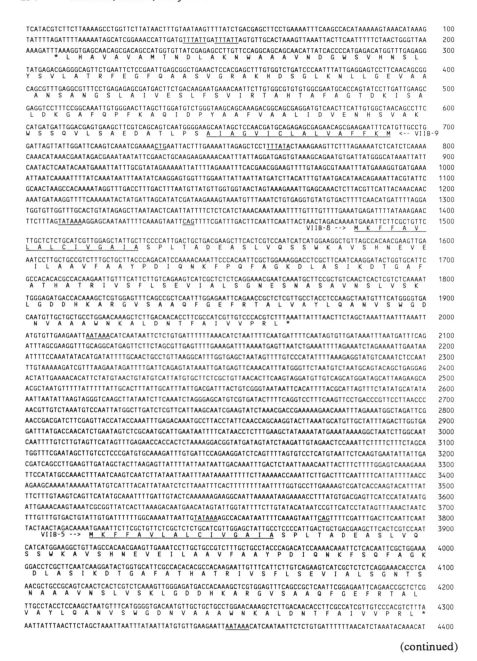

Figure 2. Nucleotide sequence of the Hb gene cluster of λpiHb1, containing nine different putative Hb genes and their intergenic regions. Gene regions are indicated in front of their start codons and their transcriptional orientation is shown by arrows.

```
ATTTTATTATTTTTTTGAAGTTACAACGTTTTCAGCTTACTTATAAATTCATCACTTCAAAGTTTTCATTAAAAAGAATATTTCGATTGATTTATCTTGA    4500
TCCTTTCTATCAGAAATAAATTTTGGAGTCTAGATGCAAATACGTTTTAATTTTTTAAATTTATTCATAAATGACGTTTATTTAAAAATGTTTTCGGATG    4600
TTTGGAACTCTTTTCTTTTAGGTCACTAAACTATTGAATTAAAAATAAATTAATAAAAATAAAAGAGACATAAAACGACAAGCTTTCTTCCTCTTCAAAA    4700
ATTGATCATTATTGATGATGACCCTTTATGGAGTTTGTTACAAATGATGACTTTATGTTTCCCTAACCCCTCTGACTTCATGTTGTGAAAAAACATATT    4800
ATAACGATCGGCACACACGAAAATTCATTGAAATGTTTGCGGACCGTCGGCCAAGTGGATAGGAGGACTAACGATTGCTCATATCACTTAAGGATGAGGG    4900
GGTTCGAAATTATGGCGATCCATTTCTAAAGTTTTTTTCATTCCATTACATTTGTTTTGCAAAATACAGTCAATTTCTCACAAAATAATGAATACAACTTA    5000
AAAGCTTACCTCAAATGATTTTGGAAAACAAGTTATTCAGATTGAAATAATTCATAAAACCAATTTAACTTCAATGTTATTACGGACACATATGTATAGT    5100
AAGAAATGAGACATTAATCTAACCCAATGCTAAAATTTTGACAAAAATAATGCATTAAAATTTTGTAATTGATGAAATTGAATCTTAAGAAATTTTTGGA    5200
TCTATATAATCCGTGTCAGTCAGAATTAATAACGCTTTGATAAGTCATATTCCAATGATTCTACCGGTTTCTTGCTTTGAAAGAACTTTCTTTCATTTGA    5300
AAATATTCAACCTTAAGTTTAAGGGATTACATAATAAATCAAAGATTCCGCCAAATAAAAAATATTTGAGTAAATCGGTGTAAATATTTAAAATTCATAA    5400
GGGGCCGTTCACTTACTGGGGGGGGGGGGGCACAAGGAAAAAGCGATACAAATTTTTTGAAAAGTTATTTTCTTTAAATATGACCGATTTGGAACGTGAT    5500
AAGTGAACGACCCCACATAATAATGTTGAAATAATAAAATATTATTGAAAAACAATTGATTTTGCTTCCCTAAACTAATGTGAATTAATTTATTTAAATT    5600
ACAATTATGCAACATATAAAACAATAATTGCGATAAGATTATATTTAAAGCGGGCTCATCTAAAATTTGATAATATTAGGTAAATTCGCTTATCAAAAAT    5700
AAAACTAGCGCTTGGATTGGTTATTGCTATACGCAATCTATTATTGTTCTGAATAATAAGAACAAATCATTCTTTTATGGCTGTATTGTGATTTTTTGGC    5800
AAAATTAGTTGTATAAAAGCCACAATAATTTTCAAAGTAATTCAGTTTTCGATTTGACTTCAATTCAATTACTAACTAGACAAAATGAAATTCTTCGCTG    5900
                       —————               ———                            VIIB-6 --> M  K  F  F  A
TTCTCGCTCTCTGCATCGTTGGAGCTATTGCTTCCCCATTGACTGCTGACGAAGCTTCACTCGTCCAATCATCATGGAAGGCTGTTAGCCACAACGAAGT    6000
 V  L  A  L  C  I  V  G  A  I  A  S  P  L  T  A  D  E  A  S  L  V  Q  S  S  W  K  A  V  S  H  N  E  V
TGAAATCCTTGCTGCCGTCTTTGCTGCCTACCCAGACATCCAAAACAAATTCTCACAATTCGCTGGAAAGGACCTCGCTTCAATCAAGGATACTGGTGCA    6100
 E  I  L  A  A  V  F  A  A  Y  P  D  I  Q  N  K  F  S  Q  F  A  G  K  D  L  A  S  I  K  D  T  G  A
TTCGCCACACACGCCACAAGAATTGTTTCATTCTTGTCAGAAGTCATCGGCTCTCTCAGGAAACACCTCAAACGCTGCCGCAGTCAACTCACTCGTCTCAA    6200
 F  A  T  H  A  T  R  I  V  S  F  L  S  E  V  I  A  L  S  G  N  T  S  N  A  A  A  V  N  S  L  V  S
AGTTGGGAGATGACCACAAAGCTCGTGGAGTTTCAGCCGCTCAATTCGGAGAATTCAGAACCGCTCTCGTTGCCTACCTCCAAGCTAATGTTTCATGGGG    6300
 K  L  G  D  D  H  K  A  R  G  V  S  A  A  Q  F  G  E  F  R  T  A  L  V  A  Y  L  Q  A  N  V  S  W  G
TGACAATGTTGCTGCTGCCTGGAACAAAGCTCTTGACAACACCTTCGCCATCGTTGTCCCACGTCTTTAAATTATTTAACTTCTAGCTAAATTAATTTAA    6400
 D  N  V  A  A  A  W  N  K  A  L  D  N  T  F  A  I  V  V  P  R  L  *
ATTATGTGTTGAAGAATTAATAAACATCAATAATTCTCTGTGATTTTTTAAACATCTCAAGTTTTGCATGTCTTTTGACTTTCTATTAATAATTATTTTTT    6500
TTCAGGAATCTTACAATTATTTTGCCTGCTCACTTTCATTTATTAAGGTCTCCAGAAGAAAGTCCACAATTTATTTTATTGTTAGCATTTTTTTTTTATT    6600
TTTTTAAAATACTGAATCTAACCGTTTAAAACATTCTAAGTACACACAAGATGCACTATGATCTCAGATAAGAACGCTAGATGTTATTTAAAGTAAACGC    6700
TCTTATCAACAATAGCTAGTTTTTGTTTATTTTGGACATAAATCATTAACCCCATTGCGTTTTATAGTTTCAAATAATCAATAAATTTTCAATGTATTGA    6800
AACTTTTTTGGACAATAGTTTAGTATAAAAGCTGAACTATTTTTCATTAGAATTCAGTTTACAATTGCATTTGAATCTACAAACTAATCCAAGCATCAAA    6900
ATGAAATTCTTCGCAGTTCTTGCACTTTGTGTCGTTGGAGCTATCGCTTCCCCATTGCTGTCGTGACGAAGCTAATCTCGTTAAGTCATCATGGGATCAAG    7000
 M  K  F  F  A  V  L  A  L  C  V  V  G  A  I  A  S  P  L  S  A  D  E  A  N  L  V  K  S  S  W  D  Q
VIIB-7 -->
TCAAGCACAATGAAGTTGACATTCTTGCTGCTGTCTTTGCTGCATACCCAGACATCCAGGCTAAGTTCCCACAATTCGCTGGAAAGGATCTCGCTTCAAT    7100
 V  K  H  N  E  V  D  I  L  A  A  V  F  A  A  Y  P  D  I  Q  A  K  F  P  Q  F  A  G  K  D  L  A  S  I
TAAGGATACTGCTGCATTCGCAACACACGCAACAAGAATTGTTTCATTCTTCACAGAAGTCATTTCTCTCTCAGGAAATCAAGCCAACCTTTCAGCCGTT    7200
 K  D  T  A  A  F  A  T  H  A  T  R  I  V  S  F  F  T  E  V  I  S  L  S  G  N  Q  A  N  L  S  A  V
TACGCACTCGTCTCAAAATTGGGAGTTGATCACAAAGCACGTGGAATCTCAGCTGCTCAATTCGGTGAATTCAGAACTGCCCTCGTTTCATACCTTCAAG    7300
 Y  A  L  V  S  K  L  G  V  D  H  K  A  R  G  I  S  A  A  Q  F  G  E  F  R  T  A  L  V  S  Y  L  Q
CTCATGTTTCATGGGGTGACAATGTTGCTGCTGCTTGGAATCACGCTTTAGACAACACTTATGCCGTTGCACTCAAGTCTCTCGAATAAACTAATAATCA    7400
 A  H  V  S  W  G  D  N  V  A  A  A  W  N  H  A  L  D  N  T  Y  A  V  A  L  K  S  L  E  *
TAAAATTAAAATTTTGAAAATGTAAAATTGAACAACTAATAAACATTTTTAATTATTTTTTTATGATTCTTAAAATTTTGTGTTATTAAAATTTAAGCACT    7500
TGAAATAATTTTAGACCTTTAAAGATTGTTCGTTCCAGATTGATGAGTTTTATATCCACCAATGTAAGTGAAATGTTCTAAAAGAATTTAAAAAAAATACA    7600
ATCTGAAATACAATAGAGCAGTAAAACAAAATAATTTCAAGTTTAAATGGACATTAAGAACTAAATCGATCTTTATTAATTCCCAATTTCACAAAATAAT    7700
AATAAATAAAAATTTTTAATATTTTTCGTTCAGGTACTTTTGTGTTTAAGCTGTGACAATTTGGAAGACAGCGGTGTAGACATTGTCAAGTGCGTGTGTC    7800
                                                  *  A  T  V  I  Q  F  V  A  T  Y  V  N  D  L  A  T
CATGCAGCTGCAACATTATCGCCCCATGAGACGTTTGATGAGATATATGAGACGAGGGCAGTTCTGAATTCATTGAATTGAGTTTGTGTAATTCCACGGT    7900
 W  A  A  A  V  N  D  G  W  S  V  N  S  S  I  Y  S  V  L  A  T  R  F  E  N  F  Q  T  Q  T  I  G  R  N
TCTTGTGGTCTGTTCCCATCTTTGAGATAAGTCCATAGATGGCTGAAAGATTGCTTTCGCTTCCTGAGAGAGCAATAAGTTCTGAAAAGAATGAGACAAT    8000
 K  H  D  T  G  M  K  S  I  L  G  Y  I  A  S  L  N  S  E  G  S  L  A  I  L  E  S  F  F  S  V  I
GCGAGTTGCATGAGTTGCGAATTGACCCACTTGTCTTAATTGAATCGAGATCCTTTCCAGCAAACTGAGGGAAACGAGCTTGGATATCTGGGTAAGCTTTG    8100
 R  T  A  H  T  A  F  Q  G  S  T  K  I  S  D  L  D  K  G  A  F  Q  P  F  R  A  Q  I  D  P  Y  A  K
AAAACAGTATAAAGGATATCAACTTCATTGTGCTTGACTTGAGCCCATGATGACTTGACGAGTGCTGCTTCGTCAGAAGTCAATGGATGAGCAATTGCTC    8200
 F  V  T  Y  L  I  D  V  E  N  H  K  V  Q  A  W  S  S  K  V  L  A  A  E  D  S  T  L  P  H  A  I  A  G
CAACAATGCAAAGAGCGAGAACAGCGAAGAATTTCATGTTGCCTAGATTTGTAAATTTGTTCAAATTCGAGTGAAAACTGATTTAATCGCAACTAATTCT    8300
 V  I  C  L  A  L  V  A  F  F  K  M  <-- Z
TCGGCTTTTATACAGGAGAAATGCTTTTGATTATTTTTTGGCTATTTTAATAAGGAACAGTTAGATCTTTACTTAATACCAAGTTCAACTTGATAATCAG    8400
CTATGCTATTTGCTGGTTCTCTTATCGCAGCACGTGCTGAAATTCTGAAATTATTGAAATAAACGAAAAAAATAATTTTATTAAGTTAAGGAATAAGCAT    8500
GATTTCAAGTTTTATTTTAAATATTATCCAATATTATCGTTATTATTCATATTTTTATGAGTATATCCGAAAACTGTCCACTTACGATAATACGAAAATT    8600
AATTTCACAATTAAGACTACTTTTTGAATCAAAAGTACAAAGATTCGAACCTTATCATCAGAAAATAATTAGCAGGAGCTTGTGTTGTGCATTTCACATT    8700
                                                                                      (continued)
```

Figure 2 (continued). The deduced amino acid sequence of coding regions is displayed in one letter code. Stop codons are marked by asterisks. The signal peptide sequences are underlined. In the 5' and 3' gene flanking sequences, putative regulatory signals (TATA boxes, cap sites, polyadenylation signals) are underlined.

```
AAAATAATCTCTCTTTTGTAATTTTGAGTTTAATATTTACTAAATATTGTAGTATATAAAGAAGTGACAAAAAATCAAAATAAATCAGTTTTCAGTTTAC   8800
ATTGAATCAACTAACAAATCCAGTCAACATGAAATTCTTCGCTGTTCTTACACTTTGCATTATTGGAGCTATTGCTCATCCATTGACTTCTGACGAAGCT   8900
                          V --> M  K  F  F  A  V  L  T  L  C  I  I  G  A  I  A  H  P  L  T  S  D  E  A

AATCTCGTTAAGTCATCATGGAACCAAGTTAAACACAATGAAGTTGACATTCTTGCTGCTGTCTTCAAAGCATATCCAGACATCCAGGCTAAGTTCCCTC   9000
N  L  V  K  S  S  W  N  Q  V  K  H  N  E  V  D  I  L  A  A  V  F  K  A  Y  P  D  I  Q  A  K  F  P
AGTTTGCTGGAAAAGATCTCGATTCAATTAAGACGAGTGGTCAATTCGCAACTCATGCAACTCGGCATTGTCTCATTCTTGTCAGAACTTATTGCTCTCTC   9100
Q  F  A  G  K  D  L  D  S  I  K  T  S  G  Q  F  A  T  H  A  T  R  I  V  S  F  L  S  E  L  I  A  L  S
AGGAAATGAAGCAAACCTTTCAGCTGTCTATGGACTCGTCAAAAAATTAGGAGTTGATCACAAGAACCGTGGAATTACACAAGGACAATTCAATGAATTC   9200
G  N  E  A  N  L  S  A  V  Y  G  L  V  K  K  L  G  V  D  H  K  N  R  G  I  T  Q  G  Q  F  N  E  F
AAGACAGCTCTTATCTCATACCTCTCAAGCCATGTCTCATGGGGTGATAATGTTGCTGCTGCATGGGAGCATGCTCTTGAAAACACATATACAGTTGCTT   9300
K  T  A  L  I  S  Y  L  S  S  H  V  S  W  G  D  N  V  A  A  A  W  E  H  A  L  E  N  T  Y  T  V  A
TTGAAGTCATTCCAGCCTAAAAAGTGTTAGATAATTTTGAATAAAATTAACTTATTTATGCTGATAATTAGAATTAATAAAGATCTGTAATTCATAATGAT   9400
F  E  V  I  P  A  *

TTTTTTAAAACTTGAAATAATTGATTTGCAGCTTCCGAAAGGCAGCTTGAAGCTAAAGTTAGACCAATTTTAATTTGGTGGGACTGTTAAGTGTAAAAAA   9500
GGCAATAAATAGTTTTAATTATATCTTTTTATTTACAACATCCTCAAATACATTATTCAAGAATAGAGTAGACAATAGCATGATGTTCATCGCCAACAAC   9600
                        *  E  L  I  S  Y  V  I  A  H  H  E  D  G  V  V
TTTCCATGCATTTTCAGTTTTCTCATCCCATGTTGTATGTGATTTAAGGTATCCCATAAATGCAATGTTAAACTTTTCAAACAATTCTTTTGTAACTCCA   9700
K  W  A  N  E  T  K  E  D  W  T  T  H  S  K  L  Y  G  M  F  A  I  N  F  K  E  F  L  E  K  T  V  G
CGAGCTTTATGATCATTTGCCATCTTCACAAGAAGTAAGTCGATAGATGATTGAACTGCTGGATTTCCAACAAGGCTTACAATTTCAGACATAAATGAAA   9800
R  A  K  H  D  N  A  M  K  V  L  L  D  I  S  S  Q  V  A  P  N  G  V  L  S  V  I  E  S  M  F  S  V
CAATTCTAGTTGAATGAACTGCAAACTCAGCAGTTTCTTTAATTGCTTCTAAATCCTTTCCAGCAAATTGAGGGAACCGATCTTGAATTTCTGGATATGC   9900
I  R  T  S  H  V  A  F  E  A  T  E  K  I  A  E  L  D  K  G  A  F  Q  P  F  R  D  Q  I  E  P  Y  A
CTTAAAGACAGTGTAGAAGAATGTCAACTTCATTATGCTTTACTTGATTCCATGATGCTTTGATGAACGGAGTTTGTCAACGTGAGCGATGGCACCAGCA   10000
K  F  V  T  Y  L  I  D  V  E  N  H  K  V  Q  N  W  S  A  K  I  F  P  A  K  D  C  H  A  I  A  G  A
ATACACAATGTAAGAATTACAAGGAACTTCATATTGATTTAATTTTCTATCATGTACTGATGATTTCATGAATGAAGCAGATGTGTTTTATATTTGACTT   10100
I  C  L  T  L  I  V  L  F  K  M  <-- W
TTGTTTTGGCAATAACATGACTTGCTAATTCAAATTTCTTTTTTATGCCCATCTATGTACCATAATCTAAATGATTAGGCAAAAGAGCTCAAAAGTCGAT   10200
TATAATGGATTTTTGTCAAAATAAAATAAGATTTTGAGACATTATTTAAACTGCTTTTGTGGATTCCCTATCAGCTTCAAACCTCCCTCCAACCTAACCT   10300
CTTTACGTGAATATCTATTTTTATGTAAAAAGATTCAAGGATCATTGAATAAAATTAAAACAAAATTTATAAATAATTTTTGATAACAGAAAGTTTTATT   10400
AATCAATCATCAAATTCTTCTATTTTTAAATGTCGACTTTCTTTTGCTGATTGTTAGCTTAAGCAACAAAAAAAAACATAATTTATTCAAGGTTAGCGTT   10500
                                                                             *  E  L  N  A  N
GATTACTTTACGGATTTCCTTTTCATTGCAGTGCCAAGCAGCATCAACATTGTCATTCCATACTGAGTGTGTGTGCAAGAAATTATGGAAGGCTTCATGG   10600
I  V  K  R  I  E  K  E  N  C  H  W  A  A  D  V  N  D  N  W  V  S  H  T  H  L  F  N  H  F  A  E  H
AATTCATCAAATTGCTTCACTGTGATGCCACGACCTTTATGATCTTTTCCCAATTTAGAGAGGAGAGACATGATTGCTGGAAGATTGTCAGGATTACCGA   10700
F  E  D  F  Q  K  V  T  I  G  R  G  K  H  D  K  G  L  K  S  L  L  S  M  I  A  P  L  N  D  P  N  G  L
GGAGTGAAATAACTTCTGTCATAAAACTGACAATTCTAGTTGCATGGACTGCAAACTCAGCAGTTCCCTTGATTGTTTCCAAGTCCTTTCCAACAAATTG   10800
L  S  I  V  E  T  M  F  S  V  I  R  T  A  H  V  A  F  E  A  T  G  K  I  T  E  L  D  K  G  V  F  Q
TGGGAATTTAGCTTGAATGTCTGGATAAGCTTTGAAAACAGTGTAGAGGATTTCAACTTCTTCATTTTTCATTGTATTCCATGCTTCCTGCATTATTTTG   10900
P  F  K  A  Q  I  D  P  Y  A  K  Y  F  T  Y  L  I  E  V  E  E  N  K  T  N  W  A  E  Q  M  I  K
AAGTCATCACATGGAGTGGCAAGTGCCCCGATGATACACAAAGCAAAAATAGCGAGAACTTTCATTTTCACTGTTTAAGGTTCAAGAATTCGAATCAAAT   11000
F  D  D  C  P  T  A  L  A  G  I  I  C  L  A  F  I  A  L  V  K  M  <-- Y
ATTGATTATAAACTGCAAGTACTCGTATTTATACACAATTTTGAAAATTTTATTTTTAGAAACTTTCAATTTGATTGTTTAAGTTCATACTTGATAACACT   11100
CACTTT                                                                                              11106
```

Figure 2 (continued).

The comparison of amino acid sequences (Figure 3) shows that the deduced proteins of Ctp HbZ and Ctp HbV are similar to the dimeric VIIB and the dimeric IIβ protein variants. It can thus be assumed that the HbV and HbZ proteins are also dimeric. HbY and HbW, however, are positioned between the protein component X, which can be either monomeric or dimeric (2), and the variant IA, which clearly is only monomeric. HbY and HbW might, therefore, occupy intermediate positions between the two lineages of the monomeric and dimeric Hb proteins in *C. th. thummi*.

These phylogenetic relationships become clearer when a dendrogram is constructed on the basis of the amino acid sequence comparisons (Figure 4). The five different Ctp HbVIIB genes are closely related, with Ctp HbVII-5 and -6 being nearly identical (no amino acid exchanges, 0.2% nucleotide substitution). These two genes are probably the result of a gene duplication event, which

Table 1. Characteristics and nomenclature of nine Hb gene coding regions found in λpiHb1.

Hb gene	coding region	deduced protein	signal peptide	'cap'-site	TATA box	poly-(A)-signal
Ctp HbVIIB-5	483	161 aa	+	− 43	− 73	+ 53
Ctp HbVIIB-6	483	161 aa	+	− 43	− 73	+ 52
Ctp HbVIIB-8	483	161 aa	+	− 43	− 73/91	+ 52
Ctp HbVIIB-9	483	161 aa	+	− 44	− 74	+ 57/65
Ctp HbVIIB-7	486	162 aa	+	− 47	− 77	+ 52
Ctp HbV	489	163 aa	+	− 44	− 74	+ 23/58
Ctp HbZ	489	163 aa	+	− 44	− 75	+ 72
Ctp HbY	480	160 aa	+	− 50	− 67	+ 86
Ctp HbW	480	160 aa	+	− 28	− 60	+ 23

happened during relatively recent evolution. Ctp HbVIIB-7 is the most distant gene of the HbVIIB complex. It is closer to HbZ and HbV. The grouping of HbV and HbZ in this dendrogram underlines the assumption that they are dimeric proteins. The proteins of Ctp HbY and Ctp HbW, which are not so closely related, are clearly positioned between the monomeric and the dimeric Hb variants. The spatial arrangement of the different Ctp Hb genes in clone λpiHb1 seems to reflect the evolution of clearly related dimeric Hb proteins (right side of the clone) from ancestors which display monomeric and dimeric properties (left end of the clone) with the 'dimeric' Ctp HbZ and HbV in between. The whole cluster might be the product of repeated gene duplication events.

Conclusions

The Hb genes of *C. th. thummi* constitute a large multi-gene family. Based upon the knowledge of 12 different protein variants (2) and an average of only three genes per variant, a minimum number of about 40 genes is expected. This, however, might prove to be an underestimate, since it could be shown that additional genes coding for undescribed proteins exist. It seems very interesting to investigate how these gene variants arose during evolution, how they are regulated within the organism, and what kind of physiological significance they have.

Another interesting feature is the possible "intron loss" of Chironomid Hb genes (3, 4). Neither of the *C. th. thummi* (or *Chironomus tentans*) Hb genes characterized thus far has introns. It has been speculated that the Hb genes of the

```
Ctp HbVIIB-5    --S----PLTADEASLVQSSWKAVSHNEVEILAAVFAAYPDIQNKFSQFAGKDLASIKDTGAFATHATRIVSFLSEVIALSGNTS
Ctp HbVIIB-6    --S----PLTADEASLVQSSWKAVSHNEVEILAAVFAAYPDIQNKFSQFAGKDLASIKDTGAFATHATRIVSFLSEVIALSGNTS
Ctp HbVIIB-8    --S----PLTADEASLVQSSWKAVSHNEVEILAAVFAAYPDIQNKFPQFAGKDLASIKDTGAFATHATRIVSFLSEVIALSGNES
CTT VII B       --S----PLTADEASLVQSSWKAVSHNEVDILAAVFAAYPDIMAKFPQFAGKDLASIKDTGAFATHATRIVSFLSEVIALMGNAS
Ctp HbVIIB-9    --S----PLTADEASLVQSSWKAVSHNEVDILAAVFAAYPDIQAKFPQFAGKDLASIKDTGAFATHATRIVSFLSEVIALSGNAS
Ctp HbVIIB-7    --S----PLSADEANLVKSSWDQVKHNEVDILAAVFAAYPDIQAKFPQFAGKDLASIKDTAAFATHATRIVSFFTEVISLSGNQA
Ctp HbZ         --H----PLTSDEAALVKSSWAQVKHNEVDILYTVFKAYPDIQARFPQFAGKDLDSIKTSGQFATHATRIVSFFSELIALSGSES
Ctp HbV         --H----PLTSDEANLVKSSWNQVKHNEVDILAAVFKAYPDIQAKFPQFAGKDLDSIKTSGQFATHATRIVSFLSELIALSGNEA
CTT II B        --A----PLSADEASLVRGSWAQVKHSEVDILYYIFKANPDIMAKFPQFAGKDLETLKGTGQFATHAGRIVGFVSEIVALMGNSA
CTT IX          --D----PVSSDEANAIRASWAGVKHNEVDILAAVFSDHPDIQARFPQFAGKDLASIKDTGAFATHAGRIVGFISEIVALVGNES
CTT VIIA        --A----PLSADQASLVKSTWAQVRNSEVEILAAVFTAYPDIQARFPQFAGKDVASIKDTGAFATHAGRIVGFVSEIIALIGNES
CTT VI          --A----VLTTEQADLVKKTWSTVKFNEVDILYAVFKAYPDIMAKFPQFAGKDLDSIKDSAAFATHATRIVSFLSEVISLAGSDA
CTT VIII        --AVT--PMSADQLALFKSSWNTVKHNEVDILYAVFKANPDIQAKFPQFAGKDLDSIKDSADFAVHSGRIVGFFSEVIGLIGNPE
CTT X *         --DPEWHTLDAHEVEQVQATWKAVSHDEVEILYTVFKAHPDIMAKFPKFAGKDLEAIKDTADFAVHASRIIGFFGEYVTLLGSSG
Ctp HbW         ---H------CDKAPFIKASWNQVKHNEVDILYTVFKAYPEIQDRFPQFAGKDLEAIKETAEFAVHSTRIVSFMSEIVSLVGNPA
Ctp HbY         --TP------CDDFKIMQEAWNTMKNEEVEILYTVFKAYPDIQAKFPQFVGKDLETIKGTAEFAVHAIRIVSFMTEVISLLGNPD
CTT I A         --GP-----SGDQIAAAKASWNTVKNNQVDILYAVFKANPDIQTAFSQFAGKDLDSIKGTPDFSKHAGRVVGLFSEVMDLLGNDA
CTT I           --GP-----SGDQIAAAKASWNTVKNNQVDILYAVFKANPDIQTAFSQFAGKDLDSIKGTPDFSKHAGRVVGLFSEVMDLLGNDA
CTT III         --------LSADQISTVQASFDKVKGDPVGILYAVFKADPSIMAKFTQFAGKDLESIKGTAPFETHANRIVGFFSKIIG------
CTT IV          --------LTADQISTVQSSFAGVKGDAVGILYAVFKADPSIQAKFTQFAGKDLDSIKGSADFSAHANKIVGFFSKIIG------
CTT III A       VATPAMPSMTDAQVAAVKGDWEKIKGSGVEILYFFLNKFPGNFPMFKKL-GNDLAAAKGTAEFKDQADKIIAFLQGVIEKLGSD-
                  .   . ..  .. *.**   .  *.    * .. *.*...*..  *  !........     ..
```

```
Ctp HbVIIB-5    NAAAVNSLVSKLGDDHKARGVSAAQFGEFRTALVAYLQANVSWGDNVAAAWNKALDNTFAIVVPRL-----
Ctp HbVIIB-6    NAAAVNSLVSKLGDDHKARGVSAAQFGEFRTALVAYLQANVSWGDNVAAAWNKALDNTFAIVVPRL-----
Ctp HbVIIB-8    NASAVNSLVSKLGDDHKARGVSAAQFGEFRTALVAYLQANVSWGDNVAAAWNKALDNTFAIVVPRL-----
CTT VII B       NAAAVQGLLDKLGDDHKARGVSAAQFGEFRTALVLQAHVSWGNNVAAAWSKALDNTFAIVVPRL-----
Ctp HbVIIB-9    NAAAVEGLLNKLGSDHKARGVSAAQFGEFRTALVSYLSNHVSWGDNVAAAWNKALDNTMAVAVAHL----
Ctp HbVIIB-7    NLSAVYALVSKLGVDHKARGISAAQFGEFRTALVSYLQHHVSWGDNVAAAWNHALDNTYAVALKSLE----
Ctp HbZ         NLSAIYGLISKMGTDHKNRGITQTQFNEFRTALVSYISSNVSWGDNVAAAWTHALDNVYTAVFQIVTA---
Ctp HbV         NLSAVYGLVKKLGVDHKNRGITQGQFNEFKTALISYLSSHVSWGDNVAAAWEHALENTYTVAFEVIPA---
CTT II B        NMPAMETLIKDMAANHKARGIPKAQFNEFRASLVSYLQSKVSWNDSLGAAWTQGLDNVFNMMFSYL-----
CTT IX          NAPAMATLINELSTSHHNRGITKGQFNEFRSSLVSYLSHASWNDATADAWTHGLDNIFGMIFAHL-----
CTT VIIA        NAPAVQTLVGQLAASHKARGISQAQFNEFRAGLVSYVSSNVAWNAAAESAWTAGLDNIFGLLFAAL-----
CTT VI          NIPAIQNLAKELATSHKPRGVSKDQFTEFRTALFTYLKAHINFDGPTETAWTLALDTTYAMLFSAMDS---
CTT VIII        NRPALKTLIDGLASSHKARGIEKAQFEEFRASLVDYLSHHLDWNDTMKSTWDLALNNMFFYILHALEVA-Q
CTT X *         NQAAIRTLLHDLGVFHKTRGITKAQFGEFRETMTAYLKGHNKWNADISHSWDDAFDKAFSVIFEVLES---
Ctp HbW         VQSSIDLLLVKMANDHKARGVTKELFEKFNIAFMGYLKSHTTWDEKTENAWKVVGDEHHAIVYSILE----
Ctp HbY         NLPAIMSLLSKLGKDHKGRGITVKQFDEFHEAFHNFLHTHSVWNDNVDAAWHCNEKEIRKVINANLE----
CTT I A         NTPTILAKAKDFGKSHKSRT-SPAQLDNFRKSLVVYLKGATKWDSAVESSWAPVLD----FVFSTLKNE-L
CTT I           NTPTILAKAKDFGKSHKSRA-SPAQLDNFRKSLVVYLKGATKWDSAVESSWAPVLD----FVFSTLKNE-L
CTT III         ELPNIEADVNTFVASHKPRGVTHDQLNNFRAGFVSYMKAHTDFAGA-EAAWGATLDTFFGMIFSKM-----
CTT IV          DLPNIDGDVTTFVASHTPRGVTHDQLNNFRAGFVSYMKAHTDFAGA-EAAWGATLDAFFGMVFAKM-----
CTT III A       -MGGAKALLNQLGTSHKAMGITKDQFDQFRQALTELL-GNLGFGGNIGA-WNATVDLMFHVIFNALDGTPV
                 .      .  *  ..    ...*. ..    ...    *  .   .
```

Figure 3. Multiple alignment of amino acid sequences derived from the gene regions of λpiHb1 ("Ctp Hb . . ."), and of the amino acid sequences of Hb protein variants, described in *C. th. thummi* ("CTT . . .") (2). The comparison allows a classification of the Ctp Hb genes. The signal peptide region is omitted. With respect to the Hb variants reported by Goodman *et al.* (2), monomeric and dimeric proteins are distinguished (monomeric ones are underlined). The CTT X component can be either monomeric or dimeric (*).

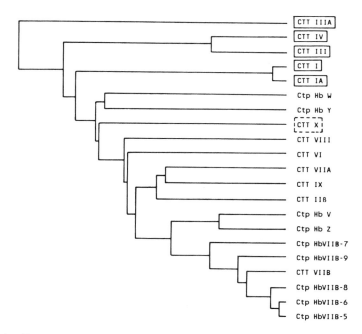

Figure 4.   Dendrogram of *C. th. piger* and *C. th. thummi* Hb variants, based on the multiple alignment of (derived) amino acid sequences (see Figure 3).   Proteins of Ctp HbVIIB-5, -6, -7, -8, -9 genes are clearly related to the dimeric CTTVIIB variant of *C. th. thummi*.   Proteins from gene Ctp HbV and Ctp HbZ are also probably dimeric, whereas the deduced proteins of Ctp HbY and Ctp HbW are grouped between the monomeric/dimeric CTT X component (dashed boxed) and the monomeric Hb variants (boxed).

midge are 'functional retrogenes,' which have lost their introns and increased their number by RNA-mediated transposition (17).   However, we have not found any signs of retro-transposition events, like poly-A-tracts or insertion site duplications in the vicinity of the genes.   More likely, according to our data, the large number of similar Hb genes is the result of successive gene duplication events, as has been pointed out by Goodman *et al.* (2).

Acknowledgements

The expert technical assistance of Miss Bettina Weich is greatly acknowledged.   We thank Miss Beate Muders for typing the manuscript, and Mrs. Ursula Boell and Mrs. Regina Gramsch for the photo-reproduction and drawing of the figures.   This work was supported by the Deutsche Forschungsgemeinschaft.

References

1. Braun, V., Chrichton, R.R. and Braunitzer. G.   (1968)   *Z. Physiol. Chem.* **349**: 197–210.
2. Goodman, M., Braunitzer, G., Kleinschmidt, T. and Aschauer, H.   (1983) *Hoppe-Seyler's Z. Physiol. Chem.* **364**: 205–217.
3. Antoine, M. and Niessing, J. (1984) *Nature* **310**: 795–798.
4. Jhiang, S.M., Garey, J.R. and Riggs, A.F. (1988) *Science* **240**: 334–336.
5. Trewitt, P.M., Saffarini, D.A. and Bergtrom, G. (1988) *Gene* **69**: 91–100.
6. Hankeln, T., Rozynek, P. and Schmidt, E.R. (1988) *Gene* **64**: 297–304.
7. Rozynek, P., Hankeln, T. and Schmidt, E.R. (1989) *Biol. Chem. Hoppe-Seyler* **370**: 533–542.
8. Schmidt, E.R., Keyl, H.-G. and Hankeln, T.   (1988)   *Chromosoma* **96**: 353–359.
9. Frischauf, A.-M., Lehrach, H., Poustka, A. and Murray, N. (1983) *J. Mol. Biol.* **170**: 827–842.
10. Yanisch-Perron, C., Vieira, J. and Messing, J. (1985) *Gene* **33**: 103–119.
11. Sanger, F., Nicklen, S. and Coulson, A.R. (1977) *Proc. Nat. Acad. Sci. U.S.A.* **74**: 5463–5467.
12. Chen, E.Y. and Seeburg, P.H. (1985) *DNA* **4**: 165–170.
13. Toneguzzo, F., Glynn, S., Levi, E., Mjolsness, S. and Hayday, A.   (1988) *BioTechniques* **6**: 460–469.
14. Higgins, D.G. and Sharp, P.M. (1988) *Gene* **73**: 237–244.
15. Saffarini, D.A., Trewitt, P.M., Castro, M., Wejksnora, P.J. and Bergtrom, G. (1985) *Biochem. Biophys. Res. Comm.* **133**: 641–647.
16. Antoine, M., Erbil, C., Muench, F., Schnell, S. and Niessing, J.   (1987) *Gene* **56**: 41–51.
17. Weiner, A.M. and Deininger, P.L. (1986) *Annu. Rev. Biochem.* **55**: 631–661.

40

The Primary Structure of Several Hemoglobin Genes from the Genome of *Chironomus tentans*

P. Rozynek, M. Broecker, T. Hankeln and E.R. Schmidt

Institute of Genetics, Johannes Gutenberg University, Becherweg 32, 6500 Mainz, Federal Republic of Germany

Introduction

In addition to *Chironomus thummi thummi*, *Chironomus tentans* is another Chironomid species in which the Hb proteins have been investigated (1) and in which the Hb genes have been localized (2). Thus, this species is an ideal candidate for the study of Hb gene structure and evolution.

C. tentans and *C. th. thummi* both have many different Hb variants and these variants can be present in the hemolymph of the larvae either as monomeric or homodimeric polypeptides (3). Although there are no amino acid sequence data available for *C. tentans* Hbs, it has been speculated from the data of *C. th. thummi* that the group of monomeric Hb variants and the group of dimeric Hb variants share one common ancestor (3).

This is an intriguing problem because it has been shown that the Hb genes of Chironomids are clustered (5, 6, 7). So far, only some of the Hb genes from *C. th. thummi* (7, 8) and *C. th. piger* (4, 5) have been analyzed. In *C. th. thummi*, all data indicate that the genes linked in one cluster are of the same type, either monomeric or dimeric Hb genes.

According to our previous data (2), *C. tentans* contains two different Hb gene clusters on one chromosome, but on separate chromosome arms. From these and other data it has been concluded that the two different loci correspond to gene clusters that contain genes for either monomeric or dimeric Hbs (3).

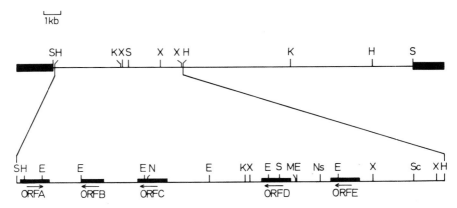

Figure 1.   Restriction map of clone *ten* Hb5.   The left-most 7 kb HindIII fragment (enlarged) has been sequenced completely.   The five open reading frames (ORFs A, B, C, D and E) with coding capacity for Hbs are shown as solid bars.   The orientations (5'-3') of the putative genes are indicated by the arrows.   S = SalI, H = HindIII, E = EcoRI, N = NcoI, K = KpnI, M = MluI, Ns = NsiI, X = XbaI, Sc = SacI.

In the present study, we have isolated two clones from a genomic DNA library of *C. tentans* using subcloned Hb VIIB genes from *C. th. piger* as probes (4, 5). From the nucleotide sequence it should be possible to decide which Hb genes are present within the cloned cluster and whether all (linked) members of the cluster belong to the subgroup of genes coding for monomeric or dimeric Hbs. Although no primary sequence data are available on the *C. tentans* Hbs, one can expect that the comparison with the established amino acid sequence data will give a clear answer to the question of which subvariant is coded by these genes.

Thus, considerable parts of the two isolated Hb gene-containing clones have been sequenced. The inferred amino acid sequences have been compared with the known sequences of *C. th. thummi* and *C. th. piger* Hbs. The results are in agreement with the suggestion that the genes for dimeric Hbs are clustered in one location. This situation probably reflects the evolutionary history of the Chironomid Hb genes, namely that the monomeric and homodimeric Hbs are derived from a common ancestor (3).

Results and Discussion

Two recombinant lambda-clones were obtained by screening approximately 40,000 clones of a genomic *C. tentans* DNA-library. These two clones, denoted as *ten* Hb5 and *ten* Hb10.1, hybridized repeatedly with a probe containing the Hb gene Ctp HbVIIB-9 from *C. th. piger* (9). On Southern-Blots, several restriction fragments hybridized readily with a Ctp HbVIIB-9 containing probe, which was a clear indication that more than one gene was present on both clones. The restriction pattern and the size of the gene-containing fragments were different, so that we have isolated two DNA fragments which do not overlap in significant parts within the genome (data not shown). We arbitrarily selected the clone *ten* Hb5 to begin the detailed analysis, which includes restriction mapping, localization of potential gene regions and nucleotide sequence determination. The results are shown in Figure 1 (restriction map) and Figure 2 (nucleotide sequence).

From the restriction analysis of *ten* Hb5, the size of the cloned region has been determined to be approximately 21 kb. According to Southern hybridizations, two HindIII fragments of approximately 7 and 2.2 kb contain Hb-gene homologous sequences. Up until now we have concentrated on the large 7 kb HindIII fragment, which represents the "left" end of the cloned region. This 7 kb HindIII fragment contains five regions with open reading frames (ORF A to ORF E) potentially coding for Hb polypeptides (Figures 1, 2). For the open reading frames (ORF) A, C, D and E, all essential regulatory sequences are present which indicates that they are putative Hb genes. ORF B, however, does not have a start codon, and is obviously truncated at the 3' end. Although a TATA box and a polyadenylation signal have been found (Figure 2), ORF B has to be considered as a pseudogene. This is the first time that a possible pseudogene has been found among the Hb gene family of Chironomids.

The inferred amino acid sequences of ORFs A, C, D and E allow a comparison to be made with the known amino acid sequences of the numerous Hb variants from *C. th. thummi* (Table 1). The results of a multiple pairwise comparison show a very high sequence identity between ORF D and CTT IX (80% aa identity), and between ORF E and CTT VIIA (84.8% aa identity). Taking into account the fact that *C. th. thummi* is only rather distantly related to *C. tentans*, the amino acid sequence similarities are surprisingly high. Therefore, we would like to conclude that *C. tentans* ORFs C, D and E represent the genes that are homologous to the genes for *C. th. thummi* Hb CTT IIβ, CTT IX and CTT VIIA. Somewhat less clear is the situation for *C. tentans* ORF A, which does not display an outstandingly high amino acid sequence identity to any of the *C. th. thummi* Hbs. The closest sequence identity between the inferred amino acid sequence of ORF A and any of the *C. th. thummi* Hb polypeptides is observed with CTT IIA (68.5%), which is, however, much less than the identity between ORF E and CTT IIA (84.8%).

Thus, ORF A clearly represents a Hb gene, but it is impossible to correlate this putative gene with one of the known Hb genes of *C. th. thummi*. It is known that there exist many more putative Hb genes than known Hb polypeptides (5, 7, 8). So, it is very possible that ORF A represents a gene for a new Hb variant in *C. tentans*.

Thus far, we have available only partial sequence data from the second clone *ten* Hb10.1. From these data (not shown), we may conclude that this DNA carries putative genes for the HbVIIB variants described for *C. th. thummi* and *C. th. piger*.

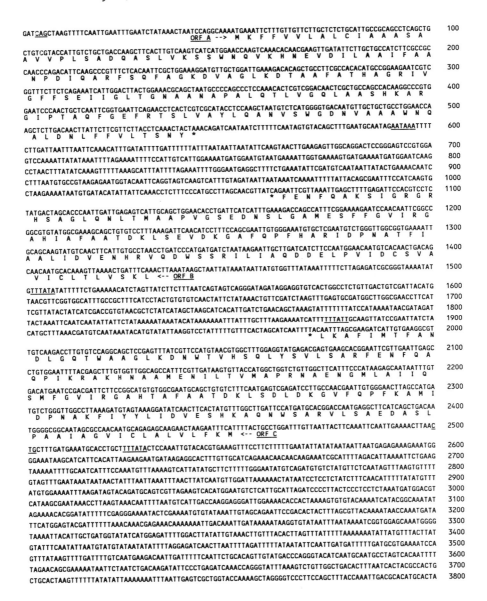

Figure 2. Nucleotide and inferred amino acid sequence of the 7 kb HindIII fragment of clone *ten* Hb5. The nucleotide sequence starts exactly at the integration site at

```
ATTGCAAGGAATTTCGCTAAATTTCTAGACCACCTAAAATGTAACAAAATTTAAAGAAAAATTCATCGACTAGAGCAAACATATCTAAAAATTAAATAAT   3900
AAAGCGTTCATGTTAATTTTTAATACGTTGACTTTTATTAAAAATGTTTCATAACATGAGGTGTTGTTTCATGTATTTGTACTTCAATCAATGCTTGATT   4000
TACAAGTGGGCGAAGATAGCGTCGAAGATGTTGTCAAGACCATGAGTCCAGGCAGCGGCTGTGGCATCATTCCATGTAGTATGGTGTGAAAGGTATGAGG   4100
*  L  H  A  F  I  A  D  F  I  N  D  L  G  H  T  W  A  A  A  A  T  A  D  N  W  T  T  H  H  S  L  Y  S  T
TCATTGAGGCACGGAATTCGTTGAATTGAGCCTTTGAGATTCCACGGTTGTGATGGTTAGTTGCCAATTCGTTGGTAAGGGTGTTCATGGCTGGGCGGTT   4200
 M  S  A  R  F  E  N  F  Q  A  K  S  I  G  R  N  H  H  N  T  A  L  E  N  T  L  T  N  M  A  P  R  N
AGCTTCGTTTCCAACAAGAGCAACAATTTCTGAGATGAATCCGACGATTCTTCCGGCGTGTGTGGCGAAAGCAGCTGTGTCCTTCAATGAGTCGACGTCC   4300
 A  E  N  G  V  L  A  V  I  E  S  I  F  G  V  I  R  G  A  H  T  A  F  A  A  T  D  K  L  S  D  V  D
TTTCCGGCGAATTGTGGGAAACGGGCTTGGATATCTGGGTGATCTTTGAAGACGGCAGCAAGGATGTCAACTTCATTGTGCTTAACTTGAGCCCATGAGC   4400
 K  G  A  F  Q  P  F  R  A  Q  I  D  P  H  D  K  F  V  A  A  L  I  D  V  E  N  H  K  V  Q  A  W  S  S
TACGGATGAGATCAGCTTGGTCAGCTGAAATTGGATCGGCGACGGCACCGACGATGCAGAGAGCGAGAACAGCGAGGAATTTCATCTTGTCTTGATTTGT   4500
 R  I  L  D  A  Q  D  A  S  I  P  D  A  V  A  G  V  I  C  L  A  L  V  A  L  F  K  M  <-- ORF D
TTGTTGCTTCAAGTTGAATTGAAAACTTAGTTTGCTTAGACAAATTTTGCTCCTTTTATACTAAAATTACGCGTGAGAATTCTAAAAGAGATTAGACATG   4600
TTACGCAAAAATAAAAAGTTTGAACATTTTAAACCTGACAAAATTGTAAACATGATAAGGATGGTTACGTTGTGAAACGATGACGAAATTATTAGTTATG   4700
TGTTTAATCGCTAGCTTATGTCAGAGTTTTCATTTACATAATAAGATTCCGCAATATTTCGGGGAGATTTCGTTTAGTTACTTATTTCGGAATTTTCAGTC   4800
ATTGAATCAATCATTAAAAAATGCTAGACAGATTTGATTTCAACAAACATTTTACTGACTGCAAACATTGTCAATCATATGCTGAAACATTGTTTTGACG   4900
CTCATGAATTTTTTTGATTTATTTTCTTTTTTAACAAATTTTCCTGTTCGAGTGAAGTATTTTATGCATTCTGTCCTCCAGCTAGCTGAACATTGTGATCG   5000
ATTGAAGCTATCTTAATATAATAAACAATAAATTTTTGATTCGTAGATTTTTATTAAAAAATCATCCAAATCATGGACTTTTACTATGCACATGCTTTGAG   5100
CTGTTAATAAATACATGTTCTTATTGATTCTATAAAATCACAATTTACAAAATTTACAAAGCGGCGAAAAGGAGTCCGTAGATGTTATCAAGTCCTTGTG   5200
                                                    *  L  A  A  F  L  L  G  Y  I  N  D  L  G  Q  T
TCCAGGCAGCAGCAACATTGTCGCCCCATGCAACATTAGCTTGGAGGTATGAGACGAGTGAGGCACGGAATTCATTGAATTGAGCTTGTGAGATTCCACG   5300
 W  A  A  A  V  N  D  G  W  A  A  N  A  Q  L  Y  S  V  L  S  A  R  F  E  N  F  Q  A  Q  S  I  G  R
GGCCTTGTGGCTGGCAGCGAGTTGTCCGACGAGTGTTTGGACAGCTGGGGCATTTGATTCGTTTCCGACGAGAGCAATGATTTCTGAGACAAATCCGACG   5400
 A  K  H  S  A  A  L  Q  G  V  L  T  Q  V  A  P  A  N  S  E  N  G  V  L  A  I  I  E  S  V  F  G  V
ATTCTTCCGGCGTGTGTGGCGAAAGCAGCTGTGTCCTTCAATGCACGAACATCCTTTCCAGCGAATTGTGGGAAACGGGCTTGGATGTCTGGGTTGGCTG   5500
 I  R  G  A  H  T  A  F  A  A  T  D  K  L  A  A  V  D  K  G  A  F  Q  P  F  R  A  Q  I  D  P  N  A  T
TGAAGACGGCGGCGAGGATGTCAACTTCGTTATTTCTAACTTGAGCCCATGTTGATTGAACAAGTGAGGCTTGGTCAGCAGTCAATGGAGTGGCGATGGC   5600
 F  V  A  A  L  I  D  V  E  N  N  R  V  Q  A  W  T  S  Q  V  L  S  A  Q  D  A  T  L  P  T  A  I  A
ACCGACGATGCAGAGGGCGAGAACAGCGAAGAATTTCATTTTGTCAGGATTAGTTTATAGATTCAAATTCAATCGAAAACTTAACTCAATTTAAAAATTC   5700
 G  V  I  C  L  A  L  V  A  F  F  K  M  <-- ORF E
CTAAACCTTTTATACAAAAAATCCGAATGAGCATTGATCGTAAGTGAGAATTAAAATTGTTGAAATAATTTAGAAAAAAGACTTTGTCATAATATAGCAT   5800
TGACCGCAATTTTGATAAGAAATTCATCTGATCTAGACAAACAAAATTTTACGATTTAACAGCATGTACGTGAGGTAGTTTTCATTTCTATAATCTAAT   5900
CTGTGTGACGAGGTTCCAGACGATTTCTTCTTATTCTTAATAATTCTCAAATTAATTGATATAAATTCTAAAAAGACCATACGCAACACGTAGCGACATT   6000
GAAGTATGAATTTTTCATTTCTTTAAAAGTGGGAATTGAACAAGGAAGAGCTGGAAAGGCTAGTAATTATTATACAATAAAAAAAAATTTTAATCTAAAT   6100
TTCACACAAAAATTAAAAAAAGAAAATGAAAATGAATTTTAAGACCTTTTTCAAACAAATATAACAAATTTGGTCATTGTTGGCCTTTTTGAAAAATTTT   6200
ATTTCACACGGTTTGAAACACACTGTTTTAGGTAGTGTTATCGCAAAGGACGATGGTGCTGACAACGAAGGATATGATAGGACTCGACTGGCCAGTGCAA   6300
ATTTCAGTCCCTTTTTTCATATGATCATTTCGTAACTAAAATTTTTCCAATCATCAAATAGCAACTCTGACTTGCAGGTGATGAAAGGTTAAATCATCAT   6400
CGCTTGCCGAGATCGACAGAAAGAAATTTGACTGGATCAGGTACACTTTAAGGAAGCATGAAACAGATCTAACCAAATAGGAACTGTTTTGATTGTCTCGA   6500
GCTCGAGAAAAAAAAACAGAATTGTAACAAATGAAGCTAGTAAAAGCAACTGTCCATATAAGAAACTGTTCAACAGCTCTTTTAGTGCCCTAAAATTCAC   6600
AGTGGATAAATAGAAAATCATTATTTTTACATAATTTTTATTTGATGTGAAATAGAATAAATTGTTCTCAATAAAGAATCTGACAGAGTCGAGGCTGCTC   6700
AGCATTGCAAAAAGCCACATACTTGAATAAAATTTGTTCGTCAAAAAAGAAAAAATAATAATGAGGAAATGAAGAAAATAATTTTCATTATTCAATAAT   6800
TTTTTTTTTAAATTACTCTAAAAAGACTTTGAAGTTATTCATAAATGACATTCACCGTTTATGGGAGAAAAGTTCTAGAATTTTGTGACAGATAATGACAT   6900
AGGGGGATGCGGATCAAAAAATCAATTTTCAATGAACTTCATTTTATAAACAACCCCTTTAGAAAAATCTGAAACATTTAGAGGGACTTTTAAAAAGCT   6997
```

the left end of the cloned genomic DNA and proceeds 6997 bp up to the HindIII site. The ORFs A, B, C, D and E are translated into amino acid sequences starting with a methionine. The directions of the possible transcriptions are indicated by arrows.

Thus, in conclusion, both gene clusters that hybridize *in situ* to one and the same chromosomal locus, contain only genes coding for dimeric Hbs. This is essentially in agreement with the hypothesis that all genes for the dimeric group of Hbs are located in one chromosomal gene cluster, while the genes for the monomeric Hbs should be located and clustered at another chromosomal locus.

Table 1. Comparison of the inferred amino acid sequences of ORFs A, C, D and E with the Hbs from *C. th. thummi.* The numbers shown are sequence identities.

	ORF A	ORF C	ORF D	ORF E
CTT IIβ	65.8%	81.4%	65.5%	66.2%
CTT IX	58.2%	64.8%	80.0%	65.5%
CTT VIIA	68.6%	64.8%	66.9%	84.8%
CTT VIII	62.6%	60.3%	61.0%	56.8%

In evolutionary terms this makes sense because the sequence data suggest that dimeric and monomeric Hbs have been separated very early in evolution, probably some 250–300 million years ago (10), while the divergence within the group of dimeric Hbs is less than that between monomeric and dimeric Hbs. Further analysis of the entire dimeric Hb gene cluster will indicate whether this hypothesis is correct. In the future, we will also be able to compare the structure and the sequence of the Hb gene cluster between the different species of *C. th. thummi, C. th. piger* and *C. tentans,* which should provide interesting insights into the evolution of a gene family with an exceptionally large number of members.

Acknowledgements

We are grateful to Miss Bettina Weich and Miss Simone Scholz for expert technical assistance, to Miss Beate Muders for typing the manuscript and to Mrs. Ursula Boell and Mrs. Regina Gramsch for the drawing and the photoreproduction of the figures. This work was supported by the Deutsche Forschungsgemeinschaft.

References

1. Tichy, H. (1975) *J. Mol. Evol.* **6**: 39–50.
2. Schmidt, E.R., Keyl, H.-G. and Hankeln, T. (1988) *Chromosoma* **96**: 353–359.
3. Goodman, M., Braunitzer, G., Kleinschmidt, T. and Aschauer, H. (1983) *Hoppe-Seyler's Z. Physiol. Chem.* **364**: 205–217.
4. Hankeln, T., Rozynek, P. and Schmidt, E.R. (1988) *Gene* **64**: 297–304.
5. Rozynek, P., Hankeln, T. and Schmidt, E.R. (1989) *Biol. Chem. Hoppe-Seyler* **370**: 533–542.
6. Antoine, M. and Niessing, J. (1984) *Nature* **310**: 795–798.
7. Trewitt, P.M., Saffarini, D.A. and Bergtrom, G. (1988) *Gene* **69**: 91–100.

8. Antoine, M., Erbil, C., Muench, E., Schnell, S. and Niessing, J.   (1987) *Gene* **56**: 41–51.
9. Hankeln, T., Luther, C., Rozynek, P. and Schmidt, E.R.   (1991)   This volume.
10. Goodman, M., Pedwaydown, J., Czelusniak, J., Suzuki, T., Gotoh, T., Moens, L., Shishikura, F., Walz, D.A. and Vinogradov, S.N.   (1988)   *J. Mol. Evol.* **27**: 236–249.

41
The Structure and Function of *Chironomus* Hemoglobins

Pawel A. Osmulski[a,b] and Wanda Leyko[b]

[a]Loomis Laboratory of Physics, University of Illinois, Urbana, IL 61801, USA
[b]Department of Biophysics, University of Łódz, 90237 Łódz, Poland

Introduction

The biological utilization of molecular O_2 has resulted in the evolution of a dynamic equilibrium between systems generating highly reactive oxygen-derived species and systems which deactivate these molecules. Sources of such potentially reactive compounds can be the result of autoxidation of low molecular weight compounds, enzymatic activity of oxidoreductases, chains of electron transport, or ions of transition metals (1). Furthermore, synthetic compounds such as pesticides, drugs or organic solvents should be added to the toxic molecules with which an organism must cope (2). The existence of an equilibrium between the physiologically important production of free radicals and their deactivation is presently a postulate since only a few elements of the system are known and a complete understanding of the process is still missing.

Heme respiratory proteins play an important role in reactions involving free radicals and in the activation and deactivation of toxic compounds, and thus provide a simple system for studying the details of these reactions. Their universal presence, usually at a high concentration, and their common reactivity as a biomolecular amplifier suggest that they can play other, very important physiological roles in addition to the storage and transport of O_2.

The scope of this article is to summarize our present knowledge of *Chironomus* Hbs and to discuss their probable role(s) in the metabolic pathways of O_2 and its active forms.

Results and Discussion

The Structural Background. Although the insects comprise, taxonomically, a large group (*ca.* 1 million species), the occurrence of Hb in this class is rare. The presence of Hb has been detected in a few representatives of Heteroptera (families Notonectidae, Corixidae and Reduviidae) (3, 4) and Diptera (families Chironomidae and Gasterophilidae) (5, 6). Chironomid Hb is the most widely studied insect respiratory protein. It is found in the following genera of Chironomidae: *Chironomus, Anatopynia, Glyptotendipes, Stictochironomus* and *Endochironomus.* Further discussion will refer mainly to *Chironomus thummi thummi* since nearly all works devoted to the Hbs of Chironomidae deal with this species.

The *C. th. thummi* Hb is freely dissolved in larval hemolymph. Its synthesis in the subepidermal fat body begins in the second instar (7). The process is controlled by juvenile hormone and ecdysterone. The juvenile hormone switches on the globin synthesis at the transcriptional level, while ecdysterone interferes with the translation of globin mRNA (8). Production of Hb ends at the pupal stage. The Hb constitutes about 80% of the total amount of hemolymph proteins. Its highest concentration is observed at the fourth instar (4–5%) (9), but also depends upon ecological conditions (10).

The most striking feature of the *C. th. thummi.* Hb is its high degree of polymorphism that depends upon the stage of development and on the population and ecological conditions. Twelve distinct Hbs, differing in amino acid sequences, have been found in a single *C. th. thummi* larva (5). The molecules form monomers (mol. wts. *ca.* 17,000) and homodimers (mol. wts. *ca.* 34,000). The formation of dimers depends upon pH, ionic strength, Hb concentration and the nature of the ligands (11). However, under physiological conditions, five components are always monomeric, six are dimeric, and one component can occur in both forms. The structural basis for the formation of the dimeric components is not clear, but some data suggest that participation of ionic bonds between residues of His or Lys to that of Glu occurs in the process (12).

The primary structures of all 12 components have been determined (5). Comparison of the sequences indicates important differences in chain lengths (136 to 151 residues) and in residue substitutions. Although there are 21 conserved residues, mainly connected with the heme cavity and maintenance of the spatial structure, the overall similarity between any two components is only 50% (15). On the basis of parsimony analysis, it appears that *C. th. thummi* Hbs have a monophyletic origin within the insecta and that the homodimeric Hbs have a monophyletic origin from monomeric forms. *C. th. thummi* Hbs were compared to 195 other eukaryotic globins, and it was estimated that the annelid *Glycera dibranchiata* was the nearest outgroup. An analysis of the transition frequency in different codon positions leads to the conclusion that *C. th. thummi* Hbs, like Mbs or cyt c, are slowly evolving proteins (5).

The tertiary structure of *C. th. thummi* Hb has been determined for component III (16, 17). Folding of the protein is typical of Mb in having eight α helical segments (about 82% of the protein) and five non-helical segments in a similar grouping. A heme *b* (a single domain arrangement) is buried between helices *F*, *G*, *E* on the proximal side, and *H* on the distal side. These helices together constitute the heme cavity. This structure is more compact than that of the Mb heme cavity, but the more important differences between the two are a 180° heme rotation around the α-τ axis and a displacement of the distal His from its usual site by an Ile residue (16). On the proximal side, the heme group is bound to His *F*8 in the usual manner.

The globin genes are located in chromosome 3. The genomic sequence of two components (CTT III and IV) is known and reveals an intron-less architecture (13). Detailed analysis of the chromosome 3 structure also suggests that the high diversity of globins has resulted from repeated duplication followed by paracentric inversion (14).

Functional Consequences. The *C. th. thummi* Hbs display the following relatively high O_2 affinities (expressed as P_{50} in mm Hg, pH 7.0, 20°C): mixture of all the components, 1.27; mixture of the monomeric components, 1.04; and mixture of the dimeric components, 1.76 (18). The shape of the O_2 affinity curve is hyperbolic and the Hill coefficient is about 1, even for the dimeric components, indicating that *C. th. thummi* Hbs bind O_2 noncooperatively. Although the O_2 affinity of *C. th. thummi* Hbs is unaffected by inorganic phosphates, high concentrations of NaCl (0.56–1.16 M) increase the P_{50} of CTT III and decrease that of CTT IV, VI, VIIB, and X (19). With the exception of CTT I and IA, the Hbs exhibit an alkaline Bohr effect; for example, the P_{50} of monomeric Hb CTT III is 0.28 and of dimeric Hb CTT VIIB is 0.94 (19).

The binding of O_2 to the heme iron produces significant rearrangements in the structure of the whole Hb molecule, particularly in the heme cavity. These changes are pH-dependent and induce two conformation states: a *t* (tense) form with a low O_2 affinity at acidic pH, and an *r* (relaxed) form with a high O_2 affinity at alkaline pH. When O_2 binds to the heme iron, the number of globin-heme contacts increases, the central atom is closer to the heme plane and the distance between the nitrogen atom of the proximal His and the iron decreases. Changes in pH affect the detailed structure of *C. th. thummi* Hb III. At low pH the structure is stabilized by a salt bridge between His *G*2 (His-95) and the *C*-terminal Met *H*22 (Met-136). The linkage leads to a stiffening of the *F* helix and makes the structure of "the cage" more compact and less accessible to all ligands. A proton dissociation from the Met residue at higher pH causes an increase in the mobility of the *F* helix and a lengthening of the Fe–His bond. As a consequence, the Hb exhibits higher ligand affinity (16, 17).

It is believed that, in general, there are at least four energetic barriers that a ligand moving towards the heme iron should overcome: a hydrophobic effect (the transfer of the ligand from solvent to the protein surface), a cage effect (its movement into the heme cavity), a heme door effect (the penetration of the distal part of the heme cavity by the ligand), and the binding to the heme iron atom (20). The data suggest that the observed high O_2 affinity arises as a result of the high accessibility of the heme (the cage effect), and the unobstructed movement of the ligand in the distal part of the heme cavity, because of the altered orientation of distal His (the heme door effect).

In order to discuss ligand affinity and its regulation, it is also necessary to take into account the detailed architecture of the heme and its surroundings. This can be described in terms of four effects: a trans effect (the interaction of proximal His through the heme plane with a ligand), a cis effect (the interactions of the bound ligand with the distal part of the heme cavity), a porphyrin effect (the interaction of the porphyrin macrocycle, especially of its side chains with globin), and a metal effect (the interaction of the central iron atom with the globin and the linked ligands) (21).

On the basis of the collective results of EPR, NMR and x-ray diffraction experiments (22, 21, 16, 17), it appears that the trans effect is the fundamental mechanism modulating both ligand affinity and the Bohr effect. The cis effect only increases the ligand affinity since the bound molecule (especially O_2) is not stabilized by any interaction with the distal His. A modification of the porphyrin side chains mainly influences the ligand affinity. Usually, an increase in affinity is observed because of steric hindrance of the modified porphyrin in the heme cavity. The Bohr effect is left unchanged if differences in the electron affinity of the heme iron and the mechanism of allosteric interactions are compensated. The heterotropic interactions also remain unaffected by the substitution of the central atom. However, the replacement of iron by a Co leads to a 250-fold decrease in O_2 affinity, probably because of its stronger σ-donor property (22).

C. th. thummi Hbs are oxidized relatively easily compared to other heme proteins. This may be due to a larger exposure of the porphyrin ring to the solvent and to the large distal heme cavity. The Hbs share a common mechanism of oxidation through a nucleophilic displacement of an oxygen-derived radical (23). From the structural point of view, the spatial organization of *C. th. thummi* metHb is similar to the deoxy Hb. The only differences found, besides the oxidized iron atom, deal with the reduced number of heme-globin contacts and the lengthened distance between the central atom and the distal His (17). Depending upon pH, the EPR and spectrophotometric data (24, 25) reveal the presence of two forms of the metHb: an acidic derivative with a water molecule bound to the heme iron and an alkaline derivative with a hydroxyl group, characterized by high and low spins of the heme iron, respectively.

C. th. thummi Hbs exhibit some interesting oxidation-reduction properties. They share with other heme respiratory proteins the property of having their hydroxymet forms much less easily reduced with ascorbate than their aquamet forms (26). However, the reduction by ascorbate is much faster in the *C. th. thummi* Hbs and is closer to the reduction rates of cyt c than that of Mb. Next, the dependence of the reduction rates of the *C. th. thummi* aquomet Hbs on pH can be divided into two distinct parts: an acidic and an alkaline range. This phenomenon was observed previously only in heterotetrameric Hbs. The imidazole (Im) derivatives of the *C. th. thummi* Hbs reduced more slowly than those of the metHbs, a feature that is again similar to cyt c but not to other Hbs (27).

C. *th. thummi* Hbs can be oxidized further to a high valent Fe^{4+} form called the ferryl derivative, which is extremely unstable, and shows an optical absorption spectrum similar to ferryl Mb, with maxima at 420 and 540 nm (27).

C. *th. thummi* Hbs demonstrate an enzymatic or pseudoenzymatic activity of *p*-hydroxylation of aromatic amines, such as aniline, in the presence of NADPH. The protein is a much more effective catalyst of the reaction than are human HbA or cyt c. It is interesting that the mixture of monomeric components and the mixture of dimeric ones show similar maximum velocities, but the Michaelis constant is about three times larger in the former and results in lowered V/K_m (27).

An analysis of these findings leads to some important observations. For *C. th. thummi* Hbs, a direct relationship between their spatial structure and their function can be seen. It seems that the presence of functional interactions between the chains restricts the reactivity of the heme group. This is clearly visible in the case of the reaction with ascorbate. On the other hand, the dimeric structure appears to be more effective in the enzymatic reaction in which the highly oxidized forms of Hb probably also take part. The role of the distal His in the stabilization of the ferryl derivative is unquestionably evidenced on the basis of the comparison of its stability with other heme proteins. The dependence of the Hb reduction by ascorbate on pH may be easily explained by the presence of two conformers of aquo-metHb with different pathways of electron transfer. The presence of such conformers was also demonstrated in the autoxidation of Mb isolated from different sources (28). The retardation in the reduction of the metHb (Im) by ascorbate is probably related to the restriction of electron transfer by the bound axial ligand that is not compensated for by the larger exposure of the heme group.

This short review of *C. th. thummi* Hb properties should help to explain possible physiological roles of the protein.

Physiological Overview. It is generally believed that Hb functions either as an oxygen store or as an oxygen transporter or both. In the case of *C. th. thummi* Hbs, the first follows from the fact that the content of the Hb is large enough to maintain the metabolic activity of larvae for up to 15 min (30). This storage function may be provided by the monomeric components which exhibit higher oxygen affinity and the alkaline Bohr effect. The second proposed function is the facilitation of O_2 diffusion and the maintenance of a proper O_2 gradient. The concentration of Hb increases during larval development as surface area and body volume decrease. Moreover, the ratio of dimeric Hbs to monomeric Hbs increases when the O_2 supply is restricted. It is on this basis that an important aspect of the O_2 transport may be related to the presence of the dimeric components.

The two possible functions of the *C. th. thummi* Hbs mentioned above explain both the quantitative and qualitative compensations in an O_2 supply for the chironomid larvae. The changes in overall concentration of the Hbs would result in a quantitative increase in O_2, while the induction of the synthesis of the specific (dimeric?) components would modulate the transport of O_2.

The description of oxidation-reduction reactions of *C. th. thummi* Hbs and the remarks in the introductory part of this work imply another possible function of the protein: the development of a resistance to toxic compounds, or, more generally, the participation of the Hbs in the metabolism of xenobiotic compounds.

C. th. thummi larvae often live in highly contaminated waters characterized by a low level of dissolved O_2 and a high level of organic substances. Under such eutrophic conditions, *C. th. thummi* can occur in masses constituting more than 80% of the total biomass (29). The ability to resist the noxious environmental conditions has been equated with the presence of Hb in the larvae.

Freshwater macroinvertebrates possessing Hb often have a much higher tolerance to polluted environments (15). It was found that *Chironomus riparius* (also Hb-rich) is highly resistant to phenol and lindane (31, 32). Of course, these phenomena may be simply explained in terms of the Hb providing a better O_2 supply to other O_2-dependent detoxification pathways such as cyt P450. On the other hand, since the hydroxylation activity of the *C. th. thummi* Hbs was demonstrated under *in vitro* conditions, a good argument can be made for the protein as an *in vivo* detoxification system. Human HbA shows a mono-oxygenase activity against many derivatives of aromatic amines, both *in vitro* and *in vivo*. HbA catalyzes *p*-hydroxylation, *O*-and *N*-demethylation in the presence of a source of electrons (NADPH or ascorbate), and, additionally, of cyt P450 reductase (33). Another important class of reactions catalyzed by HbA is the epoxidation of xenobiotics (*e.g.*, benzo[a]pyrene and its derivatives (34)), or of biologically active agents (*e.g.*, retinoic acid and derivatives of eicosatreanoic acid (35)). As a result of these activities, compounds with more pronounced hydrophilic properties are formed. Sometimes the products are more toxic (*e.g.*, the transformation of aniline into *p*-aminophenol) and/or create adducts with proteins. All of these reactions proceed with the participation of active species, mainly free radicals, and the high valent derivatives of heme respiratory proteins. In the case of *C. th. thummi* Hbs, only limited data are available, but their much greater effectiveness in comparison with other heme proteins suggests the importance of Hbs in the pathways of biotransformation.

Because of the ability of HbA to take part in the metabolism of toxic compounds and the ability of chironomids to survive in polluted environments, it is of vital importance to search for the enzymatic activity of *C. th. thummi* Hbs under *in vivo* conditions as well as to investigate their role in the biotransformation of xenobiotics.

Acknowledgements

We thank Prof. P.G. Debrunner and Dr. C.R. Vossbrinck for valuable discussion and suggestions. This work was supported in part by National Institutes of Health Grant 16406 to P.A. Osmulski and Grant RP II.1.11.0.1 from the Polish Ministry of Education.

References

1. Halliwell, B. and Gutteridge, J.M.C.   (1986)   *Trends Biochem. Sci.* **11**: 372–375.
2. Machlin, L.J. and Bendisch, A. (1987) *FASEB J.* **1**: 441–445.
3. Bergtrom, G. (1977) *Insect Biochem.* **7**: 313–316.
4. Wells, R.M.G., Hudson, M.J. and Brittain, T. (1981) *J. Comp. Physiol.* **142**: 515–522.
5. Goodman, M., Braunitzer, G., Kleinschmidt, T. and Aschauer, H. (1983) *Hoppe-Seyler's Z. Physiol. Chem.* **364**: 205–217.
6. Keilin, D. and Wang, Y.L. (1946) *Biochem. J.* **40**: 855–866.
7. Vafopoulou-Mandalos, X. and Laufer, H. (1984) *Archs. Insect Biochem. Physiol.* **1**: 191–198.
8. Vafopoulou-Mandalos, X. and Laufer, H. (1982) *Dev. Biol.* **92**: 135.
9. Tichy, H. (1980) In *Chironomidae Ecology, Systematics, Cytology and Physiology*, ed. D.A. Murray, 43–51. New York: Pergamon Press.
10. Leyko, W. and Osmulski, P.A.   (1985)   *Comp. Biochem. Physiol.* **80B**: 613–616.
11. Behlke, J., Müller, K. and Scheler, W. (1972) *Acta Biol. Med. Germ.* **28**: 1069–1072.
12. Aschauer, H., Kleinschmidt., T., Steer, W. and Braunitzer, G. (1982) *Chem. Peptide Prot.* **1**: 61–67.
13. Antoine, M. and Neissing, J. (1984) *Nature* **310**: 795–798.
14. Tichy, H. (1975) *J. Mol. Evol.* **6**: 39–50.
15. Osmulski, P.A. and Leyko, W.   (1986)   *Comp. Biochem. Physiol.* **85B**: 701–722.
16. Weber, E., Steigemann, W., Jones, T.A. and Huber, R. (1978) *J. Mol. Biol.* **120**: 327–336.
17. Steigemann, W. and Weber, E. (1979) *J. Mol. Biol.* **127**: 309–338.
18. Osmulski, P.A.   (1988)   *Acta Univ. Lodz. Folia Biochim. Biophys.* **6**: 123–143.
19. Weber, R.E., Braunitzer, G. and Kleinschmidt, T. (1985) *Comp. Biochem. Physiol.* **80B**: 747–753.
20. Dreyer, J.L. (1984) *Experientia* **40**: 653–675.
21. La Mar, G.N., Anderson, R.R., Budd, D.L., Smith, K.M., Langry, K.C., Gersonde, K. and Sick, H. (1981) *Biochemistry* **20**: 4429–4436.
22. La Mar, G.N., Anderson, R.R., Chacko, V.P. and Gersonde, K. (1983) *Eur. J. Biochem.* **136**: 161–166.
23. Shikama, K. (1985) *Experientia* **41**: 701–706.

24. Gersonde, K., Twilfer, H. and Overkamp, M. (1982) *Biophys. Struct. Mech.* **8**: 189–211.
25. Christahl, M. and Gersonde, K. (1982) *Biophys. Struct. Mech.* **8**: 271–288.
26. Tsukahara, K. and Yamamoto, Y. (1983) *J. Biochem.* **93**: 15–22.
27. Osmulski, P.A. (1989) Ph.D. Thesis, University of Lodz.
28. Suzuki, T. (1987) *Biochim. Biophys. Acta* **914**: 170–176.
29. Koehn, T. and Frank, C. (1980) In *Chironomidae Ecology, Systematics, Cytology and Physiology*, ed. D. A. Murray, 187–195. New York: Pergamon Press.
30. Fox, H.M. and Taylor, A.E.R. (1955) *Proc. Roy. Soc. London, B* **143**: 214–225.
31. Green, D.W.J., Williams, K.A. and Pascoe, D. (1985) *Archs. Hydrobiol.* **103**: 75–82.
32. Williams, K.A., Green, D.W.J. and Pascoe, D. (1985) *Archs. Hydrobiol.* **102**: 461–471.
33. Starke, D.W., Blisard, K.S. and Mieyal, J.J. (1984) *Molec. Pharmacol.* **25**: 467–475.
34. Catalano, C.E. and Ortiz de Montellano, P.R. (1987) *Biochemistry* **26**: 8373–8380.
35. Iwahashi, H., Ikeda, A., Negoro, Y. and Kido, R. (1986) *J. Biochem.* **99**: 63–71.
36. Pace-Asciak, C.R. (1984) *Biochim. Biophys. Acta* **793**: 485–488.

42
Transcriptional Control of *Vitreoscilla* Hemoglobin Synthesis

Kanak L. Dikshit, Rajendra P. Dikshit and Dale A. Webster

Department of Biology, Illinois Institute of Technology, Chicago, IL 60616, USA

Introduction

The Hb from the gram-negative bacterium *Vitreoscilla* is a homodimeric molecule that exhibits cooperative ligand binding properties (1–3). It has the most sequence homology with lupine leghemoglobin (LegHb) (4), but its function is probably quite different from that of LegHb. The latter putatively functions in symbiotic nitrogen fixation to supply O_2 to the Rhizobial respiratory chain, but is buffered at a concentration low enough to prevent inactivation of the O_2-sensitive nitrogenase (5, 6). *Vitreoscilla* Hb (*Vt*Hb), on the other hand, probably functions to facilitate respiration under hypoxic conditions by supplying O_2, buffered at a high concentration, to the terminal oxidases. Supportive evidence for this includes the dramatic rise in the intracellular concentration of *Vt*Hb in cells grown under hypoxic conditions (7), and the unusually high dissociation rate constant (k_{off}) for O_2 (3, 8). The latter confers a rather large K_d on this Hb that is consistent with its proposed function.

The gene for the globin part of *Vt*Hb (vgb) has been cloned in *E. coli* where it is highly expressed (9, 10). This has permitted the study of the control of biosynthesis of this Hb at the genetic level. *Vt*Hb can comprise 8% or more of the total protein in *E. coli*, and during respiration it is in the physiologically functional oxy form in both *Vitreoscilla* and *E. coli* (10). Its presence has been shown to facilitate the growth of *E. coli* at low O_2 concentrations (11, 12). It is now known that O_2 controls the biosynthesis of VtHb at the level of transcription. After Northern hybridization, an RNA transcript of about 500 bp, was detected using the vgb gene as a probe (10). The relative amount of vgb-specific mRNA increased in cells grown at low O_2 levels in both *E. coli* (13, 14) and *Vitreoscilla* (12, 13). Since the cloned gene fragment also included the regulatory or promoter region of the vgb gene, we decided to further characterize the regulation of this gene.   One way to facilitate this is to construct transcription fusion plasmids in which the promoter region is fused to another gene (the reporter gene), and which expresses a product that is easily assayed.

We used two such constructs: one (pKD-14) with the CAT gene which codes for chloramphenicol acetyl transferase, and the other (pKD-49) with the xylE gene which codes for catechol-2, 3-dioxygenase (CDO). The construction of these two plasmids has been described previously (12). Maps of plasmids used in this study are shown in Figure 1. The sequence of the promoter region is shown in Figure 2.

Materials and Methods

Vgb-CAT and vgb-xylE fusion plasmids, respectively, were created by fusing the HindIII-AflII fragment of pUC8 : 16 at the SmaI site of pKK232-8 and at the HindIII-EcoRI site of pVDX18.   Details are given in reference 12.   Plasmid pKD-49 is a non-self-transmissible RSF 1010 derivative and may be mobilized using helper plasmid pRK2013 (17). The construction of plasmids pDR540 : 16 and pDR540 : 17 has been described (12). Briefly, to form pDR540 : 16, the 1.4 kb HindIII-BamHI fragment of pUC8 : 16, containing the entire vgb gene including its promoter region, was ligated into pDR540, which had its tac promoter removed by the same restriction enzymes. To form pDR540 : 17, the AflII-SalI fragment of pUC8 : 16, containing the vgb gene without its promoter, was ligated into pDR540 opened with BamHI, to put vgb into correct orientation under control of the tac promoter. The maps of the plasmids are provided in Figure 1.

Results and Discussion

The expression of the two fusion plasmids (pKD-14 and pKD-49) in *E. coli* was increased in cells grown at low (5%) O_2 relative to those grown at atmospheric (20%) O_2 (Figure 3). Expression of both reporter genes was higher at all stages of the growth cycle for cells grown at 5% O_2, which indicates that O_2 does regulate the activity of the vgb promoter. Since we had difficulty trying to maintain the 20% O_2 level at late stages of the growth cycle because of the high O_2 consumption of the cells, the actual differences between the 5% and 20% O_2 levels are probably even larger than what we observed.   This is

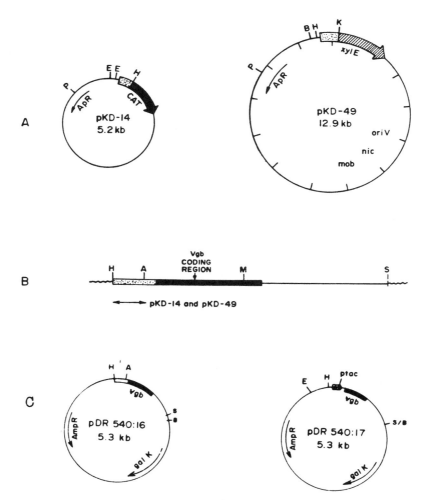

Figure 1. (*A*) Maps of recombinant plasmids pKD-14 and pKD-49. (*B*) Map of plasmid pUC8 : 16 showing cloned portion of the *Vitreoscilla* genome containing the vgb gene. The double arrow, ↔, indicates the portion of vgb that was cloned on the above promoter probe vectors. (*C*) Maps of plasmids constructed from pDR540 containing the vgb gene under control of its native promoter and the *E. coli* tac promoter: clear rectangle, promoter region of vgb; solid rectangle, vgb coding region; striped rectangle, tac promoter (*p*tac). (*Abbreviations.* Restriction enzyme sites: A, AflII; B, BamHI; E, EcoRI; H, HindIII; K, KpnI; M, MluI; P, PstI; and S, SalI. S/B designates the junction of the SalI and BamHI sites after blunt end ligation. Genetic markers: ApR (AmpR in Figure 1*C*), ampicillin resistance; CAT, gene encoding chloramphenicol acetyl transferase; gal K, gene coding for galactokinase; mob, determinant essential for plasmid mobilization; nic, relaxation nick site; oriV, origin of replication; vgb, gene coding for the globin part of Hb; xylE, gene encoding catechol 2, 3 dioxygenase.)

                         PBS1   PBS2
         -100 GTGGATTAAG TTTTAAGAGG CCAATAAAGA TTATAATAAG TGCTGCTACA

                         PBS2   S-D   -1
         -50 CCATACTGAT GTATGGCAAA ACCATAATAA TGAACTTAAG GAAGACCCTC **ATGTTAGAC**
                                                              Af

Figure 2. Sequence of vgb promoter region. From (9), confirmed in our laboratory to
−100 shown.
Af = AflII restriction enzyme site.
S-D = Shine-Dalgarno ribosome binding site *AGGAGG*.
*PBS*1 = −35 consensus sequence *TTGACA*.
*PBS*2 = −10 consensus sequence (Pribnow box) *TATAAT*.

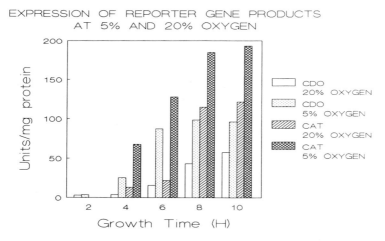

Figure 3. Expression of CAT and xylE genes fused to the vgb promoter during the
growth cycle at 20% and 5% O2. *E. coli* JM83, with fusion plasmids pKD-14 or pKD-
49, was grown in LB medium at 37°C under 20% and 5% O2. The O2 levels were
maintained by bubbling a mixture of air and N2. The assays for CAT and CDO
activities are described in references 12 and 18, respectively.

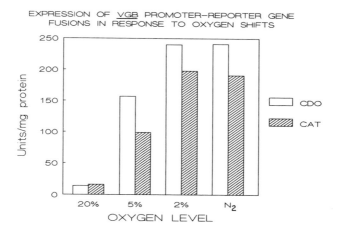

Figure 4.   Expression of vgb promoter-reporter gene fusions in response to O_2 shifts. *E. coli* JM83 harboring fusion plasmids pKD-14 or pKD-49 was grown in LB medium at 37°C with vigorous shaking (250 rpm) for 3 h, shifted to the specified O_2 level for 1 h, and then the amounts of reporter proteins were assayed. One flask was bubbled with N_2 to lower O_2 below 1%; it is unlikely to be completely anaerobic.

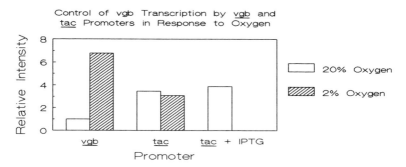

Figure 5.   Control of vgb transcription by vgb and tac promoters in response to O_2. Brief descriptions of the construction of pDR540:16 containing the native vgb promoter and pDR540:17 containing the *E. coli* tac promoter are given in the Methods section. Total RNA was isolated as described previously (13) from *E. coli* JM103 strains carrying these plasmids or control plasmid pDR540 grown under 20% or 2% O_2. RNA (50 μg) from each sample was fractionated on a 1.2% agarose gel containing 2.2 M formaldehyde (15) and immobilized on nitrocellulose paper. This was hybridized with the HindIII-MluI fragment (^{32}P labeled), which contains the vgb promoter region and part of the coding region. After developing, the films were scanned with a Model EC910 Densitometer from E-C Apparatus Corp. The concentration of IPTG used for induction of pDR540:17 (20% O_2, 2%O_2 not tested) was 5 mM. There was no hybridization observed when control cells containing pDR540 were used.

Table 1. Recognition and expression of the vgb promoter in other bacteria. Plasmid pKD-49, which has the vgb promoter fused to the reporter gene xylE, was transferred into different hosts by triparental filter mating (17) using *E. coli* HB101 to harbor the helper plasmid pRK2013. Recombinant colonies were identified visually for vgb expression (CDO activity) after spraying with 100 mM catechol in 50 mM potassium phosphate buffer, pH 7.5. Specific CDO activity was assayed as described (18).

Strains	Relative CDO activity
E. coli JM83	100[a]
Pseudomonoas putida	119
Pseudomonas aeruginosa	116
Rhizobium japonicum	22
Azotobacter vinelandii	N. D.[b]

[a]Control rate was 124 units/mg protein.
[b]Specific activity was not determined, but the intensity of the yellow color that was observed in recombinant colonies after spraying with the catechol solution was intermediate between the first three strains and *R. japonicum*.

supported by the results of the O_2 shift experiment shown in Figure 4: when cells harboring these fusion plasmids were transferred from normal, atmospheric O_2 to hypoxic growth conditions (2% O_2), the levels of the reporter proteins were 12 to 17 times larger than those growing at atmospheric O_2.

We constructed other types of transcription fusions to confirm these results. In one of these (pUC8 : 17), the vgb promoter was deleted and *E. coli* strains bearing this plasmid did not express vgb under any conditions. Two constructions were made from plasmid pDR540: in one, the vgb gene was left under the control of its native promoter; and in the other, the vgb promoter was replaced by the *E. coli* tac promoter (Figure 1). In the latter (pDR540 : 17), the gene was expressed, but the level of expression was not affected by O_2 (Figure 5). An inducer of this promoter, isopropyl-β-D-thiogalactopyranoside (IPTG), did slightly increase the level of expression of vgb (Figure 5). The rather high level of expression of vgb in the absence of an added inducer may be due to the presence of inducers in the medium.

The plasmid, pKD-49, is actually a transcriptional fusion constructed using the broad host range plasmid pVDX18 which contains the promoterless xylE gene. The HindIII-AflII fragment of pUC8 : 16 (containing the vgb promoter) was inserted into this plasmid to create pKD-49. We transferred this plasmid into different hosts by a triparental mating procedure to see if the vgb promoter could be recognized in other organisms and vgb expressed. We found that the vgb promoter was also recognized in *Pseudomonas, Azotobacter* and *Rhizobium* (Table 1).

The vgb promoter contains a region of strong homology with the CAP-binding site on the *E. coli* tac promoter, and there is evidence that a catabolite activator protein (CAP) is also involved in vgb regulation (14). We found that when glucose is used as a carbon source for *E. coli* JM83 containing plasmids pUC8 : 16 or pKD-14, the amount of *Vt*Hb or CAT produced, respectively, was about one-third to one-fifth of that produced when aerobic carbon sources like succinate, lactate, or glycerol were used (Figure 6). Our finding that glucose acts as a repressor of the vgb promoter supports previous evidence (14) that CAP has a role in the regulation of this gene.

The sequence of the promoter contains several potential transcriptional starting points (Figure 2). We used S1 nuclease mapping to identify which site or sites are used (15, 16). Starting with the 500 bp HindIII-M1uI fragment containing the vgb promoter and part of the coding sequence (Figure 7, lane *D*), we isolated a 380 bp RNA-protected fragment from *Vitreoscilla* (Figure 7, lane *B*). This localized the transcriptional starting point between the Shine-Dalgarno site and the first potential polymerase binding site (PBS2), or between about –10 and

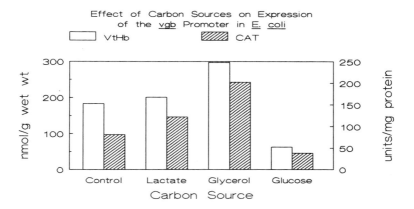

Figure 6. Effect of carbon sources on the expression of the vgb promoter in *E. coli*. Strains of *E. coli* JM83 containing plasmids pUC8 : 16 or pKD-14 were grown overnight at 37°C in 100 ml minimal medium (10) in 250 ml flasks supplemented with nitrate and 0.2% of the indicated carbon source. The O_2 concentration was lowered to obtain optimal conditions for the induction of the vgb promoter by flushing each flask, with the exception of the control, with sterile N_2 for 30 min and then by sealing each with a rubber stopper and parafilm. Results obtained using succinate (not shown) were virtually identical to those of lactate.

Lane A. DNA fragment from E. coli carrying pUC8:17.
Lane B. DNA fragment from Vitreoscilla.
Lane C. DNA fragment from E. coli carrying pUC8:16.
Lane D. Undigested control fragment.
Lane E. 123 bp DNA ladder.

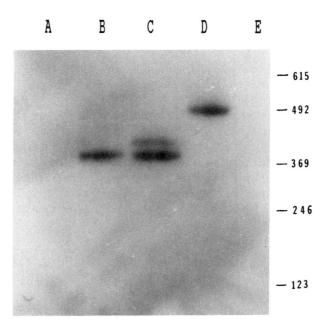

Figure 7. Nuclease S1 mapping of vgb transcripts. The 5' end labeled HindIII-MluI fragment of pUC8 : 16, which is 500 bp and contains the vgb promoter and part of the coding sequence, was protected using RNAs isolated from *Vitreoscilla* and *E. coli* carrying pUC8 : 16 and pUC8 : 17 (promoterless control). The 5' end labeled probe was prepared using polynucleotide kinase and ^{32}P-ATP, after treating with calf intestine phosphatase to remove the 5' terminal phosphate (15). After purifying, the labeled DNA probe (20,000 cpm) was mixed with 200 μg of RNA from each strain, denatured, annealed, treated with 200 units of S1 nuclease according to published procedures (15, 16), and then analyzed on a 4% polyacrylamide –7 M urea gel.

–20. The corresponding Pribnow box is at –20 to –25. A protected fragment of the same size was also identified in *E. coli*, but an additional minor fragment, about 30–35 bp larger, was also found in this organism (Figure 7, lane *C*). The latter may be an experimental artifact or a second promoter site that is recognized in *E. coli* but not in *Vitreoscilla*. Khosla and Bailey have also identified two transcriptional start sites in *E. coli* (14), but the stronger one had its Pribnow box at –64 to –69; and the weaker at –116 to –121. Whether their procedure may have missed the most proximal transcriptional starting site is not clear from the evidence presented (14). The size of the minor, larger S1 nuclease fragment that we observed in *E. coli* is consistent with the first location upstream that was identified by them, but our HindIII-M1uI fragment may not have been sufficiently long to enable detection of the transcriptional starting site that is furthest upstream. Thus, it is possible that there are actually three transcriptional starting sites that are recognized in *E. coli*.

Acknowledgements

This work was supported by National Institutes of Health grant GM27085. We are indebted to Dr. Benjamin Stark for helpful advice.

References

1. Tyree, B. and Webster, D.A. (1978) *J. Biol. Chem.* **253**: 6988–6991.
2. Webster, D.A. and Orii, Y. (1985) *J. Biol. Chem.* **260**: 15526–15529.
3. Webster, D.A. (1988) In *Advances in Inorganic Biochemistry*, eds. G.L. Eichhorn and L.G. Marzilli, vol. 7, 245–265. New York: Elsevier.
4. Wakabayashi, S., Matsubara, H. and Webster, D.A. (1986) *Nature* **322**: 481–483.
5. Appleby, C.A. (1984) *Annu. Rev. Plant Physiol.* **35**: 443–478.
6. Wittenberg, J.B. and Wittenberg, B.A. (1990) *Annu. Rev. Biophys. Biophys. Chem.* **19**: 217–241.
7. Boerman, S.J. and Webster, D.A. (1982) *J. Gen. Applied Microbiol.* **28**: 35–43.
8. Orii, Y. and Webster, D.A. (1986) *J. Biol. Chem.* **261**: 3544–3547.
9. Khosla, C. and Bailey, J.E. (1988) *Mol. Gen. Genet.* **214**: 158–161.
10. Dikshit, K.L. and Webster, D.A. (1988) *Gene* **70**: 377–386.
11. Khosla, C. and Bailey, J.E. (1988) *Nature* **331**: 633–635.
12. Dikshit, K.L., Dikshit, R.P. and Webster, D.A. (1990) *Nucleic Acids Res.* **18**: 4149–4155.
13. Dikshit, K.L., Spaulding, D., Braun, A. and Webster, D.A. (1989) *J. Gen. Microbiol.* **135**: 2601–2610.
14. Khosla, C. and Bailey, J.E. (1989) *J. Bacteriol.* **171**: 5995–6004.
15. Maniatis, T., Fritsch, E.F. and Sambrook, J. (1982) *Molecular Cloning: A Laboratory Manual*. New York: Cold Spring Harbor.
16. Berk, R.J. and Sharp, P.A. (1977) *Cell* **12**: 721–732.
17. Riess, G. and Puhler, A. (1984) In *Advanced Molecular Genetics*, eds. A. Puhler and K.N. Timmis, 51–59. Berlin: Springer-Verlag.
18. Konyecsni, W.M. and Deretic, V. (1988) *Gene* **74**: 375–386.

43
Hemoglobins of Eukaryote/Prokaryote Symbioses

Jonathan B. Wittenberg[a] and David W. Kraus[b]

[a]Department of Physiology and Biophysics, Albert Einstein College of Medicine, Bronx, New York 10461, USA
[b]Department of Biology, University of Alabama, Birmingham, Alabama 35294, USA

Introduction

Cytoplasmic Hb, developed by the host, is a constant feature of symbioses between plants and intracellular prokaryotes and between molluscs and intracellular prokaryotes. This essay attempts to define the role of Hb in each symbiosis.

The plant symbionts fix atmospheric N_2 into ammonia and supply in small or large part, the N_2 requirements of the host. The molluscan symbionts fix carbon from carbon dioxide into hexoses by way of the Calvin-Benson cycle, and supply the majority or all of the carbon nutrition of the host. The fuel of oxidative phosphorylation by plant symbionts is photosynthate, brought to root or stem nodules in the vascular system. The fuel of carbon fixation by molluscan symbionts is hydrogen sulfide, methane or other reductants, brought to each bacteriocyte (symbiont-harboring cell) of the solemyid or lucinid gill in the respiratory stream of sea water. We shall be concerned with symbioses using hydrogen sulfide.

The two symbiotic systems resemble each other in many ways. In each, O_2 is brought to the borders of the bacteriocyte. In plants, a series of air passages reaches every individual bacteriocyte. In the modified solemyid or lucinid gill, as noted, sea water bathes one face of each bacteriocyte. Hb, accordingly, functions within the boundaries of each cell and need not transport O_2 over long distances in tissue. In each, the intracellular symbiont, housed in specialized cells of specialized tissues, occupies a third of the tissue volume. In each, the symbiont is housed in peribacterial sacs set off from the host cytoplasm proper by a peribacterial membrane. The complement of proteins within the sac is different from that outside, and active transporters are located on the peribacterial membrane (1, 2). The Hb of plant symbioses is confined entirely to the plant cytoplasm proper; none is found within the peribacterial sac ((1), reviewed in (3)).

Hb occurs in non-symbiotic tissues of plants (4) and molluscs; indeed, the gills of a few clams that do not house symbionts are dark red by virtue of their contained Hb (5). We conclude that the development of Hb in symbiont-harboring tissues exploits a pre-existing ability of the host organism.

Hb in plant root nodules and clam gills may reach very high concentrations, 1.0–1.5 mM/kg wet weight tissue, perhaps 5 mM in the cytoplasmic domain to which it is confined (5, 6). The flux of nutrients and O_2 to the symbiont is large, and the return flux of metabolites to the host must be large. Animals and plants with interiorized symbionts flourish in inhospitable and otherwise inaccessible environments and reach gargantuan size and impressive population densities.

Perhaps because the mechanism of Hb action long remained unknown, most authors have regarded Hb as ancillary to tissue function, a decorative accessory easily dispensed with: not so. There is no plant/symbiont association effective in fixing N_2 without plant Hb. Carbon monoxide blocking of legHb (legume Hb) in the intact soybean root nodule largely abolishes N_2 fixation (7). Experiments in which bacteroids (the bacterium *Rhizobium* modified for symbiotic life) isolated from soybean nodules are suspended in solutions of Hb demonstrate why this is so (8). Bacterial oxidative phophorylation is dependent on and proportional to Hb-bound O_2 and uses free, dissolved O_2 to only a very limited extent (6, 8). This may not be an isolated phenomenon. Oxidative phosphorylation by mitochondria of mammalian heart muscle cells also depends in part upon Mb-bound O_2 (3, 9).

In this essay, we examine the properties of isolated Hbs, and from those properties deduce what we can about the function of the Hbs in the symbioses from which they were isolated.

Results and Discussion

The nitrogenase system of plant symbionts is highly intolerant of O_2, and, although a large flux of O_2 must reach the terminal oxidases of bacterial oxidative phosphorylation, none must penetrate to the site of N_2 fixation within the bacteroids. The free O_2 concentration within the cytoplasm is no more than 10 nM. The rate of O_2 consumption and the dimensions of the nodule are such that the diffusion of free O_2 at this concentration cannot support the desired flux.

Table 1. Kinetics and Equilibria of Reactions of Plant Hbs with O_2.

Protein	Combination	Dissociation	Equilibria
	k'_{on} x 10^{-6} $M^{-1}s^{-1}$	k_{off} s^{-1}	$K'(= k:k')$ nM
Soybean	120	5.6	48
Kidney bean	130	6.2	48
Cow pea II	140	5.5	39
Sesbania II	210	7.5	36
Green pea I	250	16	65
Green pea IV	260	16	61
Broad bean V	260	19	73
Lupin I	(540)	20	36
Lupin II	(320)	25	78
Parasponia I	165	15	89
Casuarina II	41	6	135

The concentration of legHb-bound O_2 exceeds that of free O_2 perhaps one hundred thousand-fold. We deduce that a major function of legHb is to facilitate diffusion of O_2 to the bacteroid surface.

Gibson *et al.* (10) have studied the reactions of fourteen purified plant Hbs with O_2 and other ligands. These include Hbs from legume/*Rhizobium* associations, the *Parasponia* (a tropical tree in the elm family)/*Rhizobium* association (11) and the *Casuarina/Frankia* (actinorhizal) association (12). The latter is the only Hb isolated thus far from nodules produced on the roots of diverse woody dicots (scattered among 22 families) in response to invasion by a single actinomyocyte genus, *Frankia*.

All plant Hbs studied achieve extraordinary affinity for O_2 by very rapid combination with O_2 together with moderate rates of dissociation (Table 1). The amount of O_2 transported by Hb is proportional to its saturation, and it is therefore surprising that in soybean and sweet clover nodules, the only nodules studied, the saturation of legHb was less than 20%. Although soybean legHb has a high O_2 affinity, it appears that a legHb with a significantly higher affinity would permit O_2 to be transported more efficiently. Inspection of Table 1, however, shows that the wide range of plant Hbs studied all have affinities very close to that of soybean legHb. The plant Hbs share the three dimensional structure of Mb and retain the conserved amino acid residues of the heme pocket required for reversible O_2 binding. One may speculate that no higher affinity can be achieved by modifying the Mb pattern further, bearing in mind that the rate of O_2 binding to several of the plant Hbs is within an order of magnitude of that expected for the diffusion of O_2 into a sink. There may also be a lower limit to the rate of dissociation from a Mb type of molecule, since, where the structure is known, the lowest rate thus far reported is 2 s^{-1} for soybean legHb at pH 5. Despite similar O_2 combination and dissociation rates, geminate reactions of plant Hb vary widely. We conclude that selection has forced plant Hb into a common pattern of very high O_2 affinity achieved by nearly diffusion-limited combination with O_2.

In contrast to the sameness of plant Hbs, Hbs from the mollusc/bacteria symbioses present a thought-provoking diversity. We begin with observations of the living, symbiont-harboring gill of the small East Coast clam, *Solemya velum* (13) collected from marine sediments. The ambient pO_2 required to half-saturate the Hb of respiring gill filaments was 6 torr, substantially less than the probable pO_2 in the burrow, say 60–80 torr. Accordingly, in nature the pO_2 in the bacteriocyte cytoplasm, although unknown and probably limited by a balance of consumption and rates of entry through anatomical barriers, probably exceeds the pressure required to saturate the Hb (P_{50} *Lucina* Hb = 0.1–0.2 torr).

Introduction of a small concentration of hydrogen sulfide into the ambient oxygenated sea water rapidly and reversibly brings about conversion of about half the Hb of the living gill to a new spectral species, later unequivocally identified as ferric Hb sulfide; that is, ferric Hb with hydrogen sulfide or with an hydrosulfide anion (HS^-) ligated to the heme Fe in the distal position. To our knowledge, the rapid and reversible formation of ferric Hb sulfide is unique to the clam gill; it was not observed in 12 other representative vertebrate and invertebrate tissues.

Two Hbs may be isolated in roughly equal amounts from the *S. velum* gill. One, HbI, reacts quantitatively with hydrogen sulfide and O_2 to form ferric Hb sulfide; it may be called "sulfide-reactive." The other, HbII, remains oxygenated and may be called "oxygen-reactive." We assume that only HbI reacts with sulfide in the living gill, HbII remaining oxygenated.

The concentration of hydrogen sulfide required in ambient sea water for half-maximal formation of ferric Hb sulfide in the living *S. velum* gill, 200–250 μM (13) is of the same order as that estimated in the environmental pore water, 30–1000 μM, and is comparable to the 200 mM required to support sulfide-dependent carbon fixation. These numbers exceed roughly 10,000-fold the probable P_{50} for the reaction of ferric Hb with hydrogen sulfide to form ferric Hb sulfide (P_{50} purified ferric *Lucina* Hb, H_2S = 3nM). Since ferrous and ferric Hb are interconverted in the living gill independent of the presence of sulfide (see reference 13, experiments using cyanide as a trapping agent), the steady state level of ferric Hb sulfide is directly related to the cytoplasmic hydrogen sulfide activity. The latter must be in the nanomolar range under the chosen conditions (ferric Hb sulfide formation half-maximal). Accordingly, we envision a steady state in the living gill in which inflowing hydrogen sulfide is balanced by removal, and the concentration of free hydrogen sulfide remains low, perhaps in the nanomolar range. This concentration of free hydrogen sulfide may not be sufficient to support the flux of hydrogen sulfide to the symbiont, and we suggest that HbI may facilitate the diffusion of hydrogen sulfide through the cytoplasm.

In the steady state envisioned, the concentration of ferric Hb sulfide (say 0.25 mM) exceeds by roughly 100,000-fold the concentration of free hydrogen sulfide (say 3nM). This implies that the symbionts act as a sink for Hb-bound hydrogen sulfide, constantly removing it to support oxidative phosphorylation without prior dissociation of the bound ligand.

To explore the properties of gill Hbs, we turn to the large and easily accessible Puerto Rican clam, *Lucina pectinata*. The sulfide reactive Hb, HbI, isolated from the symbiont harboring gill of *L. pectinata* (14, 15) is a rather small (M_r = 14,443), monomeric protohemeprotein of moderate O_2 affinity (P_{50} = 0.18 torr equivalent to 325 nM dissolved O_2). An electron paramagnetic resonance (EPR) and optical spectral study shows that the proximal ligand to the heme Fe is histidyl imidazole. Acid and alkaline ferric forms, and the pK of their interconversion, are similar to those of Mb. Oxygen combination is conspicuously fast; O_2 dissociation is moderately fast (Table 2), but similar to those of other molluscan tissue Hbs. These properties are consonant with a Mb-like structure, but would also accord with a structure in which a residue such as Gly, Leu, or Ile replaces His in the position distal to the heme.

The reactions with sulfide are extraordinary. Very rapid combination of the ferric protein with hydrogen sulfide and very slow dissociation underlie an extraordinary ligand affinity, K' = 3.4 nM, 4000-fold greater than that of ferric Mb, Table 3. Equally unusual is the speed of the reaction in which hydrogen sulfide converts oxyHb I to ferric Hb sulfide: $t_{1/2}$ = 350 – 1000 seconds, three orders of magnitude more rapid than comparable reactions of other proteins. A probable reaction mechanism is nucleophilic displacement of bound superoxide anion from oxyHb, Hb (heme $d_{1/2}^5$).O_2^-, by the hydrosulfide anion, HS^-. However, interconversion of ferrous and ferric Hb in living tissue proceeds ($t_{1/2}$ = 63 s^{-1}) independently of the presence of hydrogen sulfide and may be somewhat faster than the *in vitro* reaction of *Lucina* HbI (13); perhaps it is enzyme assisted.

Dissociation of hydrogen sulfide from ferric Hb sulfide is very slow, k_{off} = 0.00022 s^{-1}, implying a turnover time of 5000 seconds. It seems improbable that delivery of sulfide could be achieved by the simple dissociation of ligated sulfide. We note that purified ferric Hb I sulfide can accept electrons without prior dissociation of the ligand, liberating ferrous Hb and hydrogen sulfide in a rapid reaction. We suggest that, in the living cell, the reduction of ferric Hb sulfide near the peribacterial membrane surface may precede the delivery of the ligand.

The O_2 reactive Hbs of *Lucina* gill are two very similar proteins, HbII (M_r = 16,128) and HbIII (M_r = 17,762), which, when mixed at millimolar concentrations, form a tetramer. Perhaps (HbII)$_2$(HbIII)$_2$ is the cytosolic species.

Table 2. Kinetics and equilibria of reactions of *Lucina* Hbs with O_2 compared to those of legHb and Mb. P_{50} is related to K' by the relation 1 torr equivalent to 1800 nM at 20°C. All data were obtained at pH 7.5 and 20°C.

Proteins	Kinetics		Equilibria	
	k'_{on} x 10^{-6} M^{-1} s^{-1}	k_{off} s^{-1}	$K'(= k{:}k')$ nM	P_{50} torr
HbI	100–200[a]	61.1	–	0.18
HbII	0.390	0.11	282	0.16
HbIII	0.288	0.075	260	0.14
Whale Mb	19	10.0	526	0.3
Soybean Lb	116	5.55	48	0.03

[a]Estimated from k_{off} and P_{50}.

Table 3. Reactions of ferric *Lucina* Hbs with hydrogen sulfide. Data obtained at pH 7.5 and 20° C.

Protein	k'_{on} x 10^{-3} $M^{-1}s^{-1}$		k_{off} s^{-1}	K' (=$k_{off}{:}k'_{on}$) nM
	Acid limit	Alkaline limit		
HbI	226	0.066	0.22	0.0034
HbII	11.3	0.032	17	14.5
HbIII	41.7	0.12	16	16
whale Mb	8.8	0.052	48	18.5

If so, it is a non-interactive tetramer, because the Hill coefficient, h, is always unity and the O_2 affinities of each purified Hb and its mixture are independent of protein concentration. Oxygen combination and dissociation are both extremely slow, although their balance, the O_2 affinity, is about the same as that of HbI, Table 2. The proximal ligand to the heme Fe is histidyl imidazole. A Tyr probably occupies the distal position, as it does in *Dicrocoelium* Hb and in HbM's Boston and Saskatoon. At alkaline pH, tyrosyl competes with hydroxyl as a ligand to the heme Fe. At acid pH, the tyrosyl is assumed to be protonated and water ligates to the ferric Fe. Binding of tyrosinate to the heme Fe is weaker than in HbM Saskatoon and much weaker than in HbM Boston, permitting the formation of ferrous and of oxyHbsII and III. We suggest that the tyrosyl (or a nearby residue) may remain in the heme pocket of ferrous HbII and III, where it may interact with the bound O_2 molecule and contribute to the very slow rates of O_2 dissociation from these proteins.

Since HbI in the gills of clams in their natural enironment is pre-empted as ferric Hb sulfide; the burden of O_2 transport must fall on HbII and III. However, the rate of O_2 dissociation from these proteins (*Lucina* HbII and HbIII) is among the slowest known, and may be insufficient to support O_2 delivery. We suggest, as a working hypothesis, that cytoplasmic oxyHbs may accept electrons originating from the symbiont, in that fashion assuming the character of a terminal oxidase (3).

Acknowledgements

Many of the ideas presented in this essay were developed in collaboration with Drs. Cyril A. Appleby and Quentin H. Gibson, whom we thank.

References

1. Day, D.A., Price, G.D. and Udvardi, M.K. (1989) *Aust. J. Plant Physiol.* **16**: 69–84.
2. Werner, D., Morschel, E., Garbers, C., Bassarab, S. and Mellor, R.B. (1988) *Planta* **174**: 263–270.
3. Wittenberg, J.B. and Wittenberg, B.A. (1990) *Annu. Rev. Biophys. Biophys. Chem.* **19**: 217–241.
4. Appleby, C.A., Dennis, E.S. and Peacock, W.J. (1990) *Aust. J. Systematic Botany* **3**: 81–89.
5. Wittenberg, J.B. (1985) *Bull. Biol. Soc. Washington* **6**: 301–310.
6. Appleby, C.A. (1984) *Annu. Rev. Plant Physiol.* **35**: 433–478.
7. Bergersen, F.J., Turner, G.L. and Appleby, C.A. (1973) *Biochim. Biophys. Acta* **292**: 271–282.
8. Wittenberg, J.B., Bergersen, F.J., Appleby, C.A. and Turner. G.L. (1974) *J. Biol. Chem.* **249**: 4057–4066.
9. Wittenberg, B.A. and Wittenberg, J.B. (1987) *Proc. Nat. Acad. Sci. U.S.A.* **84**: 7503–7507.
10. Gibson, Q.H., Wittenberg, J.B., Wittenberg, B.A., Bogusz, D. and Appleby, C.A. (1988) *J. Biol. Chem.* **264**: 100–107.
11. Wittenberg, J.B., Wittenberg, B.A., Gibson, Q.H., Trinick, M.J. and Appleby, C.A. (1986) *J. Biol. Chem.* **261**: 13624–13631.
12. Fleming, A.I., Wittenberg, J.B., Wittenberg, B.A., Dudman, W.F. and Appleby, C.A. (1987) *Biochim. Biophys. Acta* **911**: 209–220.
13. Doeller, J.E., Kraus, D.W., Colacino, J.M. and Wittenberg, J.B. (1988) *Biol. Bull.* **175**: 388–396.

14. Kraus, D.W. and Wittenberg, J.B.   (1990) *J. Biol. Chem.* **265**: 16043–16053.
15. Kraus, D. W., Wittenberg, J. B., Lu, J. F. and Peisach. J.   (1990) *J. Biol. Chem.* **265**: 16054–16059.

44
Modulation of Oxygen Binding in Squid Blood

Bruno Giardina,[a] Saverio G. Condo[a] and Ole Brix[b]

[a]Department of Experimental Medicine and Biochemical Sciences, II University of Rome, via Orazio Raimondo, 00173 Rome, Italy
[b]Zoological Laboratory, University of Bergen, Allegaten 41, 5007 Bergen, Norway

Introduction

An extensive set of data related to the binding of O_2 by Hc from the squid *Todarodes sagittatus* has been collected under various experimental conditions. The results obtained show that, within the range of physiological pH, the concentration of protons affects mainly the high-affinity state of the molecule without significantly affecting the low-affinity state. As far as the effect of temperature is concerned, the data show a characteristic feature which is very similar to that previously described in the case of Hbs from arctic mammals such as reindeer and musk ox (1–4). The shape of the O_2 equilibrium curve shows strong temperature dependence, since the overall heat of binding of O_2 to the low-affinity state of the molecule is strongly exothermic, while that to the high-affinity state is very close to zero.

The results provide an outline of the molecular compromise that optimizes the loading and unloading of O_2 under the various environmental conditions experienced by this species of squid (5–7) through the interplay of temperature and protons.

The example described here could be generally applicable to all those animals in which the temperature is allowed to change throughout the entire body.

Results and Discussion

Figure 1 shows a series of Hill plots obtained at different values of pH and temperature. As is evident from the data reported, the distribution of the Bohr protons all along the binding curve is strongly dependent upon pH. In particular, in going from pH 7.4 to more acidic values, the concentration of protons acts mainly on the upper asymptote, representative of the high-affinity state of the molecule, and leaves unchanged the lower asymptote, representative of the low-affinity state of the molecule. This differential effect of pH on the upper and lower asymptotes is evidenced in a decrease in the free energy of interaction between the O_2 binding sites, and results in a lowering of the overall cooperativity of ligand binding.

In addition to the effect of pH described above, Figure 2 outlines the effect of temperature, as observed within the same range of pH values at a constant concentration of protons (see also Figure 1). The data, presented in the form of a Hill plot, show a characteristic feature which is very similar to that previously

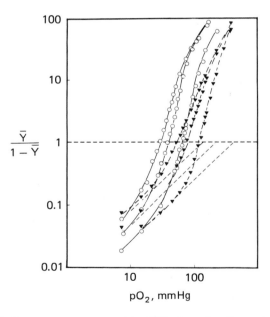

Figure 1. O_2 binding curves, expressed by Hill plots, for *T. sagittatus* blood, as a function of temperature and pH. (O) pH 7.4 and 6, 10 and 20°C (from the left to the right); (▼) pH 7.19 and 6, 10 and 20°C (from the left to the right). The O_2 equilibrium experiments were performed in a thermostated chamber fed with equilibrating gas from serially connected Wosthoff gas mixing pumps.

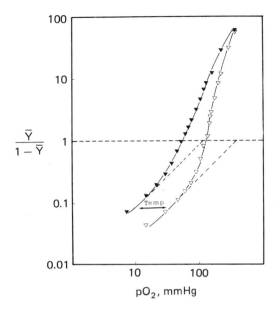

Figure 2. The differential effect of temperature at pH 7.15 on the low- and the high-affinity state of Hc from *T. sagittatus* blood. (Δ) 20°C; (▼) 6°C.

described in the case of Hbs from arctic mammals like reindeer and musk ox (1–4). As in the case of those Hbs, the shape of the binding curve shows a strong temperature dependence, since an increase in temperature induces a great decrease in the association constant for the binding of O_2 to the low-affinity state of the molecule, without having any significant influence on the association constant for binding to the high-affinity state. Thus, while the overall reaction of O_2 binding to the low-affinity state is strongly exothermic, the heat of reaction relative to the high-affinity state is very close to zero, and the reaction may even absorb heat if a correction is made for the heat of solubilization of O_2. As is indicated by the unchanged position of the upper asymptote and the concomitant, marked shift of the lower one, cooperativity of O_2 binding increases substantially as the temperature is increased.

It should be emphasized that the different thermodynamic parameters that characterize the low- and the high-affinity states of this squid Hc appear to be strongly dependent upon pH. This dependence is shown in Figure 3 where the overall heat of oxygenation, measured at three different degrees of saturation, is reported as a function of pH within the range of pH 7.7 to pH 7.2. At alkaline pH values (about 7.7), the overall reaction of O_2 binding is strongly exothermic and is very similar for the three degrees of saturation (from 13 to 15 Kcals/mol of O_2). This observation implies that, at this pH, the binding curves are shifted, almost without any change in shape, with increasing temperature, as is often observed with other respiratory proteins (8, 9). The same applies at pH 7.4 where a smaller heat of binding is observed (from -7 to -9 Kcals/mol of O_2). Towards more acidic pH values, while the enthalpy of O_2 binding at 25% and 50% saturation is seen to level off, that at 95% continues to decrease, tending to about -1 Kcals/mol of O_2 at pH 7.2. At this pH value, as outlined in Figure 2, the dependence of the effect of temperature on the degree of saturation of the protein with O_2 is dramatic. The results reported here show that the functional behavior of the Hc from *T. sagittatus* displays peculiar features which, like those previously observed with Hbs from reindeer and musk ox, make this protein a very interesting example of molecular adaptation to specific physiological requirements. These characteristic features result from the unusual interplay of protons and temperature (Figures 2, 3). In fact, the strong dependence of the effect of temperature on the degree of O_2 saturation observed at pH 7.2 (Figure 3) may be explained in large part by the differential effect of protons between states with very high and very low saturation by the ligand.

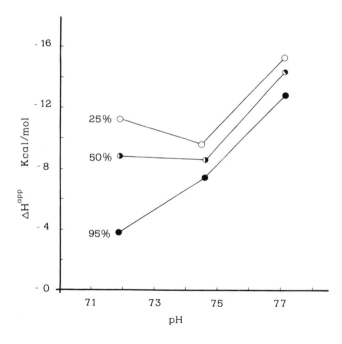

Figure 3. Overall heat of oxygenation (ΔH_{app}) for Hc from *T. sagittatus* as a function of pH, at three different levels of saturation (as indicated). The reported values are not corrected for the heat of solubilization of O_2 (-3.0 Kcals/mol) and are expressed in Kcals per mole of O_2.

Thus, while the Bohr effect in the case of the low-affinity state of the protein is almost absent at pH values lower than 7.4, that of the high-affinity state is operative over the entire range of pH values examined. Therefore, for pH values lower than 7.4, most of the Bohr protons are released with the binding of the last molecules of O_2. In other words, below pH 7.4, the overall enthalpy of O_2 binding to the low-affinity state is no longer influenced by the endothermic release of the O_2-linked protons and levels off at about −10 Kcals/mol of O_2; by contrast, the enthalpy of oxygenation of the high-affinity state is more and more diminished by the endothermic contribution of the Bohr effect that is still operative within this pH range. The net result of this interplay is the peculiar dependence of the shape of the binding curve on temperature, which is clearly discernible at pH 7.2 (Figure 2).

These functional properties seem to fit very well with the physiological requirements of *T. sagittatus*. This animal, like most cephalopods, is characterized by a high aerobic metabolism which has to be satisfied by very efficient respiration. In this connection it should be recalled that, for a functional Haldane coefficient of about 1.0, the Bohr shift would not come into operation by means of an arterial-venous difference but only by means of a change in pH brought about by temperature at the level of both arterial and venous blood (5, 6). The particular physico-chemical properties of the blood of *T. sagittatus* seem, in this respect, very well-designed to optimize the transport of O_2 when the temperature is allowed to change throughout the animal's body. Thus, the particular differential and opposite effects of protons and temperature will minimize the handicap of the unloading of O_2 that is induced by a decrease in temperature and a subsequent increase in pH. In fact, as far as the concentration of protons is concerned, the position of the lower asymptote is fixed and is allowed to change only as a result of a change in temperature. This change will increase the O_2 affinity of the lower asymptote of the binding curve, but not to the extent that would have been observed in the presence of a significant Bohr effect at these levels of saturation. As a consequence, and on the basis of the insensitivity of the high-affinity state to changes in temperature, a decreased cooperativity should have been observed; however, this effect is almost completely counterbalanced by the Bohr effect which is operative only at high levels of saturation (below pH 7.4). Therefore, at the level of the high-affinity state, a minimalization of the effect of concomitant changes in temperature and pH is also observed. At this level, the increase in the O_2 affinity is solely determined by the change in pH and the effect of temperature is negligible. Moreover, the increased affinity of the upper asymptote, brought about by pH would increase the cooperativity and counteract the effect of temperature at the level of the lower asymptote. The results reported outline very clearly the intramolecular compromise that optimizes O_2 loading and unloading under the various environmental conditions experienced by these animals through the interplay between temperature and protons. The example described here could be of general applicability and validity for all those animals in which temperature is allowed to change over the entire body.

References

1. Giardina, B., Brix, O., Nuutinen, M., El-Sherbini, S., Bardgard, A., Lazzarino, G. and Condo, S.G. (1989) *FEBS Lett.* 247: 135–138.
2. Giardina, B., Condo, S.G., El-Sherbini, S., Mathisen, S., Tyler, N., Nuutinen, M., Bardgard, A. and Brix, O. (1989) *Comp. Biochem. Physiol.* 94B: 129–133.
3. Brix, O., Bardgard, A., Mathisen, S., El-Sherbini, S., Condo, S.G. and Giardina, B. (1989) *Comp. Biochem. Physiol.* 94B: 135–138.
4. Brix, O., Condo, S.G., Lazzarino, G., Clementi, M.E., Scatena, R. and Giardina, B. (1989) *Comp. Biochem. Physiol.* 94B: 139–142.
5. Brix, O., Bardgard, A., Cau, A., Colosimo, A., Condo, S.G. and Giardina, B. (1989) *J. Exp. Zool.* 252: 34–42.
6. Brix, O., Lykkeboe, G. and Johansen, K. (1981) *Respir. Physiol.* 44: 177–186.
7. Reeves, R.B. (1980) *Respir. Physiol.* 42: 317–328.
8. Brunori, M., Giardina, B. and Bannister, J.V. (1979) *Inorg. Biochem.* 3: 126–131.

45
The Role of Heme Compounds in Sulfide Tolerance in the Echiuran Worm *Urechis caupo*

Alissa J. Arp

Biology Department, San Francisco State University, San Francisco, CA 94132, USA

Introduction

Since the discovery of the hydrogen sulfide-rich hydrothermal vent environments in the late 1970s, much research effort has been directed at understanding the abilities of metazoan animals to tolerate high sulfide environments. A variety of sulfide-rich habitats have been reinvestigated along with the deep ocean vents (such as sewer outfalls, marshes, and mudflats), and animals from diverse phyla have been shown to tolerate toxic sulfide and/or to capitalize on its high energy state to fuel metabolism (1–5). In many cases this tolerance and/or exploitation depends upon the central role played by a Hb or heme-containing molecule (5–13).

U. caupo, the "fat innkeeper worm," is an echiuran which inhabits *U*-shaped burrows in intertidal mudflats off the coast of California (14). During high tide, fresh, oxygenated seawater is pumped through these burrows by peristaltic contractions of the worms' body walls (15, 16). With daily and seasonal tidal fluctuations, however, these mudflats are subject to recurrent and sometimes lengthy periods when no fresh seawater is present. Burrow water O_2 concentration may decrease during low tide due to animal consumption, and toxic sulfides may accumulate due to bacterial reduction of sulfates and by the decay of organic matter (17, 18, 19).

When confronted with low O_2 levels, *U. caupo* has been shown to maintain aerobic respiration by using a large pool of coelomic fluid. This coelomic fluid contains an erythrocytic, tetrameric Hb capable of storing large quantities of O_2 (15, 20–22). The presence of even nanomolar levels of sulfide during low tide could seriously impede or destroy this ability to maintain aerobic respiration due to the binding of sulfide to cytochrome c oxidase, as well as to the heme sites of the coelomic fluid Hb molecules (23).

In order for *U. caupo* to survive in such an environment, it must be able to protect its tissues against the toxic effects of sulfide. Possible mechanisms include complete anaerobiosis, complete exclusion of sulfide at the body wall and hindgut (the proposed respiratory organ), and/or some type of sulfide detoxification system involving a chemical conversion of sulfide (H_2S, HS^- or $S^=$) to non-toxic forms such as sulfate, sulfite or thiosulfate ($SO_4^=$, $SO_3^=$ or $S_2O_3^=$). Laboratory studies have shown that *U. caupo* does not reduce aerobic respiration with sulfide exposure, but significantly increases its rate of O_2 consumption (5, 24). Furthermore, *in vitro* studies have demonstrated that the body wall and hindgut of *U. caupo* are freely permeable to sulfide (25, 26). Clearly, *U. caupo* is not undergoing periods of anaerobiosis during intermittent sulfide exposure, nor is it able to exclude sulfide at the epithelial surfaces.

The presence of an unusual brown heme compound contained in the coelomocytes (Hb-containing erythrocytes within the coelomic fluid) has been noted in *U. caupo* (13, 27, 28). This brown pigment is hematin, an oxidized heme compound which, unlike Hb, is not associated with a globin moiety. Hematin is contained in granules which co-occur with Hb in the coelomocytes. Powell and Arp (13) report a correlation between sulfide oxidation activity and hematin content of the coelomic fluid and propose the use of hematin as a sulfide detoxification mechanism in *U. caupo,* which catalyzes the oxidation of sulfide to non-toxic forms (13).

Materials and Methods

In order to investigate the use of a heme-based sulfide tolerance mechanism in *U. caupo*, the heme compound concentrations and coelomic fluid characteristics of organisms from three separate northern California environments have been analyzed over a period greater than one year, and have been compared on the basis of age, sex, season, population site and sulfide exposure. Total heme was assayed spectrophotometrically by acid-acetone extraction, Hb was assayed spectrophotometrically using the extinction coefficient of human Hb, hematin was calculated as the difference (13), and standard measures of the coelomic fluid were taken with hemacytometer and hematocrit techniques (29). All data are presented in the text as mean values plus the standard error of the mean, followed by the number of experimental analyses in parentheses. In all cases, data are considered significant if $p < 0.05$.

Results and Discussion

The major population sites for *U. caupo* in northern California are Elkhorn Slough in Monterey County, Princeton Harbor in Half Moon Bay, San Mateo County, and Bodega Bay, in Sonoma County. Animal distributions, sediment type and concentration of sulfide vary between these sites. At Elkhorn Slough burrow water sulfide concentration during tidal exposure in summer was 10.6 ± 1.4 μM (n = 20) with only intermittent occurrences of sulfide during other months. At Bodega Bay burrow water sulfide concentration in summer was 29.3 ± 7.3 μM (n = 14). Burrows have not been analyzed at Princeton Harbor. The mean O_2 concentration of water samples from exposed Elkhorn Slough and Bodega Bay burrows was 3.08 ± 0.13 ml O_2/l (n = 45) and 2.91 ± 0.21 ml O_2/l (n = 17), respectively (29, 30). These values are between 52% and 46% of the O_2 concentration of saturated sea water under the environmental conditions of 15°C and 30 ppt.

Heme Compound Analyses. Results of heme analyses of animals from these three sites show that the concentration of Hb and hematin in the coelomic fluid increased significantly with the size of the animals at both the Elkhorn Slough and Princeton Harbor sites (Figures 1A –D). Juvenile animals of 20 g or less in weight had very low heme concentrations, whereas animals 30 g and greater had higher, but variable, heme concentrations, suggesting that older animals may accumulate more heme compounds. This relationship was not apparent with the Bodega Bay animals, as all animals collected from this site were 28 g or greater in size. For this reason, further comparisons of heme concentrations in the coelomic fluid between animals collected at the three sites were made only among animals greater than 30 g.

The pooled data for heme compound concentration for all animals greater than 30 g from each site showed that mean Hb and hematin concentration was highest in the Bodega Bay animals, followed by the Elkhorn Slough animals and finally the Princeton Harbor animals (Table 1). In many cases there are significant differences in the concentration of heme compounds between the environments. High variability in hematin concentrations could be due to the seasonal differences in this compound, as the pooled data are from animals collected throughout the year.

Heme compound concentrations showed a great deal of seasonal variability. Hb concentrations ranged from 0.12 to 2.0 mM and hematin concentrations ranged from 0 to 5.4 mM in animals greater than 30 g over a one year period. Additionally, the variations in heme compound concentrations are independent of the sex of the animal (Table 1B). It has been reported that hematin occurs only in the male coelomocytes of *U. caupo* (27). The data in Table 1B show that this is not true for analyzed specimens from Elkhorn Slough, and that females had a greater mean concentration of hematin than males, although there is no statistically significant difference in heme compound concentration between the two sexes.

A.

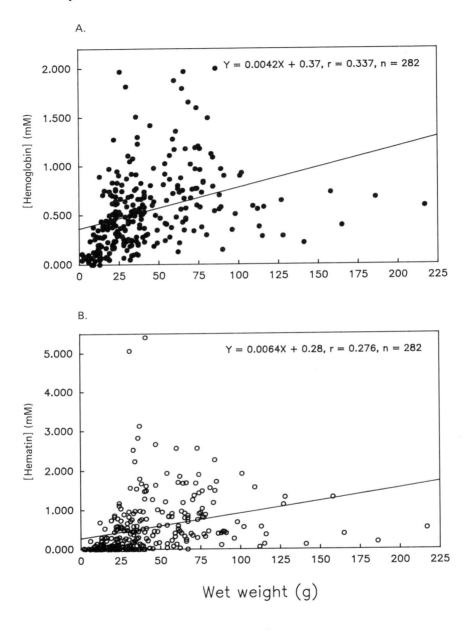

Figure 1*A, B*.   The concentration of Hb and hematin in the coelomic fluid of specimens of *U. caupo* from two environments as a function of animal wet weight. Regression equations appear on the graphs. Data are from Arp *et al.* (29). (*A*) Hb concentration of the coelomic fluid of animals from Elkhorn Slough. (*B*) Hematin concentration of the coelomic fluid of animals from Elkhorn Slough.

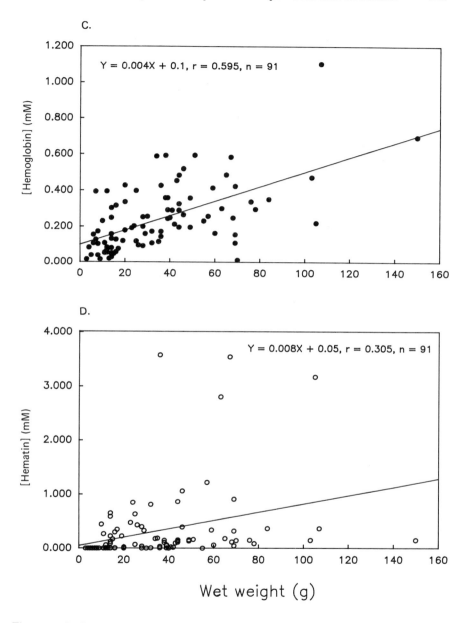

Figure 1C, D. (C) Hb concentration of the coelomic fluid of animals from Princeton Harbor. (D) Hematin concentration of the coelomic fluid of animals from Princeton Harbor.

Table 1. Characteristics of the coelomic fluid heme compounds of *U. caupo*. All values are means plus the standard error of the mean, followed by the number of experimental observations in parentheses. Data are from Arp *et al.* (29).

	Total heme (mM)	Hb (mM)	Hematin (mM)
A. Site[a]			
Princeton Harbor	0.85 ± 0.15 (44) **	0.34 ± 0.03 (44) **	0.51 ± 0.14 (44) *B
Elkhorn Slough	1.48 ± 0.08 (159) *P	0.67 ± 0.03 (159) *P	0.82 ± 0.07 (159)
Bodega Bay	1.79 ± 0.17 (60) *P	0.70 ± 0.04 (60) *P	1.08 ± 0.16 (60) *P
B. Sex[b]			
Male	1.26 ± 0.13 (39)	0.71 ± 0.05 (39)	0.55 ± 0.10 (39)
Female	1.43 ± 0.16 (32)	0.60 ± 0.06 (32)	0.83 ± 0.13 (32)

[a]Concentration of heme compounds for animals greater than 30 g wet weight from three population sites. One asterisk (*) indicates that the value is significantly different from only one other population site, this site is indicated by the first letter of the name of the site. Two asterisks (**) indicate that the value is significantly different from both of the other sites.
[b]Concentration of heme compounds from male and female animals greater than 30 g wet weight.

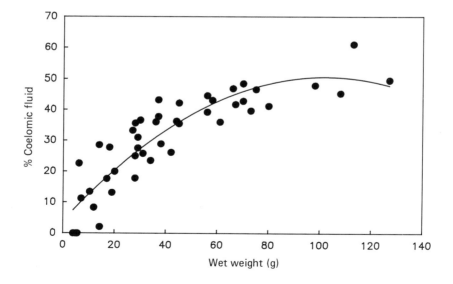

Figure 2. The relationship of coelomic fluid volume to the wet weight of *U. caupo* specimens. Data are from Arp *et al.* (29).

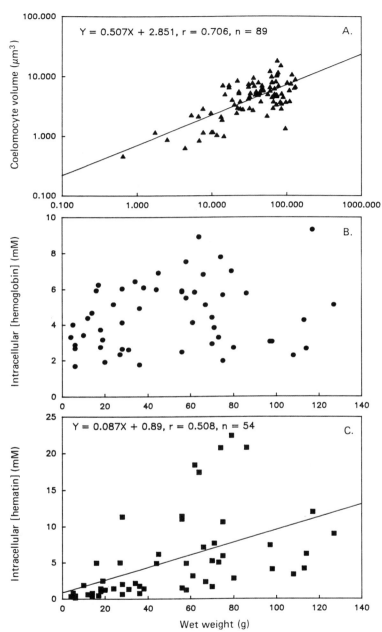

Figure 3. Coelomocyte characteristics of *U. caupo*. Regression equations appear on the graphs. Data are from Arp *et al.* (29). (*A*) The relationship of coelomocyte volume to the wet weight of animals. (*B*) The relationship of intracellular Hb concentration to the wet weight of animals. (*C*) The relationship of intracellular hematin concentration to the wet weight of animals.

Coelomic Fluid Characteristics. The volume of coelomic fluid of sacrificed specimens from Elkhorn Slough had a mean value of 0.34 ± 0.02 ml/g ($n = 41$) in animals that ranged in wet weight from 10 to 127 g. This value was calculated for animals greater than 10 g, as smaller animals had negligible amounts of coelomic fluid. The volume of coelomic fluid varied significantly with the size of the animal, as large animals have proportionally greater coelomic fluid volumes (Figure 2).

The mean hematocrit for 41 analyzed specimens was 13.2 ± 0.9 %, and the volume of the average individual cell was 8185 ± 547 μm^3 ($n = 32$), in animals ranging from 10 to 127 g. The cell volume varied significantly with the size of the animal, as large animals have proportionally greater coelomocyte volumes (Figure 3A). These data suggest that larger animals may synthesize larger coelomocytes, and/or that coelomocytes may be retained by animals and grow as the animal grows. Analysis of the coelomocytes also showed a significant increase in the intracellular hematin concentrations with body size (Figure 3C). Juvenile animals have low intracellular heme concentrations, but adults have variable and often high intracellular heme concentrations, suggesting that these compounds may increase with animal age, thus providing an additional explanation for the variation in heme compounds with size reported above.

In Vivo Observations. The pO_2 in the coelomic fluid of specimens maintained in air saturated seawater in the laboratory for greater than one week were highly variable and low. Coelomic fluid pO_2s measured using a blood gas analyzer ranged from 4 to 61 mm Hg, with a mean value of 27 ± 18 mm Hg for 10 animals. The pH values of the coelomic fluid of these same specimens varied from 7.7 to 7.9, with a mean value of 7.8 ± 0.1. Animals exposed to 70 μM sulfide and 25–80 mm Hg O_2 in the laboratory show no accumulation of sulfide in the coelomic fluid after 20 hours and pO_2s are maintained at 1 to 5 mm Hg (31). Although these *in vivo* studies show coelomic fluid pO_2s of less than 5 mm Hg, the absence of sulfide accumulation suggests that sulfide oxidation does, in fact, occur.

The prevalent heme compounds in the abundant coelomic fluid of *U. caupo* appear to be important for sulfide tolerance. Hematin and Hb may act together to allow for continued aerobic respiration in the presence of sulfide, as Hb binds and stores O_2 that can be utilized for sulfide oxidation as well as for aerobic respiration, and hematin acts as a catalyst for the oxidation of sulfide into nontoxic sulfur-containing compounds. Older animals (as indicated by greater weight) have variable but greater Hb and hematin concentrations, larger coelomocytes with higher intracellular hematin concentrations and a greater volume of coelomic fluid. The variability in heme compound concentrations in the coelomic fluid of larger animals may reflect their sulfide exposure history, with animals from high-sulfide areas, such as Bodega Bay having significantly higher concentrations than those from lower-sulfide environments, such as Elkhorn Slough or Princeton Harbor. Inhabitants of high sulfide environments may accumulate heme compounds as coelomocytes persist and grow, or they may produce new, larger coelomocytes with higher heme levels at all stages of their life cycle. These data support the hypothesis that *U. caupo* tolerates high sulfide environments by detoxification of sulfide with abundant heme compounds.

Acknowledgements

This research was supported by Research Corporation grant C-2455 and National Science Foundation grant DCB-891776. I thank the diligent co-workers and collaborators whose names appear on the initial manuscripts, including; M. Powell, D. Julian, B. Hansen, R. Eaton and S. Mansour.

References

1. Hand, S.C. and Somero, G.N. (1983) *Biol. Bull.* **165**: 167–181.
2. Felbeck, H. (1983) *J. Comp. Physiol.* **153**: 3–11.
3. Vetter, R.D., Wells, M.E., Kurtsman, A.L. and Somero, G.N. (1987) *Physiol. Zool.* **60**: 121–137.
4. Arp, A.J., Powell, M.A. and Hansen, B.M. (1987) *Am. Zool.* **27**: 32A.
5. Arp, A.J., Hansen, B.M. and Julian, D. (1989) *Am. Zool.* **29**: 69A.
6. Arp, A.J. and Childress, J.J. (1981) *Science* **213**: 342–344.
7. Arp, A.J. and Childress, J.J. (1983) *Science* **219**: 295–297.
8. Powell, M.A. and Somero, G.N. (1983) *Science* **219**: 297–299.
9. Arp, A.J., Childress, J.J. and Fisher, Jr., C.R. (1984) *Physiol. Zool.* **57**: 648–662.
10. Arp, A.J., Childress, J.J. and Vetter, R.D. (1987) *J. Exp. Biol.* **128**: 139–158.
11. Dando, P.R., Southward, A.J., Southward, E.C., Terwilliger, N.B. and Terwilliger, R.C. (1985) *Mar. Ecol. Prog. Ser.* **23**: 85–98.
12. Doeller, J.E., Kraus, D.W., Colacino, J.M. and Wittenberg, J.B. (1988) *Biol. Bull.* **175**: 388–396.
13. Powell, M.A. and Arp, A.J. (1989) *J. Exp. Zool.* **249**: 121–132.
14. Fisher, W.K. and Mac Ginitie, G.E. (1928) *Annu. Mag. Nat. Hist.* **10**: 204–213.
15. Garey, J.R. and Riggs, A.F. (1984) *Arch. Biochem. Biophys.* **228**: 320–331.
16. Lawry, J.V., Jr. (1966) *J. Exp. Biol.* **45**: 343–353.
17. Theede, H., Ponat, A., Hiroki, K. and Schlieper, C. (1969) *Mar. Biol.* **2**: 325–337.
18. Fenchel, T.M. and Riedl, R.J. (1970) *Mar. Biol.* **7**: 255–268.
19. Aller, R.C. and Yingst, J.Y. (1978) *J. Mar. Res.* **36**: 201–254.
20. Hall, R.E., Terwilliger, R.C. and Terwilliger, N.B. (1981) *Comp. Biochem. Physiol.* **70**B: 353–357.
21. Mangum, C.P., Terwilliger, R.C. and Hall, R. (1983) *Comp. Biochem. Physiol.* **76**A: 253–257.
22. Pritchard, A. and White, F.N. (1981) *Physiol. Zool.* **54**: 44–54.

23. *Hydrogen Sulfide.* (1979) The Subcommittee on Hydrogen Sulfide, Committee on Medical and Biological Effects of Environmental Pollutants. Division of Medical Sciences, Assembly of Life Sciences, National Research Council. Baltimore: University Park Press.
24. Eaton, R.A. and Arp, A.J. (1990) *Am. Zool.* **30**: 69A.
25. Julian, D. (1989) *Am. Zool.* **29**: 69A.
26. Julian, D. and Arp, A.J. (1991) In preparation.
27. Baumberger, J.P. and Michaelis, L. (1931) *Biol. Bull.* **61**: 417–421.
28. Terwilliger, R.C., Terwilliger, N.B. and Schabtach, E. (1985) In *Blood Cells of Marine Invertebrates: Experimental Systems in Cell Biology and Comparative Physiology*, ed. W.D. Cohen, 193–225. New York: Alan R. Liss, Inc.
29. Arp, A.J., Hansen, B.M. and Julian, D. (1991) In preparation.
30. Hansen, B.M. (1989) M. Sc. Thesis, San Francisco State University.
31. Julian, D, Mansour, S. and Arp, A.J. (1990) *Am. Zool.* **30**: 69A.

Author Index

Subject Index